HOW TO BUILD
OUTDOOR
STRUCTURES

By DEBORAH MORGAN and NICK ENGLER

Popular
Science
Books

Sterling Publishing Co., Inc.   New York

Designed by Linda Watts, Bookworks, Inc.

Copyright © 1987 by Deborah Morgan and Nick Engler
First published in paperback by Sterling Publishing Co., Inc.
387 Park Avenue South, New York, N.Y. 10016
Distributed in Canada by Sterling Publishing
% Canadian Manda Group, P.O. Box 920, Station U
Toronto, Ontario, Canada M8Z 5P9
Distributed in Great Britain and Europe by Cassell PLC
Artillery House, Artillery Row, London SW1P 1RT, England
Distributed in Australia by Capricorn Ltd.
P.O. Box 665, Lane Cove, NSW 2066
Manufactured in the United States of America
All rights reserved

Sterling ISBN 0-8069-6742-0 Paper

Acknowledgments

Ｔhe authors would like to acknowledge and give special thanks to Thomas A. Fister for his help, advice, and expertise in preparing this book. Tom is a member of the Oregon Group Architects, of Dayton, Ohio. He is also a Building Official, Electrical Safety Inspector, and a Fire Safety Official, certified by the State of Ohio.

We would also like to thank:

- **Brimark Industries, Inc.,** Grand Rapids, Michigan
- **84 Lumber Company,** Springfield, Ohio
- **Furrow Building Supplies,** Springfield, Ohio
- **Koppers Company,** Pittsburgh, Pennsylvania
- **Forrest M. Morgan, The Martin Electric Company,** Dayton, Ohio
- **Shopsmith, Inc.,** Dayton, Ohio
- **Teco, Inc.,** Chevy Chase, Maryland
- **Wickes Lumber,** Springfield, Ohio

Contents

Tailoring Our Plans to Your Backyard

There are instructions for more than three dozen outdoor structure projects in this book. That's a lot of projects to choose from; but then, "outdoor structures" is a broad topic. Even with a good selection of plans and instructions, we realize that you may not find just the right deck or just the right storage building for your backyard. To solve this dilemma, we've done some things differently in this book.

We've arranged the materials to make it easy for you to modify the designs you see in the following chapters. Many outdoor structures, while they may differ in appearance, share many of the same features and components. For example, all storage buildings have foundations, walls, and roofs. You can set your storage building on a pier-and-beam foundation, a pole foundation, a concrete pad, and a few others. You can top this building with a gabled roof, slant roof, or a gambrel (barn) roof. A storage building with almost any style of roof will sit on almost any type of foundation. The great variety of designs for outdoor structures comes from 'mixing and matching' these components.

If you don't see your 'perfect' outdoor building project in this book, design your own by substituting components and features from structures in other chapters. We carefully chose the projects in this book so that there is an enormous selection of components and features included in the project plans. Here's an example: If you want to use pole construction for your enclosed garage, you can do it with the information in this book, even though we show our enclosed garage with a concrete footer-and-wall foundation. Just turn to several of the buildings we show with pole foundations, such as the *Stand-Alone Carport* or *Slant-Roof Shed,* and study how they're put together. Then revise the plans for the *Enclosed Garage* accordingly. Instead of pouring a concrete foundation, set posts in the ground. Instead of framing in the walls, attach cleats and a top plate to the posts. The steps for building a gabled roof remain the same. However, if you would rather have a gambrel roof, take a look at the *Storage Barn* plans, and adjust the shape of your roof trusses accordingly. No matter what sort of structure you want to build, you can 'mix and match' foundations, walls, roofs, supports, trim, dimensions, and many other features until you get what you want.

There are some other places we broke with tradition in putting together this book, to help you do more with the information. Notice that the instructions are divided into easy-to-follow steps. Each of these steps is subtitled with a short sentence that sums up the procedures. This makes it easy for you to locate and follow the instructions you need to complete a particular feature or component.

The working drawings are also presented somewhat differently than what you may be used to. First of all, the drawings you need to complete each step are right there with the instructions—

there is little need to flip pages back and forth. (Some of the drawings are repeated several times throughout the book, just so you won't have to hunt through the book to find a reference.) Each drawing presents just the information you need. You won't have to search for a dimension or a detail that's part of an all-inclusive elevation or view. And the drawings for each major component or feature are presented separately, making it easier for you to imagine them incorporated in different projects. All of this helps you to modify our designs—or do your own.

However, before you start to modify a plan or draw up a new one, there are a few precautions you should take. First of all, remember that if you change the dimensions of any one component, you may also have to change the materials and perhaps some of the individual parts. For example, you can't build a two-car garage on a 4" thick concrete pad with 12" footings. You have to dig the footers 24"-36" deep and increase the thickness of the pad to 6". Likewise, you can't simply take our plan for a gambrel truss that fits over a mid-size storage shed and blow it up to span that same two-car garage. You may have to beef up some of the truss members—particularly the joist—and add bracework.

Also, be sure that all the different components you want to use work well together. In particular, the type of foundation you decide to use must be compatible with the type of construction. For example, a pole foundation cannot be easily used with frame construction because it doesn't provide the necessary support for the 2 x 4 framework. If you wish to use a pole foundation, you have to use pole construction: Use cleats and a top plate to hang the siding and support the roof. Pier-and-beam, footer-and-wall, and concrete pad foundations all work well with frame construction. They can be made to work with pole construction, but you have to anchor and brace the posts properly.

Finally, consider your neighbors. Many communities have strict ordinances governing the types of foundations and construction that you may use. If you live in an historical district, there may be codes governing the style of architecture. While all of these may seem like restrictions on your freedom of choice, they ultimately work for everyone's benefit. They keep the property value up and help create a harmony of architectural design in your community.

If you have any doubts about whether your final plan for your 'perfect' outdoor structure will work perfectly, consult an architect, a civil engineer, or your local building inspector. Any one of these people will be able to tell you whether your plan is structurally sound and will meet the building codes and any other local ordinances. They may also be able to suggest further modifications that will increase the usefulness or durability of the structure.

Storage Buildings

If your shop and garden tools are crowding the family car out of the garage, it may be time to put up a storage shed in the back-yard. These handy structures are easy to build from scratch; you can make one in practically no time. By following the steps outlined in this chapter, you can build a shed that is custom-made to suit your storage needs.

Remember, there are several different types of foundations and roofs from which you can choose for your structure. Although we have, for example, paired a gabled roof with a pier and beam foundation, you may prefer a gambrel roof on your shed. In that case, just flip the pages to the chapter featuring gambrel roof construction and follow the techniques listed there. Mix and match the plans included in this chapter until you're satisfied that your shed is adequate for your storage needs—and that it is an attractive addition to your backyard.

Before You Begin

The storage buildings featured here can be constructed in a week-end, but take some time before you begin to determine which design will best meet your needs. Consider the amount of time, money, and effort you want to put into your project. A small-scale project will use fewer materials, but will it handle your storage needs?

Before you buy that first piece of lumber, take an inventory of all the items you plan to store in your shed. The size and weight of all those tools and lawn equipment will determine not only the design of your structure, but the type of foundation, floor, and roof as well.

Next, consider the location of your shed. Since you want it to be handy, choose a location that is convenient to the areas where the stored items will be used. And, since the structure's orientation to the sun will affect how much heat and light it takes in, you'll want to consider how you'll position it in the yard. Face the long side south to keep the interior temperature down.

Select a site that is flat, firm, and dry so your shed will sit squarely, settle evenly, and drain naturally. Although pier-and-beam or wall-and-footer foundations can be used on a hilly site, putting up your shed will be much simpler if it's on even ground. Avoid soft ground because it could cause excessive or uneven settling. And be sure to select a well-drained spot, or grade it so that any runoff will flow away from the building. Moisture from standing water can be harmful to both your shed and its contents.

Finally, check your local building codes to be sure your building meets the regulations for your area. You may also need to get a build-ing permit before you begin construction.

Gabled-Roof Shed

This small shed can be used for storage, a work area, and play area—or any combination of the three. You might even build one of these in some remote spot for a hunting or fishing cabin! Next to your home, the gables will echo the architecture better than other sorts of sheds. You can even roof and paint the shed to match your house exactly.

A pier-and-beam foundation provides solid footing for both hilly and flat locations—the piers can be cast to compensate for the slope of the ground. The wooden floor keeps your tools and equipment off the ground and away from the ground moisture. As designed, the doors are in the center of one of the long walls, but they can be easily moved to another location. You can also add windows on any wall, as you wish.

10

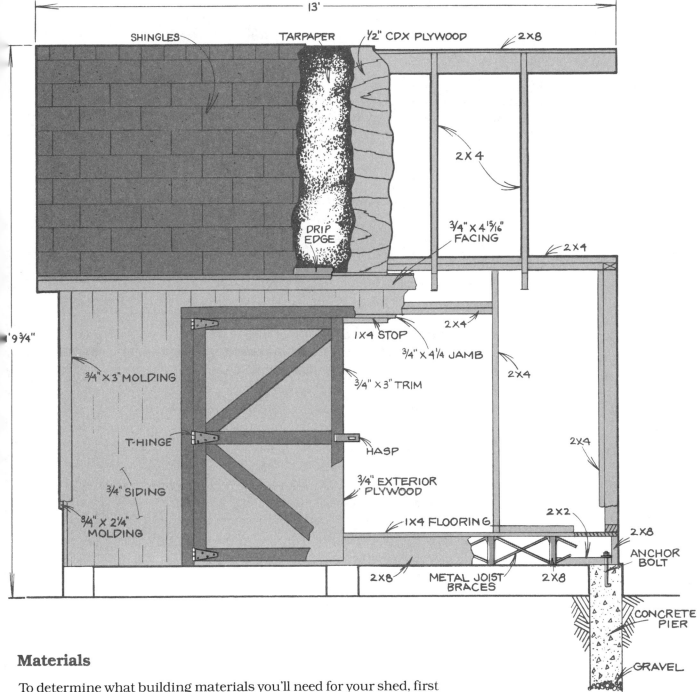

13'

SHINGLES   TARPAPER   ½" CDX PLYWOOD   2×8

2×4

DRIP EDGE

¾" X 4¹⁵⁄₁₆" FACING

2×4

9¾"

¾" X 3" MOLDING

2×4

1X4 STOP

¾" X 4¼ JAMB

2×4

¾" X 3" TRIM

T-HINGE

HASP

2×4

¾" SIDING

¾" EXTERIOR PLYWOOD

¾" X 2¼" MOLDING

1X4 FLOORING

2×2

2×8

ANCHOR BOLT

2×8   METAL JOIST BRACES   2×8

CONCRETE PIER

GRAVEL

SIDE ELEVATION

Materials

To determine what building materials you'll need for your shed, first evaluate what you will be storing in it. The size of the floor joists and the type of flooring you need depends upon the weight of the items you'll be storing. Check the charts on the next page to determine the size of the beams you'll need for your floor joists and the type of flooring your project requires.

As shown, the shed is designed for medium duty. Use pressure-treated 2 x 8 joists and 1 x 4 flooring. (The joists must be chemically treated to be moisture resistant.) You'll also need 2 x 2's for cleats, 2 x 4's for the framing, a 2 x 6 for the ridgeboard, 1 x 6 stock for the facing and molding, and ¾" exterior plywood for the siding and doors.

Of course, you should buy concrete for the piers. The roofing materials needed include drip edge, tarpaper, and shingles. Assemble all these materials with galvanized nails and either brass or stainless steel screws—ordinary nails and screws will rust. As for hardware, purchase some metal floor braces, aluminum vents, 'T' hinges, a hasp, and anchor bolts. If your shed will have windows, purchase glass.

> **TIP** It's best to purchase the window glass *after* the windows have been framed in. Then you'll know exactly how big the window panes should be.

Size of Floor Joists

Size of Beam	Duty	Equipment Stored
2" x 6"	Light	Lawnmower, garden tools
2" x 8"	Medium	Small riding mower, lawn furniture
2" x 10"	Heavy	Large garden implements, lumber

Types of Flooring

Duty	Flooring Type
Light	¾" Exterior plywood
Medium	1 x 4 Tongue-and-groove flooring
Heavy	¾" Exterior plywood over shiplap subflooring (install the subflooring on a diagonal)

Before You Begin

1 Adjust the size of the shed.

As shown in the working drawings, the shed is 8' wide, approximately 12' high (at the peak), and stretches 12' long. However, this may be larger or smaller than you need. If you just want a small storage shed for a few garden implements, reduce all three dimensions. If your intention is to build a workroom or a fishing cabin, make the walls 1'-2' taller, so you don't have to stoop to get through the door. The design can be elongated in either direction simply by adding more studs to the wall frames and stretching the rafters. You can raise the roof by using longer wall studs.

Note: If you stretch the rafters beyond 8' long, use 2 x 6's for the rafters and a 2 x 8 for the ridgeboard. Or: build 2 x 4 trusses to frame the roof.

Because this shed is attached to a permanent foundation, it will likely be affected by building codes. Check your local codes and pay particular attention to regulations governing the location of out-buildings. You'll probably find that, unless you apply for a variance, you have to locate this shed several yards back from your property line. If needed, secure a building permit before you start work.

2 **Check the building codes.**

Setting the Piers

Use stakes and a string to lay out the locations for your piers. (See Figure 1.) Locate the center on each pier along the 12′ walls exactly 69¾″ apart, as shown in the *Pier Layout* drawing. Mark the locations with stakes, then remove the string while you dig the holes.

3 **Lay out the piers.**

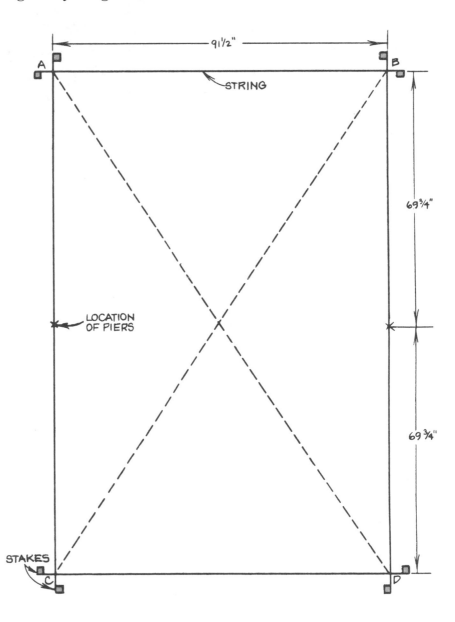

Figure 1. Use string and stakes to lay out the foundation of your shed, then mark the locations of each pier by measuring along the string and driving more stakes where the center of each pier will be. Check that the foundation is square by measuring diagonally from corner to corner. AD must equal BC.

4   Pour the piers.

Dig the holes for the piers 24″ to 36″ deep or below the frost line for your area—you can find out where the frost line is from your local building inspections department. Set 8″ diameter round cardboard forms in the holes. (You can also use 8″ stovepipe for forms.) The tops of the forms should be at least 8″ above ground level. Put the string back up, and level the string with a string level. Then adjust the tops of the forms flush to the string, so that the piers will be level with each other. Once you're sure the forms are level, mix concrete and pour it into the form.

TIP   Before you pour the concrete, throw 2″-3″ of gravel into the bottom of each form. This will help to drain ground water away from the piers.

5   Set the anchor bolts.

Before the concrete cures, position anchor bolts in the center of each pier. (See Figure 2.) Use the string to locate the precise location of the bolts. The tops of the bolts should protrude 2½″. Wait at least 24 hours for the concrete to cure, then remove the stovepipe forms.

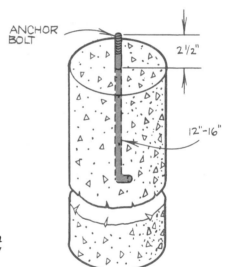

Figure 2. Set the anchor bolt in the pier so that it sticks up 2½″ from the pier.

TIP   To hold the anchor bolts at the proper height, drive two small stakes on either side of the piers. Wrap a wire around the end of the bolt, then wrap the ends of the wire around the stakes so that the bolt is suspended in the wet concrete.

Building the Floor

6   Construct the floor frame.

Cut the floor frame beams and joists from 2 x 8 stock. Nail the beams (the outside frame members) together first with 16d nails, then assemble the floor joists (the inside members) to the frame. Place the joists exactly 16″ on center, as shown in the *Floor Frame Layout* drawing.

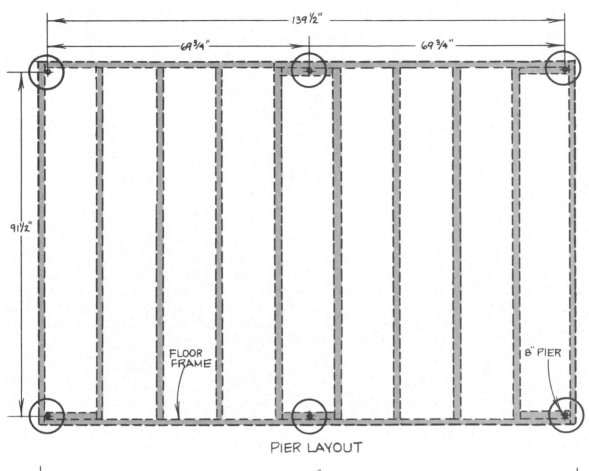

139½"

69¾" 69¾"

91½"

FLOOR FRAME

8" PIER

PIER LAYOUT

144"

16" 16" 16" 16" 16" 16" 16" 16" 16"

2×8

96"

CLEATS

FLOOR FRAME LAYOUT

7 Attach the floor frame to the piers.

Cut 2 x 2's for cleats and drill slightly oversized holes in the center of each cleat so they will fit over the anchor bolts with some slop. This slop will let you adjust the cleat position slightly. Put the cleats in place on the piers. Then set the floor frame in place around the cleats. Nail the cleats to the frame with 16d nails. (See Figure 3.) Check the frame to be sure it is level and square. Secure the cleats to the piers with fender washers and nuts.

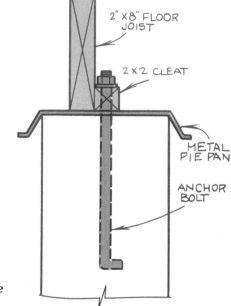

Figure 3. The floor frame is attached to the pier by cleats. Bolt the cleats to piers first, then nail them to the floor frame. If you wish, place an aluminum pie pan over the anchor bolts, in between the pier and the frame, to protect the shed from moisture and pests.

TIP To protect the floor frame from moisture and pests, drill a hole in the center of an aluminum pie pan and slip it upside down over the anchor bolt, between the cleat and the pier.

8 Add bracing to strengthen the floor.

Install metal braces diagonally between the joists to evenly distribute the load over the entire floor. Attach the braces so they form an X pattern between the joists. (See Figure 4.) Nail the bracing to the joists using 6d nails.

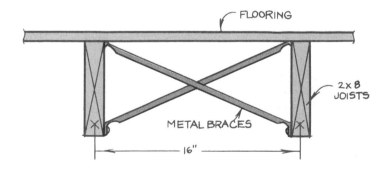

Figure 4. Use metal braces to evenly distribute the load over all the floor joists. Run them between the joists in an X pattern and nail them in place.

Lay 1 x 4 tongue-and-groove flooring across the floor frame, perpendicular to the joists. Beginning at one end of the frame, nail each board to the joists with 6d nails, then fit the groove of the next board over the tongue of the board you just nailed to the frame. Be sure the tongue-and-groove joint closes completely, all along the length of the floor board, before you nail each successive board in place. Continue until you have covered the entire floor.

9   **Attach the flooring.**

Putting Up the Walls

Use 2 x 4's to frame the walls. Cut the studs, sole plate and top plates and assemble them as shown in the *Side Wall Frame Layout, Back Wall Frame Layout,* and *Front Wall Frame Layout* drawings. If you want to add a window in the side walls, use the *Side Wall Frame Layout with Optional Window* drawing as a guide. Assemble the studs and plates with 16d nails.

Note: You need only set wall studs every 4' in this small building. The plywood siding will strengthen the walls and adequately support the weight of the roof. However, if you make the building any larger than what we show, or if local codes dictate, you need to place wall studs on 24" centers. The same considerations apply to the window and door openings. It's not necessary to use headers and trimmers to strengthen the frame around each opening—unless you enlarge the building or the codes require it.

10   **Build the wall frames.**

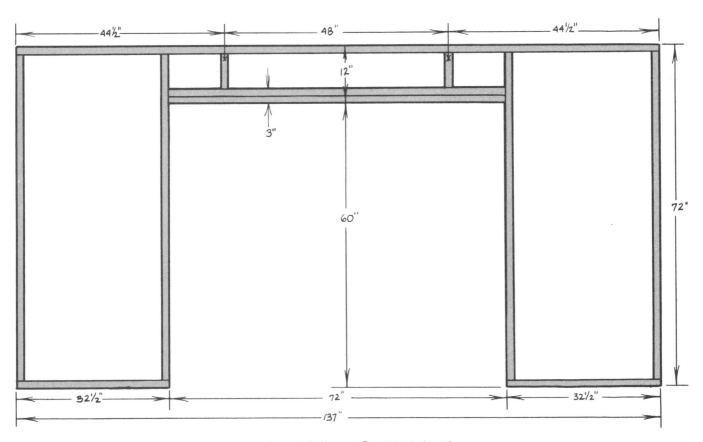

FRONT WALL FRAME LAYOUT

11 Attach the wall frames to the floor.

With the help of a friend, raise the walls into place as shown in the *Floor Layout* drawing. Temporarily brace them upright with scrap lumber. Nail the bottom plates into the floor frame with 16d nails spaced every 16″ to connect the sole plates to the joists. Connect the walls to each other at the corners with 16d nails spaced every 24″ on center. Check with your carpenter's level to be sure the walls are plumb.

12 Attach the top plates.

When all four walls are up, nail another 2 x 4 top plate flat against the existing top plate to tie the walls together. This is sometimes called the 'cap' plate. (See Figure 5.) Be sure the ends of the cap plate overlap those of the top plate.

Figure 5. Nail a second top plate (sometimes called a cap plate) over the first to tie the walls together. Make sure that the ends of the cap plates overlap the joints between the top plates.

BACK WALL FRAME LAYOUT

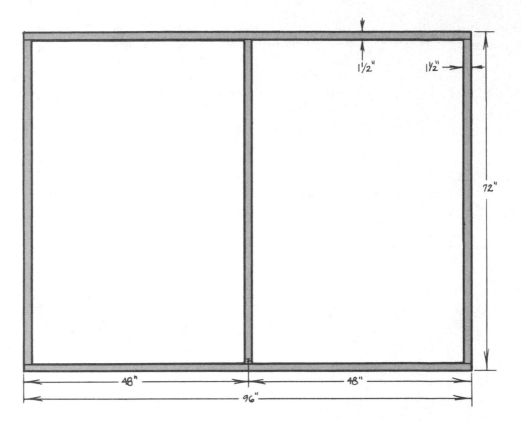

SIDE WALL FRAME LAYOUT

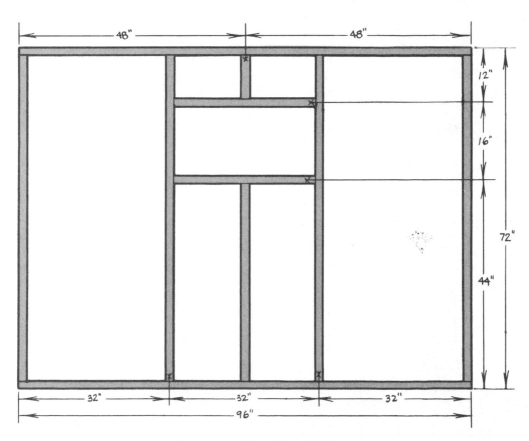

SIDE WALL FRAME LAYOUT
WITH OPTIONAL WINDOW

13 **Attach the siding to the front and back walls.**

Attach siding to the front and back walls *only,* using 6d nails spaced every 12″. This will save you some work later on—you won't have to notch the siding around the rafters. However, before you can put siding on the side walls, you'll have to put up the roof frame. The bottom edge of the siding should be flush with the bottom edge of the floor frame.

TIP Save time and money by using ready-to-finish structural siding such as square edge rough sawn pine that will act as both a wall sheathing and siding. It is available in 4′ x 8′ sheets and will save you the time and expense of installing sheathing and siding.

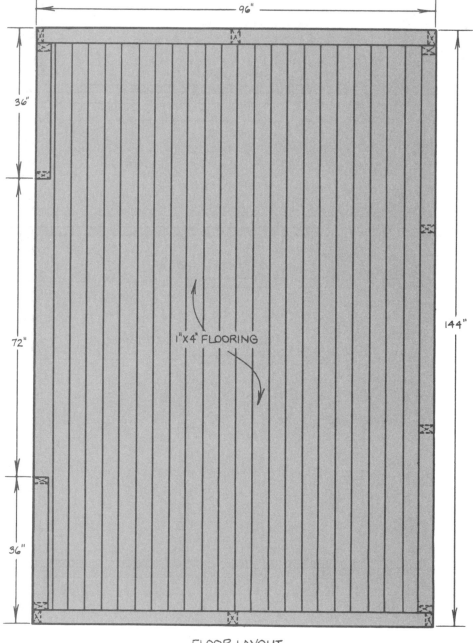

FLOOR LAYOUT

Adding the Roof

Use 2 x 4 stock for the rafters and cut them to the desired length, as shown in the *Rafter Layout* drawing. After you have cut the rafters, notch them to fit over the top plates and miter the ends. Bevel the top edge of the ridgeboard as shown in the *Roof Frame, Side View* drawing.

14 **Cut the rafters and ridgeboard.**

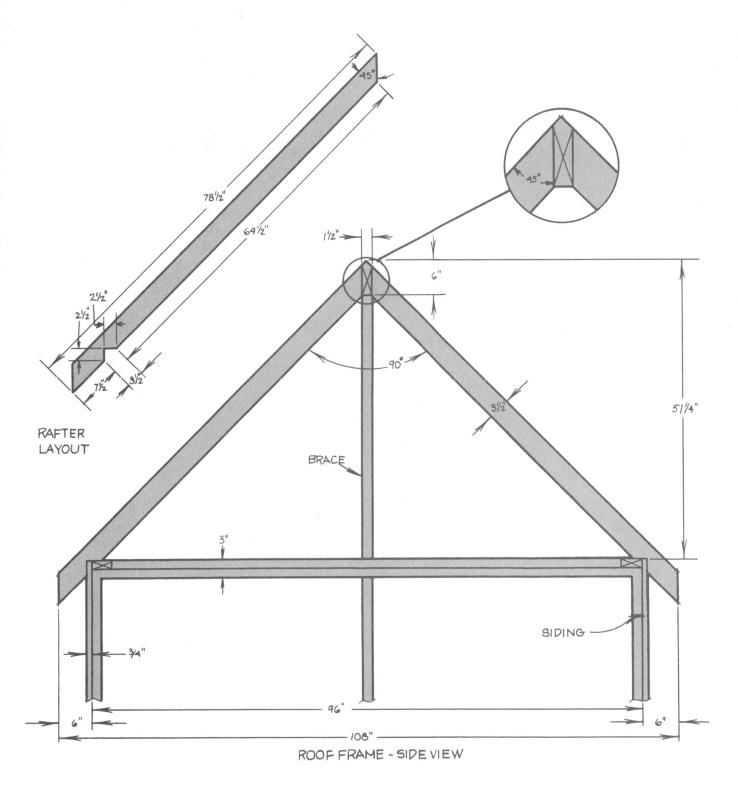

RAFTER LAYOUT

ROOF FRAME - SIDE VIEW

15   Put up the roof frame.

Temporarily, support the ridgeboard above the wall frame. Make temporary supports by nailing 2 x 4's to the side wall frames so that they stick up about 5' above the frames. Clamp (don't nail) the ridge board to these supports. The clamps will allow you to adjust the height of the board, if you need to. Nail the rafters to the ridgeboard and the top plate, every 2' on center, using 16d nails as shown in the *Roof Frame, Front View* drawing. After you attach the rafters, toenail a permanent 2 x 4 support at each gable end, between the ridgeboard and the top plate, as shown in the drawings. Remove the temporary supports.

    Note: Unlike the wall studs, the rafters must be spaced every 2', no matter how small the shed is. If they are spaced too far apart, the roof sheathing will sag.

TIP   Save the time and trouble of notching rafters by using metal rafter ties to secure the rafters in place.

16   Attach the front and back facing strips.

Cut the front and back facing strips and bevel the top edge. (See Figure 6.) Cut the facing strips 13' long—a foot longer than the length of the building. These strips overhang the building 6" on each side. Nail the facing strips to the ends of the rafters with 6d nails.

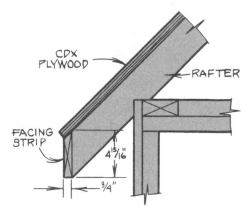

Figure 6. Attach CDX plywood roof sheathing over the rafters, then nail the facing strips along the ends of the rafters. Bevel the top edge of the facing strips at 45°.

17   Install the roof sheathing.

Place ½" CDX plywood roof sheathing on the rafters, making sure that it extends 6" from the front and back of the shed, as shown in the *Roof Frame, Front View* drawing. Nail the sheathing to the rafters using 6d nails spaced every 6" along the edges and 12" on center in the sheets.

18   Attach the side facing strips.

Cut the side facing strip pieces. Miter the top of each strip, as shown in the *Side Facing Detail* drawing. Connect the facing strips together at the peak with a metal truss plate attached on the inside. Then put the strips in place and screw them to the overhanging sheathing with #12 x 1½" flathead wood screws. (The screws will draw the facing flush to the underside of the sheathing.) Then screw the ends of the front and back facing to the side facing.

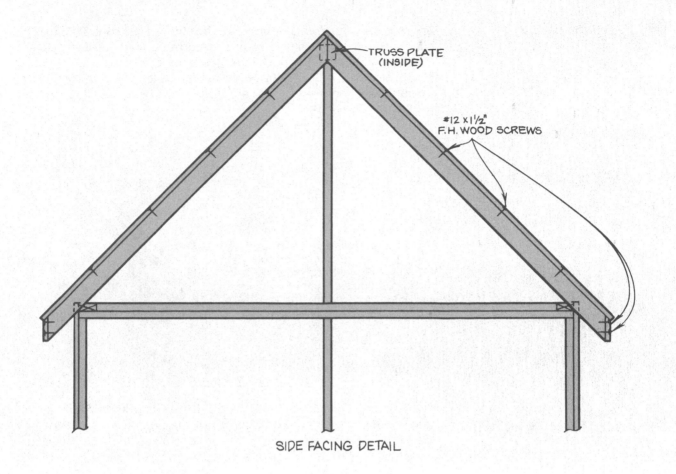

TRUSS PLATE
(INSIDE)

#12 x 1½"
F.H. WOOD SCREWS

SIDE FACING DETAIL

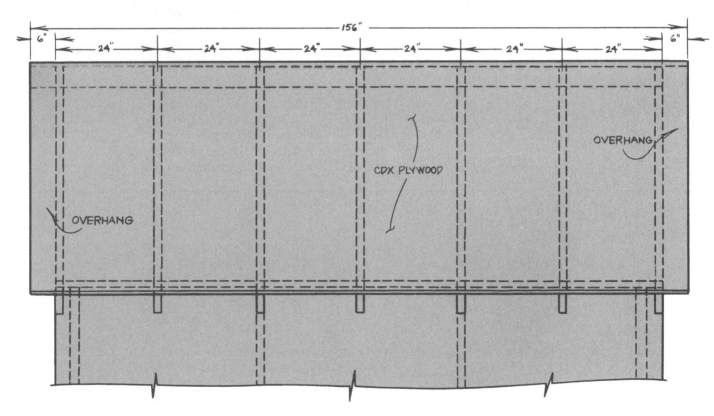

156"

6" 24" 24" 24" 24" 24" 24" 6"

OVERHANG

CDX PLYWOOD

OVERHANG

ROOF FRAME, FRONT VIEW

19 **Install the roofing materials.**

Run the metal drip edge around all the sides of the roof. This drip edge will keep the rain water from collecting under the shingles at the edge of the roof and possibly rotting out the sheathing. Then cover the entire roof with tarpaper and shingles. (See Figure 7.) Put a double layer of tarpaper and shingles at the peak to keep your roof from leaking.

Figure 7. Attach metal drip edge to the roof sheathing. Then cover the sheathing with tarpaper (adding a double layer at the eaves and top) and install the shingles.

SHINGLES

TARPAPER

DRIP EDGE
FACING

SIDING

Finishing Up

20 **Attach the siding to the side walls.**

Nail plywood siding to the side walls using 6d nails placed 6″ on center along the edges and 12″ on center in the sheets. Make sure the siding goes down to the base of the floor frame. Attach the siding at the gable ends to the rafters, the gable supports, and the ends of the ridgeboard.

Note: As shown in the *Corner Molding Detail* drawing, the siding does *not* overlap at the corners.

21 **Attach the corner molding.**

Rip the corner molding stock to size. Nail this molding to the corners, as shown in the *Corner Molding Detail* drawing, with 6d nails. Miter and notch the top ends of the corner molding to fit around the rafters and butt flush against the roof sheathing.

Option: You may want to forego the corner molding altogether, and install a simple 1 x 1 corner block at each corner.

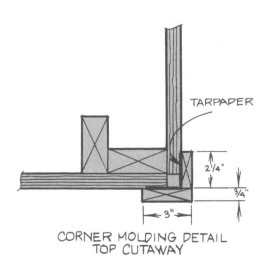

TARPAPER

2¼"

¾"

3"

CORNER MOLDING DETAIL
TOP CUTAWAY

Rip the door jamb, door molding, and door stop stock to size. Staple a 6″ wide strip of tarpaper to the frame around the door opening, lapping the siding. This will help prevent the siding from being damaged by moisture that might collect under the jamb. Install the top part of the door jamb first, then the sides, using 6d nails. This jamb must cover the door frame studs *and* the siding, as shown in the *Door Jamb Detail* drawing. Nail the door stop to the jamb, and nail the door molding to the siding so that it overlaps the jamb. Caulk all around the molding.

Option: You may want to build a short ramp for smooth access into the door of your shed. Do this by strapping planks of 2 X 4's together and nailing across their undersides.

22 **Install the door jamb and door molding.**

If your shed has one or more windows, rip the stock for the window casing, window sill, and window molding. Staple a 6″ wide strip of tarpaper to the frame around the window opening, lapping the siding. Install the window sill over the tarpaper with 6d nails. This sill overlaps the window frame studs and the siding, and protrudes 1¼″ beyond the siding, as shown in the *Window Casing Detail* drawing. Next install the top part of the window casing, then the sides parts. Like the door jamb, the window casing overlaps the siding, but does not protrude. Finally trim around the window with molding, as shown in the *Window Molding Detail* drawing. Caulk all around the molding.

23 **Install the window casing and window molding.**

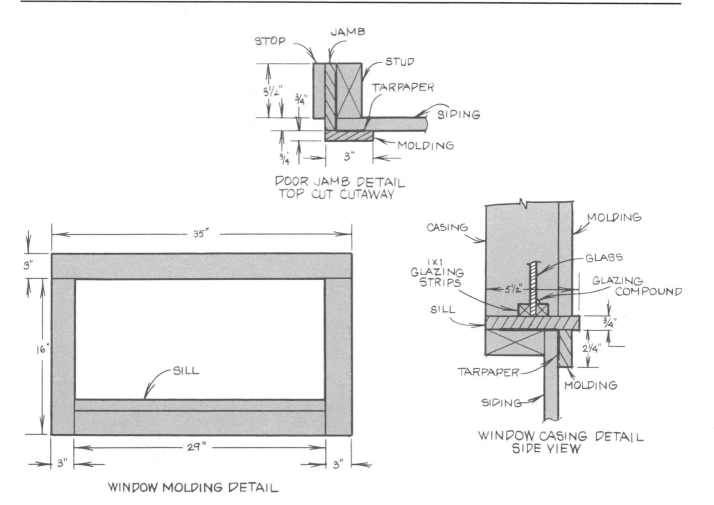

DOOR JAMB DETAIL
TOP CUT CUTAWAY

WINDOW MOLDING DETAIL

WINDOW CASING DETAIL
SIDE VIEW

24 Build the doors.

To build the doors, cut two pieces of plywood siding 59″ long and 35⅛″ wide, as shown in the *Door Layout* drawing. Attach the ¾″ x 3″ trim to the outside face of the plywood using #12 x 1¼″ flathead wood screws. Drive these screws in from the *back* side of the door. Create a 'barn door' pattern with the trim, as shown in the drawing.

---

25 Hang the doors.

Mount three T-hinges on each door, bolting the strap part of the 'T' to the door trim. Use bolts rather than screws to attach the hinges, to prevent a thief from dismounting the doors. These bolts should pass through the doors, with the washers and nuts on the inside. After you tighten the nuts, mash the threads of the ends of the bolts with a hammer. This will make it impossible to remove the hinges from the outside. After you mount the hinges to the doors, flop the butt parts of the 'T' over so that the hinges turn the corner around the edge of the door. With a friend, hold the doors in place so that the butt part of the hinges is flat against the door jamb. Mark the location of each hinge, and set the doors aside. Chisel out shallow recesses in the door jambs to accommodate the pins, then put the doors back in place, opening them part way so you can reach the butt part of the hinges. Screw the hinges to the door jamb with 2″ long flathead wood screws. (See Figure 8.) After hanging the doors, install a hasp so you can lock them.

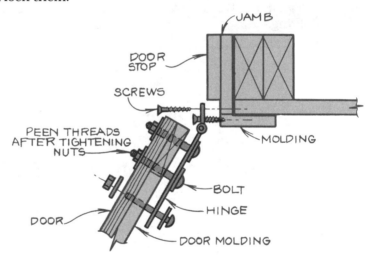

Figure 8. Hang the doors on T-hinges, as shown. Bolt the strap part of the hinge to the outside of the door, and mash the threads so you can back the nuts off. Screw the butt part of the hinge to the door jamb, so that the screw heads will be hidden when the door is closed. These precautions will help to thwart intruders.

---

26 Install the windows.

If your shed has a window or two, nail a row of 1 x 1 glazing strips all around the inside of the window casing with 4d finishing nails. Put the glass in place against these strips, and add a second row of glazing strips on the other side of the glass. Caulk all around the outside of the window with glazing compound to waterproof the seam between the glass and the outside glazing strip.

Install aluminum vents near the peak of the gable ends and near the bottom of the sides. This will keep air moving through the shed and prevent moisture from condensing on the contents. To install a vent, cut an opening in the plywood siding with a saber saw. Be careful to work around the 2 x 4 frame; don't cut through it. Put some caulk around the opening, and press the vent in place. Screw the flanges of the vent to the plywood siding.

27 **Install the vents.**

Paint or stain the shed to match—or contrast with—your home. Prime the raw wood first, then cover with two coats of paint. As designed, the shed should require about 2 gallons each of primer and paint.

28 **Paint the exposed wood surfaces.**

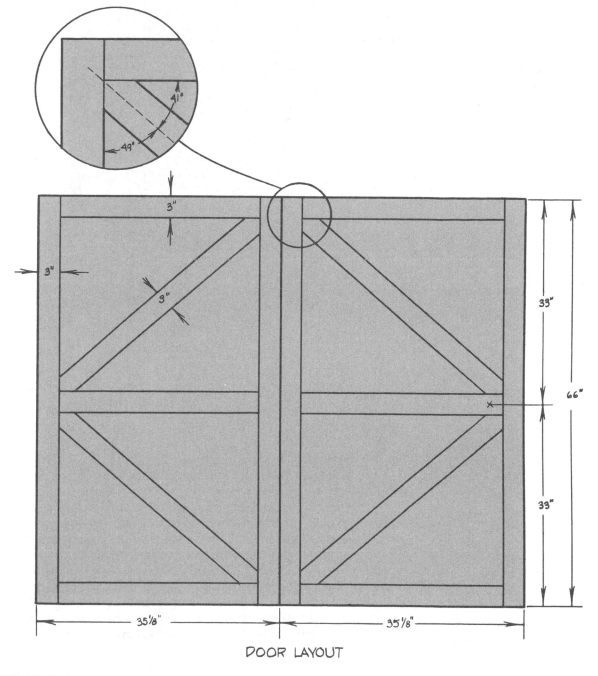

DOOR LAYOUT

Slant-Roof Shed

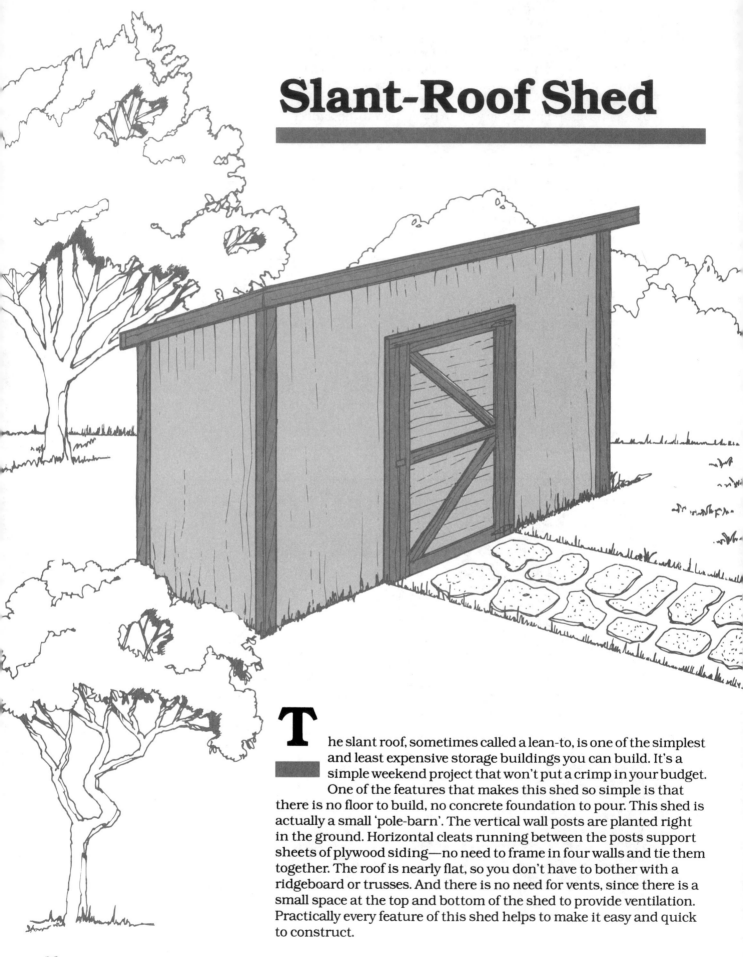

The slant roof, sometimes called a lean-to, is one of the simplest and least expensive storage buildings you can build. It's a simple weekend project that won't put a crimp in your budget. One of the features that makes this shed so simple is that there is no floor to build, no concrete foundation to pour. This shed is actually a small 'pole-barn'. The vertical wall posts are planted right in the ground. Horizontal cleats running between the posts support sheets of plywood siding—no need to frame in four walls and tie them together. The roof is nearly flat, so you don't have to bother with a ridgeboard or trusses. And there is no need for vents, since there is a small space at the top and bottom of the shed to provide ventilation. Practically every feature of this shed helps to make it easy and quick to construct.

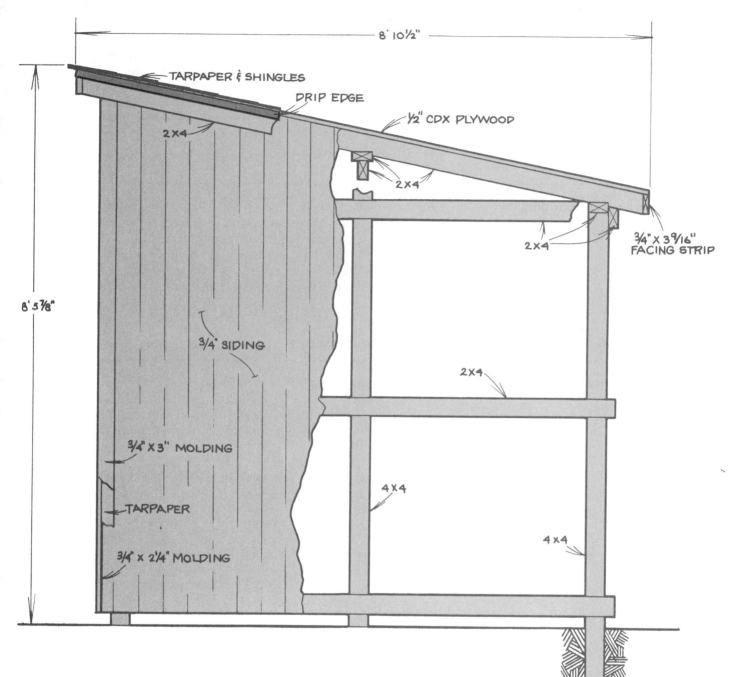

TARPAPER & SHINGLES

DRIP EDGE

½" CDX PLYWOOD

2×4

2×4

2×4

¾" X 3⁹⁄₁₆" FACING STRIP

8' 10½"

8' 5⅞"

¾" SIDING

2×4

¾" X 3" MOLDING

4 X 4

TARPAPER

4 X 4

¾" X 2¼" MOLDING

SIDE ELEVATION

Materials

Use pressure-treated 4 x 4 posts for the vertical wall members. Buy these posts at least 1′ longer than you will actually need, to allow yourself extra stock to 'level' them once they have been set.

Use 2 x 4's for the horizontal frame members and rafters, ¾″ thick plywood siding (in 4′ x 8′ sheets) to cover the walls, and ½″ thick CDX plywood to sheath the roof. You'll also need some 'one-by' (¾″ thick) stock to make the facing strips, corner moldings, door jamb and door trim. Since this shed has a dirt floor, the inside will probably be wetter than other designs. Buy pressure-treated or rot-resistant lumber for all frame parts.

In addition to these materials, you'll have to purchase galvanized nails, roofing nails, tarpaper, shingles or roofing felt, roofing cement, and drip edge. For the doors, you'll need T-hinges and a hasp.

Before You Begin

1 **Adjust the size of the shed.**

As shown in the working drawings, the shed is 8' wide, approximately 9' high, and stretches 12' long. However, this may be larger or smaller than you need. You can raise the roof by lengthening the wall posts. The length and width can be adjusted by adding or subtracting posts. However, if you make the building more than 8' wide, use 2 x 6's for the rafters. If you make it more than 12', use 2 x 8's. Otherwise, the roof may sag.

2 **Check the building codes.**

Because this shed is planted permanently in the ground, it will likely be affected by building codes. Check your local codes and pay particular attention to regulations governing the location of outbuildings. You'll probably find that, unless you apply for a variance, you'll have to locate this shed several yards back from your property line. If needed, secure a building permit before you start work.

Setting the Posts

3 **Lay out the posts.**

Use stakes and string to mark the locations of your posts. For this shed, you'll need to plant 10 posts, one every 4'. (See Figure 1.) Follow the *Post Layout* drawing to determine the exact location of your posts. If you have changed the dimensions of the shed, remember the posts must still be spaced every 4' to give the structure adequate support. Mark the location of the posts with stakes, then dig the holes 24"-36" deep. The holes must be deeper than the frost line for your area, and at least twice as wide as your posts. This will allow space for packing gravel and dirt around the posts.

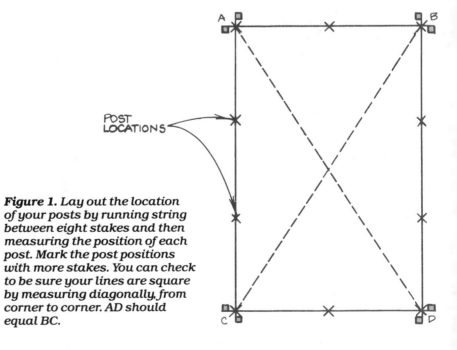

Figure 1. Lay out the location of your posts by running string between eight stakes and then measuring the position of each post. Mark the post positions with more stakes. You can check to be sure your lines are square by measuring diagonally, from corner to corner. AD should equal BC.

Place a large rock in the base of each hole to keep the posts from settling. These rocks should be twice the diameter of the posts—at least 8″ in diameter. (See Figure 2.) With the help of a friend, raise the posts in place, then shovel gravel in the hole to a depth of 12″. This will help drain the ground water away from the posts. Finally, fill the rest of the hole with dirt and tamp lightly around the post.

4 **Set the posts in the ground.**

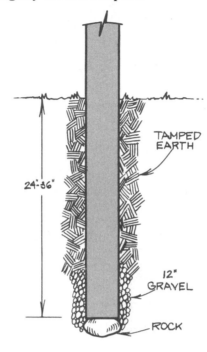

Figure 2. Set the posts at least 2′ into the ground or below the frost line in your area. Plant the posts on a large rock and surrounded by a gravel base to provide drainage.

TIP Don't tamp the earth completely until after you have aligned the posts with a level and braced them so that they are perfectly straight up and down.

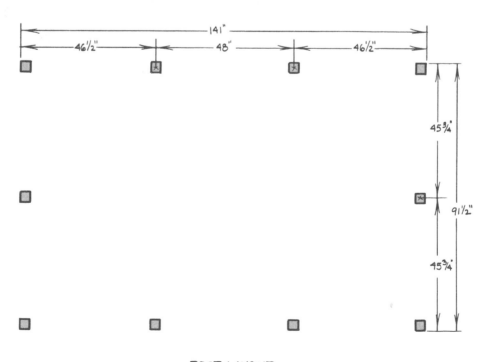

POST LAYOUT

5 **Cut the posts to the proper height.**

Once the posts are in the ground, you should brace them upright by driving stakes about 3' away and nailing scrap lumber from the post to the stakes. You must use at least two braces per post, and these must be at right angles to each other. Use a carpenter's level to make sure they are plumb. When the posts have been braced, cut them to the proper height. To do this, measure one corner post on the front side of the building and mark the top 94½" above the ground. Mark a back corner post 76½" above the ground. Using these marks as references, stretch a string along the front and back walls. With a string level, find the tops of the other front and back posts. Then stretch string between the front and back corner posts to find the tops of the middle posts in the side walls. (See Figure 3.) Remove the strings and cut the posts off with a handsaw. Check again that the posts are still plumb, and tamp the dirt down as tight as you can.

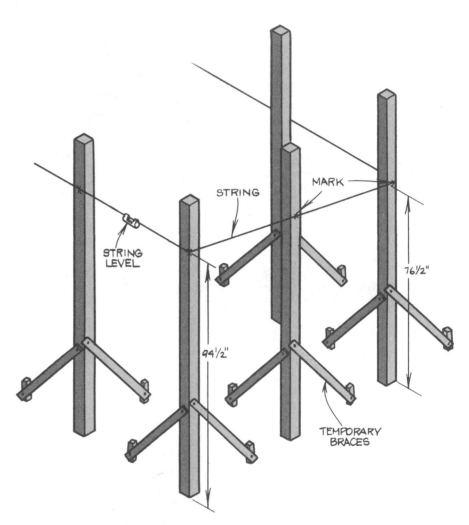

Figure 3. Use a string and string level to mark the cutting position for each post. Remember, the front posts will be higher than the back to allow for the slant of the roof.

TIP   Be sure that the lowest edge of the roof is oriented toward the prevailing weather so that rain and snow will drain off the roof faster. This will lengthen the life of your shed.

Building the Slant-Roof

Cut top plates from 2 x 4 stock to stretch across the tops of the wall posts, from side to side. The top plates must overhang the outside post by 1½″ on each end. The front and back top plates will be supported in the middle of their run by wall posts. However, the middle top plate doesn't have any support along the run. To keep it from sagging, nail the flat side of the middle top plate to the edge of another 2 x 4 to make a 'T' shaped beam, as shown in the *Middle Top Plate Detail* drawing. This reinforcing 2 x 4 should be 10″ shorter than the top plate and centered so that the top plate overhangs it 5″ on each side. That way, the reinforcing member will butt up against the wall posts when the top plate is nailed in place. Attach the top plates to the wall posts with 16d nails, as shown in the *Roof Frame, Side View* drawing.

6 **Attach the top plates.**

Cut the 2 x 4 rafters as shown in the *Rafter Layout* drawing. Cut nine rafters, but only notch seven. The two un-notched rafters will be used as side facing strips. You can save time by cutting and notching one rafter and then checking its fit to the top plates. If it fits properly, use it as a template for cutting the others.

7 **Cut the rafters.**

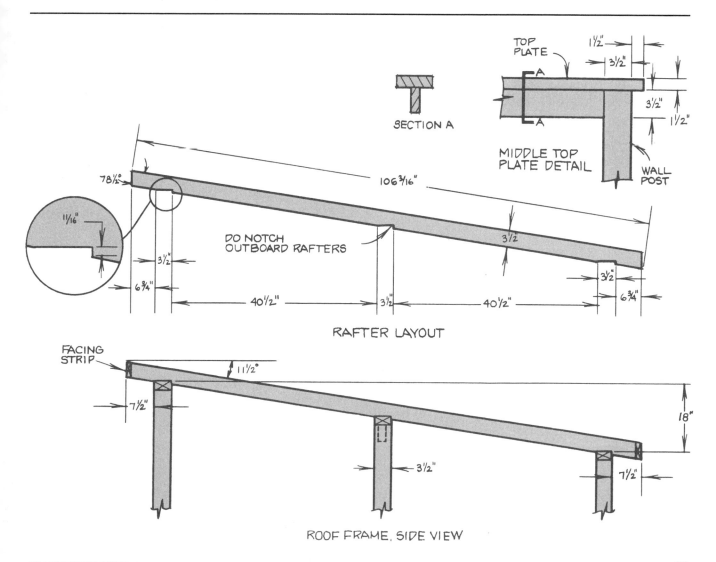

SECTION A

MIDDLE TOP PLATE DETAIL

RAFTER LAYOUT

ROOF FRAME, SIDE VIEW

8 **Attach the rafters to the top plates.**

Carefully measure along the top plates, marking where the rafters will go. They should be spaced every 24″ on center as shown in the *Roof Frame, Top View* drawing. Nail the rafters in place using 16d nails.

　　Option: Instead of notching the rafters, use metal rafter ties to secure the rafters in place.

---

9 **Attach the facing strips.**

Cut front and back facing strips and rip-bevel the edges. (See Figure 4.) Attach these facing strips to the ends of the rafters with 6d nails. The front and back facing strips should be 12″ longer than the top plates, and they should protrude beyond the end rafters 6″ on each side, as shown in the drawings. After you've attached the front and back facing strips, nail the side facing strips to them.

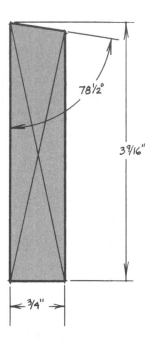

Figure 4. *Bevel the top edges of the front and back facing strips to match the slope of the roof.*

TIP　Drill the nail holes in the facing strips to prevent the strips from splitting.

---

10 **Cover the roof frame with sheathing.**

Attach sheets of ½″ CDX (exterior) plywood sheathing to the roof frame with 6d nails. If you're not going to install a ceiling, be sure to turn the good side down.

Nail metal drip edge to the eaves. If you use drip edge around the entire roof, seal it with roofing cement to keep the wind from blowing in under it. Next, cover the entire roof with a double layer of tarpaper. Install the shingles according to the manufacturer's instructions. (See Figure 5.) Seal around the entire edge of the roof with roofing cement so that no water can penetrate between the drip edge, tarpaper, or the shingles.

Option: If your weather often comes from more than one direction, you may want to forego the drip edge on the highest eave. Instead cover the eave, including the facing board, with tarpaper and shingles. Double wrap the shingles as if the eave were a roof peak.

11 **Install the roofing materials.**

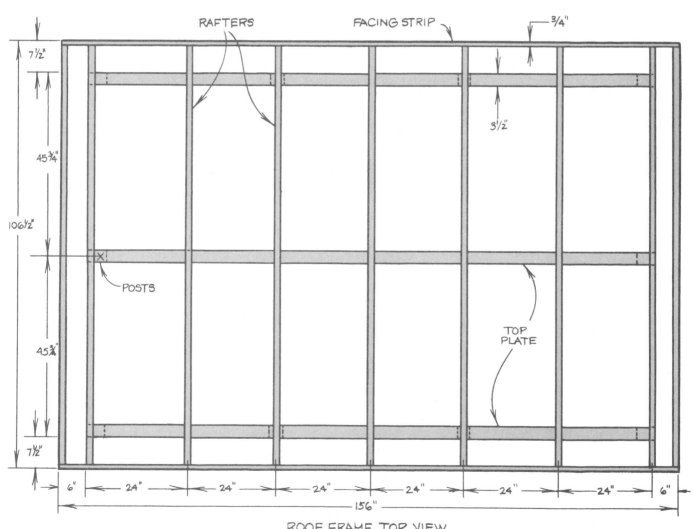

SHINGLES

TARPAPER

DRIP EDGE

Figure 5. *Attach metal drip edge to the roof sheathing on all four sides. Then cover the sheathing with a double layer of tarpaper and apply the shingles.*

RAFTERS   FACING STRIP   3/4"

7 1/2"

3 1/2"

45 3/4"

POSTS

106 1/2"

45 3/4"

TOP PLATE

7 1/2"

6"   24"   24"   24"   24"   24"   24"   6"

156"

ROOF FRAME, TOP VIEW

12 **Attach the cleats to the posts.**

Cut 2 x 4 stock to make the horizontal cleats that will support the wall siding. Place the cleats as shown on the *Front Wall Frame, Back Wall Frame,* and *Side Wall Frame* drawings. Make sure the bottom cleats are at least 2″ above the ground so they won't mildew and rot. Nail 2 x 4's to the posts with 16d nails.

13 **Put up the siding.**

Nail the plywood siding vertically to the cleats with 6d nails spaced every 6″ along the edges and 12″ in the field. Make sure the bottom edge of the siding is flush with the bottom edge of the bottom cleat, but that it doesn't touch the ground. Moisture from the ground can cause the wood to rot. On the front and back walls, the siding should butt up against the rafters, but it shouldn't be notched around them. On the side walls, the siding laps the end rafters and is flush to the roof. (See Figure 6.) This arrangement provides vents at the top and bottom of the shed.

Option: You can also side this shed with aluminum siding. Follow the manufacturer's directions to install it.

Figure 6. Don't notch the siding to fit against the rafters on the front and back walls, instead leave an open space to provide ventilation for your shed. Fit the siding flush against the roof on the sides of the shed.

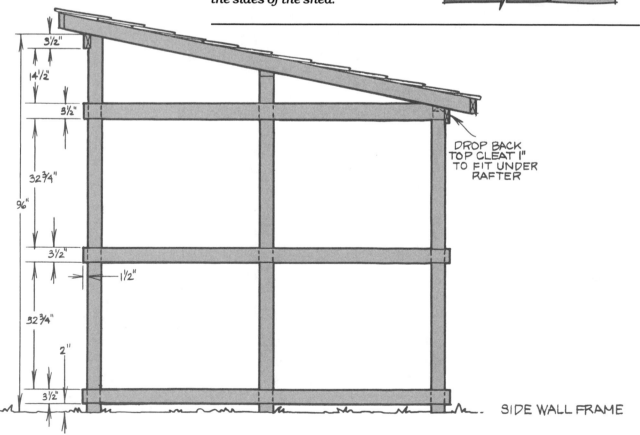

SIDE WALL FRAME

STORAGE BUILDINGS

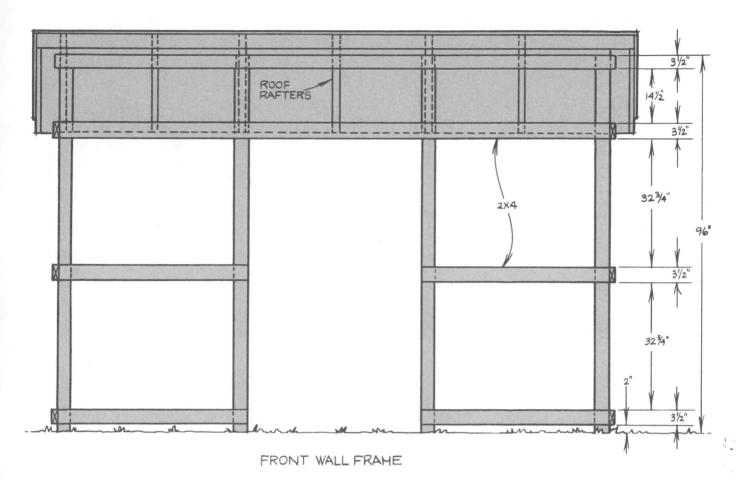

ROOF RAFTERS

2×4

3½"

14½"

3½"

32¾"

3½"

32¾"

2"

3½"

96"

FRONT WALL FRAME

2×4

3½"

1⅛"

32¹⁄₁₆"

3½"

32¾"

2"

3½"

78"

BACK WALL FRAME

14 Attach the corner moldings.

Cut the molding from 'one-by' stock. Bevel and miter the upper ends of the corner molding so they fit flush against the roof sheathing, and notch them to fit around the rafters. (See Figure 7.) Attach the molding as shown in the *Corner Joinery Detail* drawing, using 6d nails.

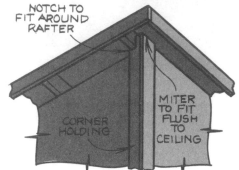

Figure 7. Miter, bevel, and notch the ends of the corner molding to fit flush to the ceiling and to fit around the end rafters.

Finishing Up

15 Install the door jamb and door molding.

Rip the door jamb, door molding, and door stop stock to size. Staple a 6″ wide strip of tarpaper to the frame around the door opening, lapping the siding. This will help prevent the siding from being damaged by moisture that might collect under the jamb. Install the top part of the door jamb first, then the sides, using 6d nails. This jamb must cover the door frame studs *and* the siding, as shown in the *Door Jamb Detail* drawing. Nail the door stop to the jamb, and nail the door molding to the siding so that it overlaps the jamb. Caulk all around the molding.

16 Build the door.

To make the door, cut a piece of plywood siding 72⅜″ long and 42¾″ wide, as shown in the *Door Layout* drawing. Attach the 1″ x 3″ trim to the outside face of the plywood using #12 x 1¼″ flathead wood screws. Drive these screws in from the *back* side of the door. Create a 'barn door' pattern with the trim, as shown in the drawing.
  Option: Make two doors, each 21¼″ wide. The smaller doors don't get in the way as much and they put less strain on the hinges.

TIP You can use both glue and screws to attach the trim pieces to your door so they won't work loose as the years go by. Use an exterior construction adhesive.

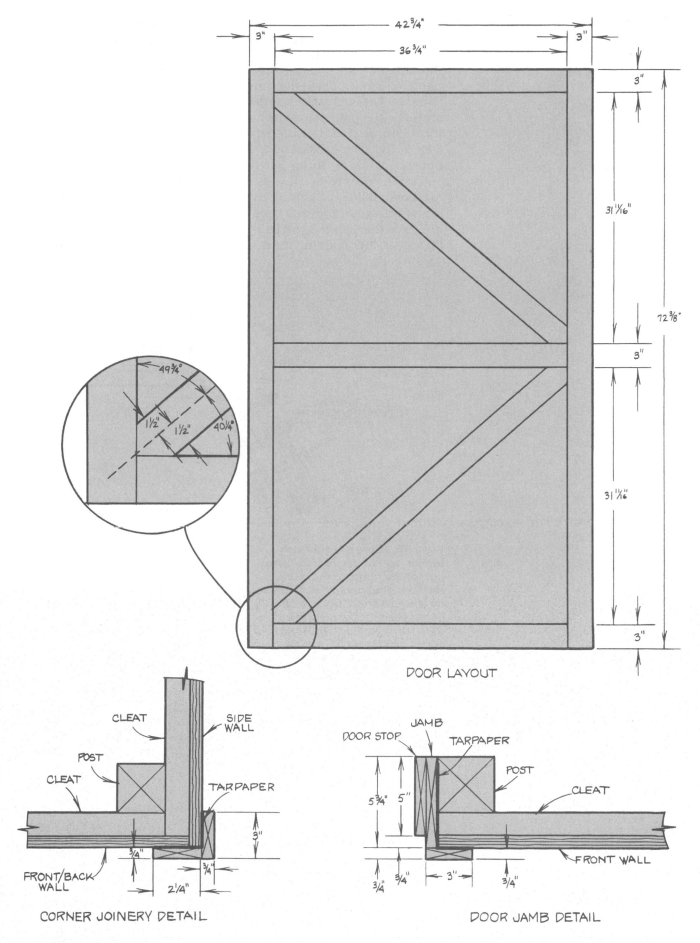

42¾"

36¾"

3"

3"

3"

31¹¹⁄₁₆"

72⅜"

3"

31¹¹⁄₁₆"

3"

49¾°

1½"

1½"

40¼°

DOOR LAYOUT

CLEAT

SIDE WALL

POST

CLEAT

TARPAPER

3"

¾"

¾"

FRONT/BACK WALL

2¼"

CORNER JOINERY DETAIL

DOOR STOP

JAMB

TARPAPER

POST

CLEAT

5¾"

5"

¾"

¾"

3"

¾"

FRONT WALL

DOOR JAMB DETAIL

17 Hang the door.

Mount three T-hinges on the door, bolting the strap part of the 'T' to the door trim. Use bolts rather than screws to attach the hinges, to prevent a thief from dismounting the door. These bolts should pass through the door, with the washers and nuts on the inside. After you tighten the nuts, mash the threads of the ends of the bolts with a hammer. This will make it impossible to remove the hinges from the outside. After you mount the hinges to the door, flop the butt parts of the 'T' over so that the hinges turn the corner around the edge of the door. With a friend, hold the door in place so that the butt part of the hinges is flat against the door jamb. Mark the location of each hinge, and set the door aside. Chisel out shallow recesses in the door jambs to accommodate the pins, then put the door back in place, opening it part way so you can reach the butt part of the hinges. Screw the hinges to the door jamb with 2″ long flathead wood screws. (See Figure 8.) After hanging the door, install a hasp so you can lock it.

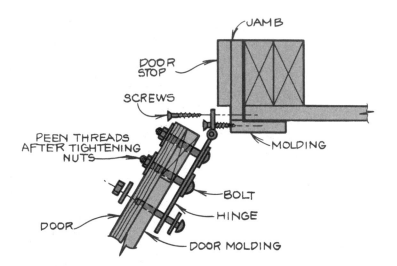

Figure 8. Hang the doors on T-hinges, as shown. Bolt the strap part of the hinge to the outside of the door, and mash the threads so you can back the nuts off. Screw the butt part of the hinge to the door jamb, so that the screw heads will be hidden when the door is closed. These precautions will help to thwart intruders.

Paint or stain the shed to match your home, or to contrast with it. Use exterior paint or stain to protect the wood from rotting.

18 **Paint or stain all exposed wood surfaces.**

Grade and level the floor as you would in preparing the site for a concrete slab. Remove all debris and rocks. If you want to lay a gravel floor, after you have graded and leveled the area, spread a base of crushed rock or stone, checking it with a string level as you go. Then spread small-size "pea" gravel over the base and tamp it down until the surface is well compacted.

19 **Grade and level the floor.**

TIP To keep critters out of your storage building, dig a trench between the poles and put aluminum screen between the ground and the siding.

Storage Barn

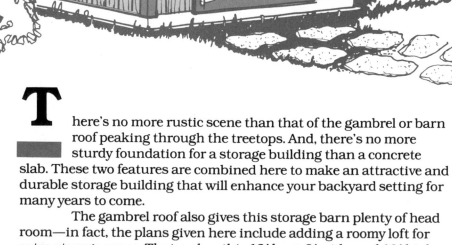

There's no more rustic scene than that of the gambrel or barn roof peaking through the treetops. And, there's no more sturdy foundation for a storage building than a concrete slab. These two features are combined here to make an attractive and durable storage building that will enhance your backyard setting for many years to come.

The gambrel roof also gives this storage barn plenty of head room—in fact, the plans given here include adding a roomy loft for extra storage space. That makes this 12′ long, 8′ wide, and 12′ high shed large enough and sturdy enough to store garden tractors, lumber, mowers, patio furniture, and much more. Paint the wood and roof the shed to match—or contrast with—your home.

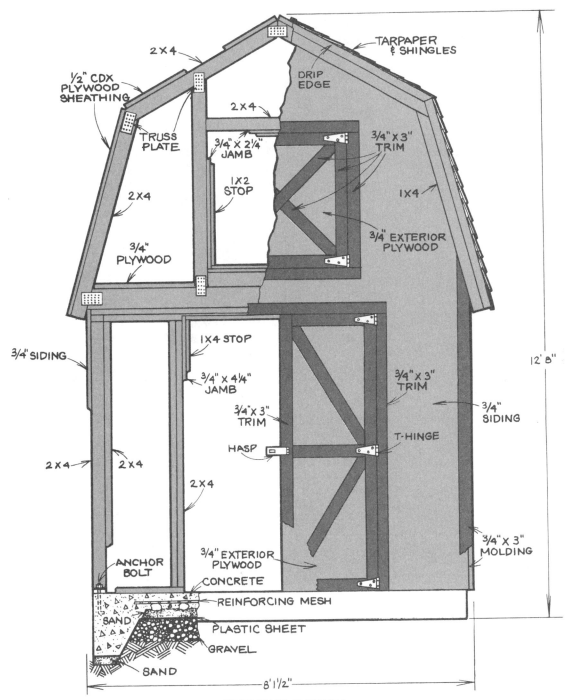

FRONT ELEVATION

Materials

To build the storage barn as shown, use 2 x 4's for the studs, sole plates, top plates, and rafters; and 2 x 6's for the ceiling joists. Cover the roof with ½" CDX plywood, and cut the facing strips from 1 x 4 stock. Cover the walls with ¾" exterior plywood siding, or another ¾" thick siding of your choice. If you add a loft to your shed, you'll need ¾" plywood for the flooring.

For the concrete slab, you'll need 2 x 8's for the forms, pea-size gravel, plastic sheeting (for the vapor barrier), sand, 6" x 6" reinforcing mesh, anchor bolts, and—of course—concrete. As shown, this 4" thick, 8' x 12' slab, with footings 12" below grade, will require 2⅓ yards of concrete.

In addition to these materials, you'll also have to purchase galvanized nails, brass or stainless steel screws, tarpaper, shingles, drip edge, T-hinges, and a hasp.

Before You Begin

1 Determine the thickness of the slab your need.

Determine how thick your concrete slab should be and whether you'll need footers by first considering the equipment you plan to store on it. If your inventory includes a light lawnmower and small garden tools, a 4″ slab will be ample for your needs. However, if you own, or plan to own, large garden equipment, you'll need the more solid base that a 6″ slab and footers will provide. This chart can help you decide how thick to pour your slab and whether you'll need footers.

Thickness of Concrete Slab

Floor Duty	Equipment Stored	Footers	Slab Thickness
Light	Lawnmower, small garden tools	No	4″
Medium	Small riding mower, patio furniture	No	4″
Heavy	Small tractor, large garden implements	Yes	6″

Note: You should never pour a foundation slab less than 4 inches thick because it will crack just from settling.

2 Adjust the size of the barn.

As shown in the working drawings, the barn is 8′ wide, approximately 12′ high (at the peak), and stretches 12′ long. However, this may be larger or smaller than you need. If you just want a small storage barn for a few garden implements, reduce all three dimensions. If you need a barn for a tractor or a horse, make the walls 1′-2′ taller and raise the height of the door. The design can be elongated in either direction simply by adding more studs to the wall frames and stretching the trusses. You can raise the roof by using longer wall studs.

Note: If you stretch the width of the barn past 8′, use 2 x 8's for the ceiling joists and 2 x 6's for the rafters. If you stretch it past 12′, use 2 x 10's for the joists.

3 Check the building codes.

Because this barn is attached to a permanent slab foundation, it will likely be affected by building codes. Check your local codes and pay particular attention to regulations governing the location of outbuildings. You'll probably find that, unless you apply for a variance, you'll have to locate this barn several yards back from your property line. If needed, secure a building permit before you start work.

Pouring the Slab Foundation

Use stakes and a string to lay out the location of your slab. Place the stakes outside of the foundation lines and fasten the string between them so that the points where the horizontal and vertical strings cross mark the exact location of the corners of your shed. (See Figure 1.) Measure carefully—if your slab isn't square your building won't be either.

4 Stake out the foundation.

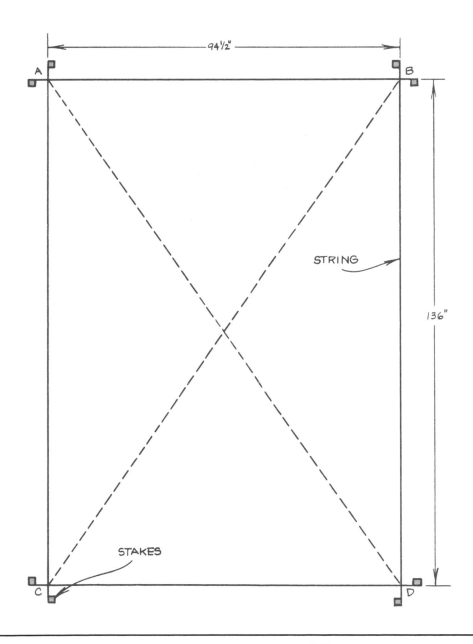

Figure 1. Use stakes and strings to mark the foundation lines for your shed. Measure diagonally from corner to corner to be sure your layout for the foundation is square. AD should equal BC.

Peel up the sod and remove any large rocks, debris, or plants from the area where you will pour the slab. Stake 2 x 8's around the outside edge of the slab as marked by your string. Be sure the forms are level and that all four corners are square. (Check by measuring the diagonals.) Use stakes and scrap lumber to brace the forms. Since the edges of the slab will serve as the footings for the shed walls, dig a trench inside the forms that is 8″ wide and 12″-14″ deep. Level the bed inside the trench with a flat shovel; then tamp the bed down. *It's*

5 Set the forms and prepare the bed.

important that the top of the bed be 6″-8″ below the tops of the forms. Build up the center of the bed with 4″-6″ of gravel (for drainage) and cover with plastic sheeting to make a vapor barrier. Next, cover the sheeting with 2″ of sand; put a little sand down in the trenches to hold the edges of the vapor barrier down. When you've finished spreading the sand, there should be 4″ between the top of the sand layer and the tops of the forms. Finally, put down steel reinforcing mesh. Hold the mesh 1″-2″ above the sand with racks or scrap wood. (See Figure 2.)

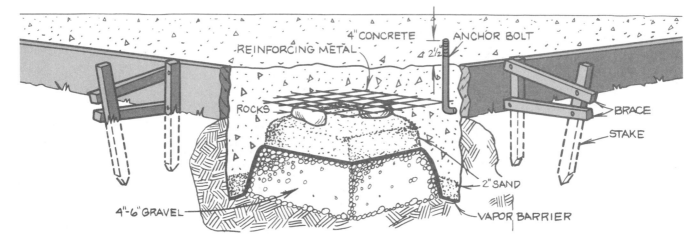

Figure 2. *To make forms for the slab, stake 2 x 8's to the ground. Dig a footing trench and prepare the bed with 4″-6″ of gravel (for drainage), a sheet of plastic (for a vapor barrier), 2″ of sand, and wire mesh held 1″-2″ off the sand (for reinforcement). Pour the concrete over the bed and set the anchor bolts in place.*

6 **Pour the concrete.**

For this foundation, you'll need about 2⅓ yards of concrete, so you'll probably want to order a ready-mix delivery. Arrange for extra wheelbarrows or a pump truck if the concrete truck cannot back up to the site. Once the concrete is poured, 'screed' it off level with the tops of the forms by dragging a 10-foot 2 x 4 back and forth with a sawing motion across the tops of the form boards.

7 **Set the anchor bolts.**

Set three anchor bolts along each 12′ side. To do this, measure 2½″ from each end and place the anchor bolts on center 1¾″ from the edge of the slab, as shown in the *Slab Layout* drawing. Then place the center of the middle anchor bolt 65½″ on center from each end bolt, and 1¾″ from the edge. Repeat the procedure for the other side of the slab. The bolts should protrude 2½″ from the top of the slab. Smooth out the concrete with a 'darby' or a trowel, then let the concrete cure for at least 24 hours before you continue with any further construction.

Building the Wall Frame

Cut the sole plates from pressure-treated or rot-resistant lumber, because they will draw moisture from the slab. Line up the sole plate along the edge of the foundation and mark the position of the anchor bolts. Then drill the holes about ¼″ larger than the diameter of the bolts so the wood will fit easily over the bolt but still be snug enough to prevent shifting. For example, drill a ¾″ hole to fit over a ½″ bolt. Lay the sole plate in place to make sure the edges are straight and the holes fit over the bolts.

8   Cut and drill the sole plates.

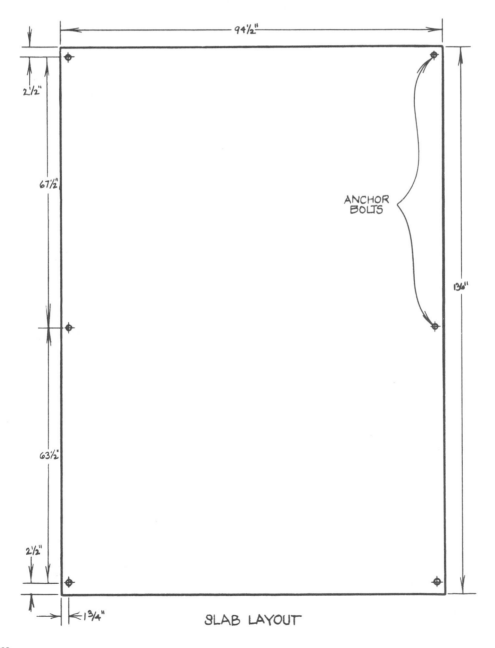

SLAB LAYOUT

9 **Build the wall frames.**

Cut the frame parts to length. Put the top plate next to the sole plate, bottom side up. Using a framing square, measure where each stud will be placed and draw a pencil line across both plates. Use the *Side Wall Frame, Back Wall Frame, Front Wall Frame,* and *Stud Layout* drawings to determine the location of each stud. Once you have marked the location of each stud, lay out the parts for each wall in turn on the concrete pad, and nail them together with 16d nails.

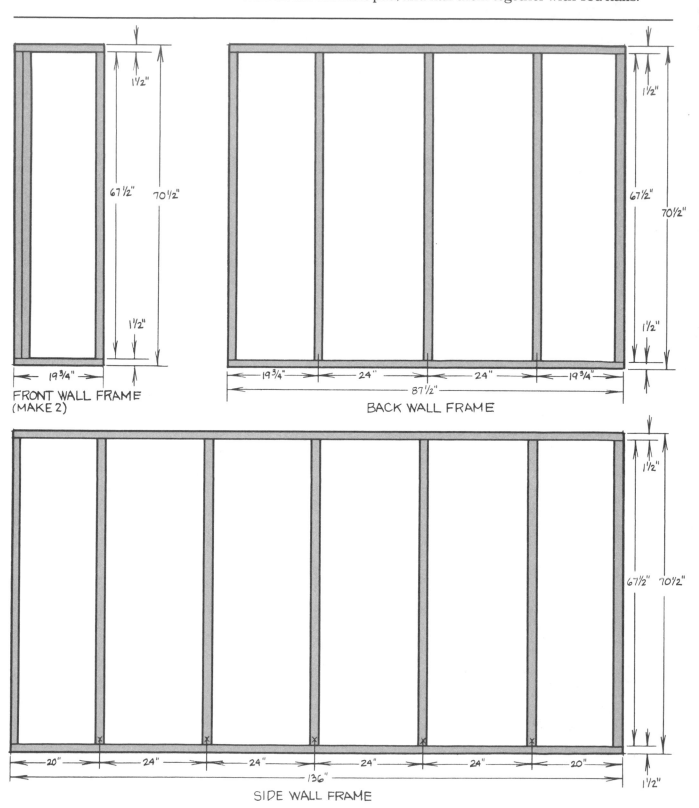

FRONT WALL FRAME
(MAKE 2)

BACK WALL FRAME

SIDE WALL FRAME

Put down a 6″ wide strip of tarpaper or plastic sheeting all around the edge of the pad, where the sole plates will rest. This will help prevent the wood from being damaged by the moisture in the concrete slab. With the help of a friend positioned at one end of the wall frame and you at the other, gently lift each wall into place. Temporarily brace them upright as you go. Then fasten the sole plates to the pad with the anchor bolts. Connect the walls to each other at the corners using 16d nails spaced every 24″. Use a carpenter's level to make sure all the studs are plumb vertically and the top plates are level horizontally.

10 **Raise the walls.**

TIP To correct the level or squareness of your wall frames, drive a wooden wedge under the sole plate until the wall is properly positioned.

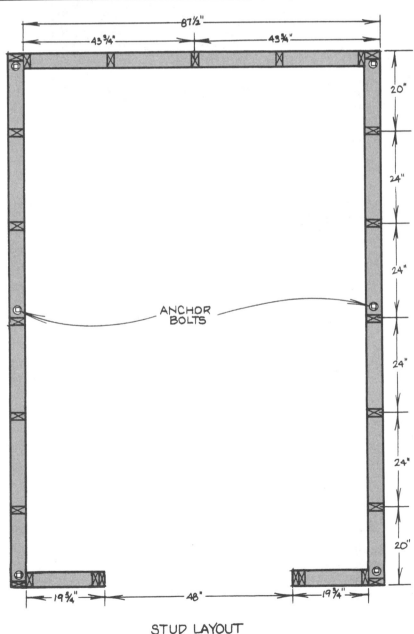

STUD LAYOUT

11 **Attach the top (cap) plate.**

To tie the entire assembly together, nail a second top (or 'cap') plate down onto the first top plate with 16d nails. Be sure the ends of the cap plate lap the joint between the top plate members. (See Figure 3.) Nail the top plate across the two front wall frames, leaving an opening for the door. (See Figure 4.)

Note: In a large building, the door would require a header. However, in this small storage barn, the 2 x 6 joist on the end truss will serve adequately as a header.

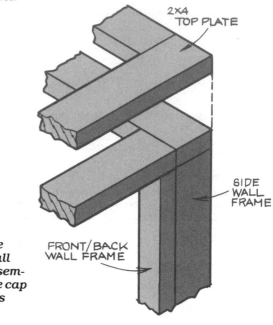

Figure 3. *The top (cap) plate is nailed to the top of the wall frames and ties the wall assembly together. The ends of the cap plate must overlap the joints between the wall frames.*

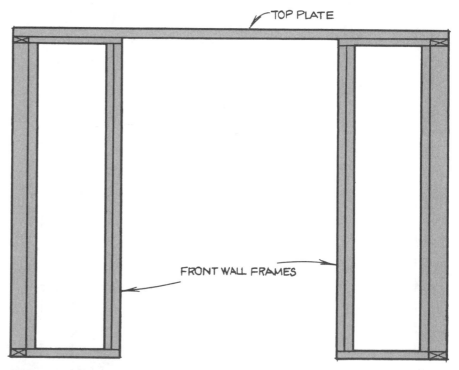

Figure 4. *Join the two front wall frames together with the top (cap) plate. Because this is a small building, the door opening needs no special header. The joist in the front end truss will serve as a header.*

Attach the siding to the side walls *before* putting up the roof trusses. This will give some rigidity to the structure. Attach the plywood siding with 6d nails.

12 Attach the siding to the side walls.

Constructing the Gambrel Roof

Cut the truss parts—rafters, braces, and joists—for seven trusses from 2 x 4 and 2 x 6 stock. Assemble these parts with truss plates, as shown in the *Roof Truss Layout* drawing. The end trusses each get a vertical brace to provide additional support for the peak and a place to nail the siding.

13 Build the roof trusses.

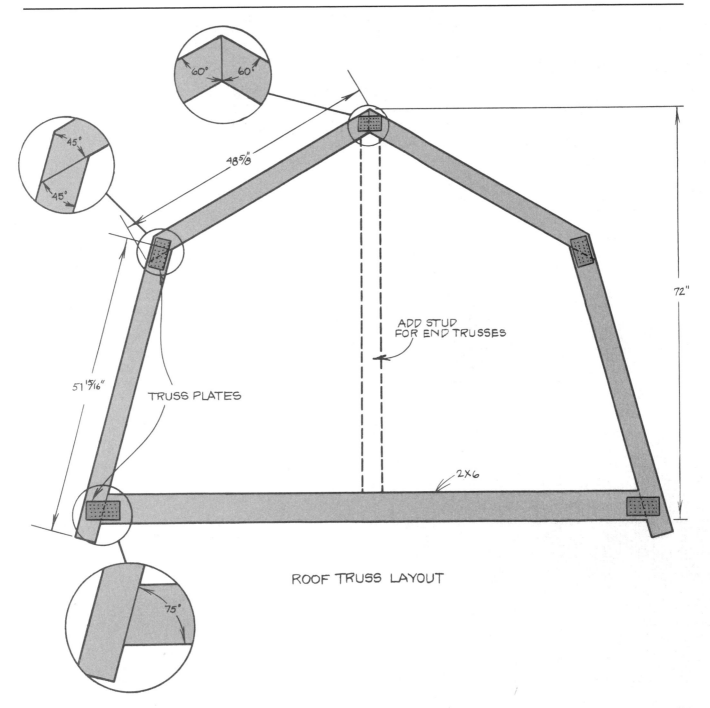

ROOF TRUSS LAYOUT

Option: You may wish to put a loft door in your barn, to make it easier to get things up and down from the loft. If that's the case, cut two braces for the front truss and assemble them to the rafters and joists 36″ apart. Add two horizontal headers, also 36″ apart, between these braces. (See Figure 5.) This will complete the frame for the loft door.

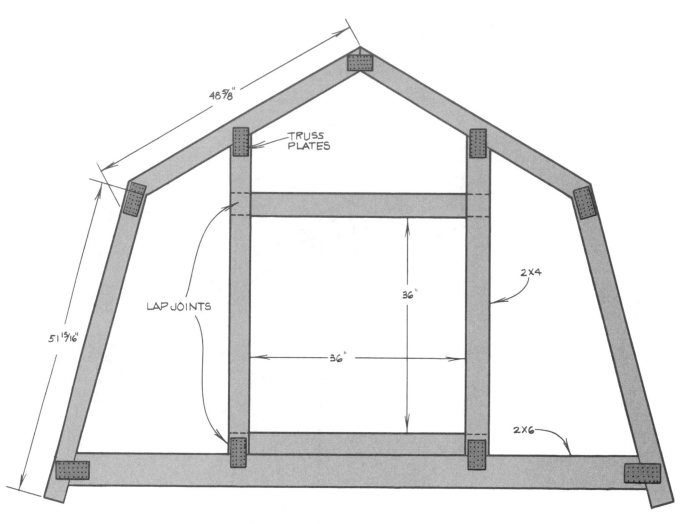

Figure 5. *To frame an opening for a loft door, use two braces and two horizontal headers, assembled to the truss as shown.*

TIP Build one truss, then lay out and build the rest of the trusses on top of the first to be sure that all trusses are exactly the same.

Toenail the joists to the top plate with 16d nails. (See Figure 6.) Space the trusses as shown in the *Roof Frame, Side View* drawing. Temporarily brace them upright with scrap lumber.

Note: With the trusses spaced every 24", you can store light objects in the loft—boxes of Christmas ornaments, portable patio furniture, flower pots, and stuff like that. However, if you plan to store anything heavy in the loft, make ten trusses and space them every 16". You might also want to make the joists from 2 x 8 stock, and install metal joist braces between them.

14 **Attach the trusses and joists to the top plate.**

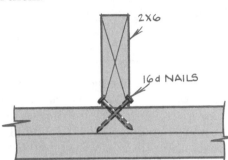

Figure 6. *Toenail the joists of the trusses to the top plates with 16d nails, as shown. You can also use rafter ties.*

Nail ½" CDX (exterior) plywood sheathing to the roof frame using 6d nails spaced every 6" along the edges and 12" in the field. Overhang the sheathing 4" on the front and back, as shown in the drawings.

15 **Install the roof sheathing.**

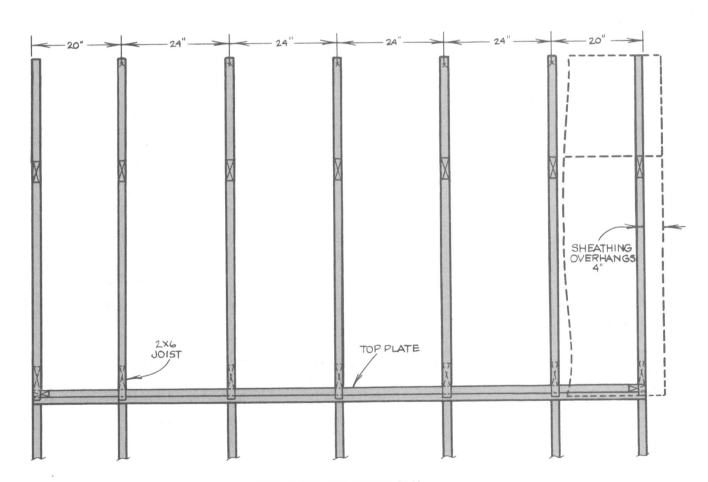

ROOF FRAME SIDE VIEW

16 **Attach the front and back facing strips.**

Cut the front and back facing strips from 1 x 4 stock, mitering the ends as shown in the *Front/Back Facing Strip Detail* drawing. Nail the strips together with truss plates on the *inside* faces, then attach the facing assemblies to the overhang on either end of the bar. Use #12 x 1¼" flathead wood screws to attach the facing to the sheathing.

TIP Cut and assemble the facing *before* you put up the trusses. Use a truss as a jig to nail the facing strips together. That way, you'll be sure they match the roof angles precisely.

17 **Install the roofing materials.**

Install metal drip edge around the edges of the roof. To fit the drip edge over the peaks and bends in the roof, snip the bottom flange as shown in Figure 7. Fit the drip edge to the roof, bending the top flange. Then nail it in place. When the drip edge has been installed, cover the entire roof with a double layer of tarpaper. Install the shingles according to the manufacturer's instructions. (See Figure 8.)

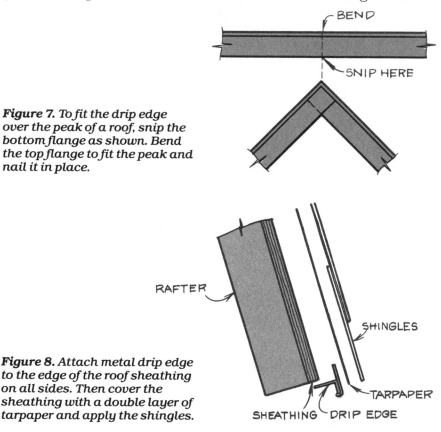

Figure 7. To fit the drip edge over the peak of a roof, snip the bottom flange as shown. Bend the top flange to fit the peak and nail it in place.

Figure 8. Attach metal drip edge to the edge of the roof sheathing on all sides. Then cover the sheathing with a double layer of tarpaper and apply the shingles.

Installing the Loft and Front/Back Walls

18 **Build the loft.**

Cover the ceiling joists with ¾" thick plywood, and nail it in place with 6d nails. Leave an opening so that you can get up to and down from the loft, as shown in the *Loft Layout* drawing.

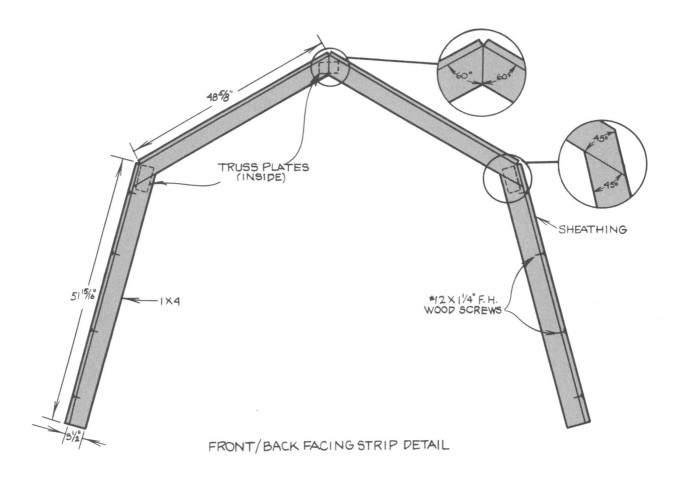

48⅝"

60°   60°

TRUSS PLATES
(INSIDE)

51¹⁵⁄₁₆"

1 X 4

45°

45°

SHEATHING

#12 X 1¼" F.H.
WOOD SCREWS

3½"

FRONT/BACK FACING STRIP DETAIL

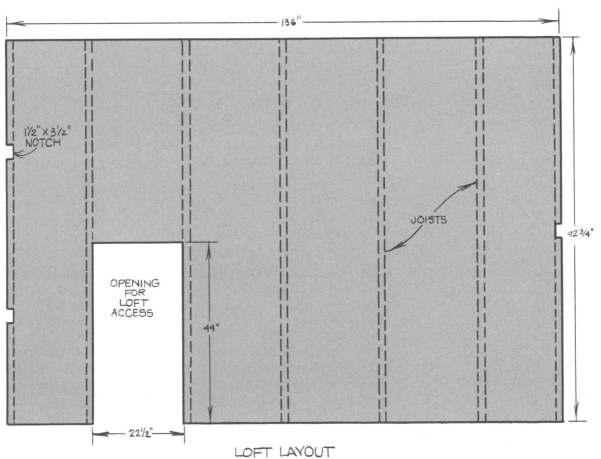

136"

1½" X 3½"
NOTCH

JOISTS

92¾"

OPENING
FOR
LOFT
ACCESS

44"

22½"

LOFT LAYOUT

19 **Install the front and back walls.**

Cover the front and back walls with ¾" plywood siding, as you did the side walls. Cut the siding so that it fits flush against the roof sheathing.

20 **Attach the corner molding.**

Cut the molding from 'one-by' stock. Bevel and miter the upper ends of the corner molding so that it fits flush to the roof sheathing. (See Figure 9.) Attach the molding as shown in the *Corner Joinery Detail* drawing, using 6d nails.

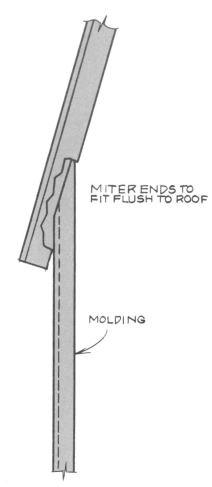

MITER ENDS TO
FIT FLUSH TO ROOF

MOLDING

Figure 9. Miter and bevel the ends of the corner molding to fit flush to the roof sheathing.

Finishing Up

21 **Install the door jamb and door molding.**

Rip the door jamb, door molding, and door stop stock to size. Staple a 6" wide strip of tarpaper to the frame around the door opening, lapping the siding. This will help prevent the siding from being damaged by moisture that might collect under the jamb. Install the top part of the door jamb first, then the sides, using 6d nails. This jamb must cover the door frame studs *and* the siding, as shown in the *Door Jamb Detail* drawing. Nail the door stop to the jamb, and nail the door molding to the siding so that it overlaps the jamb. Caulk all around the molding.

To make the doors, cut two pieces of plywood siding 71″ long and 23⅛″ wide, as shown in the *Door Layout* drawing. Attach the 1 x 3″ trim to the outside faces of the plywood using #12 x 1¼″ flathead wood screws. Drive these screws in from the *back* side of the door. Create a 'barn door' pattern with the trim, as shown in the drawing.

22 **Build the doors.**

TIP You can use both glue and screws to attach the trim pieces to your door so they won't work loose as the years go by. Use an exterior construction adhesive.

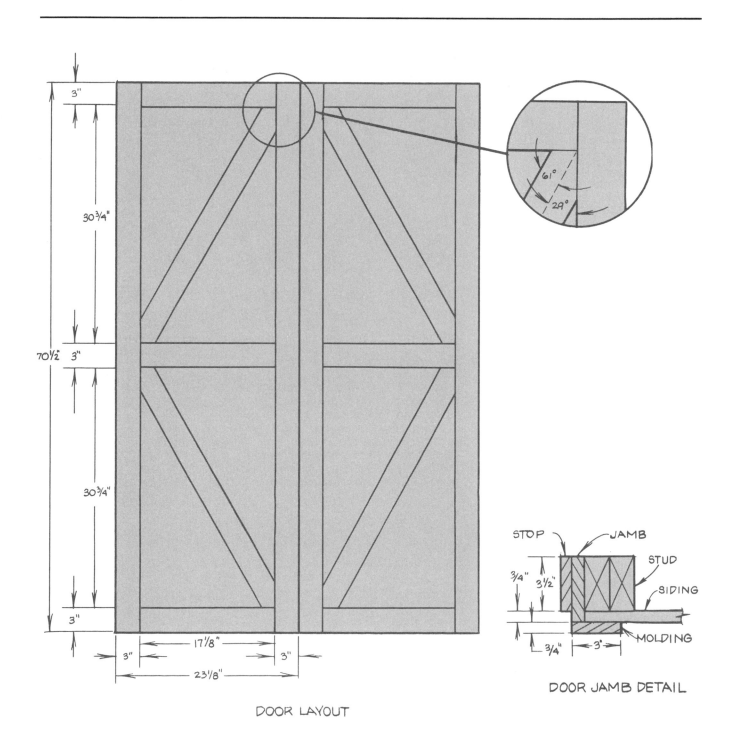

DOOR LAYOUT

DOOR JAMB DETAIL

23 **Hang the doors.**

Mount three T-hinges on each door, bolting the strap part of the 'T' to the door trim. Use bolts rather than screws to attach the hinges, to prevent a thief from dismounting the doors. These bolts should pass through the doors, with the washers and nuts on the inside. After you tighten the nuts, mash the threads of the ends of the bolts with a hammer. This will make it impossible to remove the hinges from the outside. After you mount the hinges to the doors, flop the butt parts of the 'T' over so that the hinges turn the corner around the edge of the door. With a friend, hold the doors in place so that the butt part of the hinges is flat against the door jamb. Mark the location of each hinge, and set the doors aside. Chisel out shallow recesses in the door jambs to accommodate the pins, then put the doors back in place, opening them part way so you can reach the butt part of the hinges. Screw the hinges to the door jamb with 2″ long flathead wood screws. (See Figure 10.) After hanging the doors, install a hasp so you can lock them.

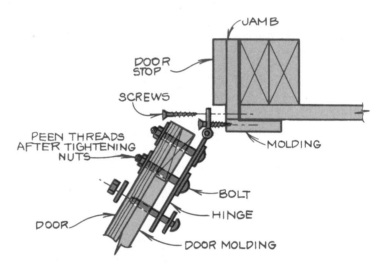

Figure 10. *Hang the doors on T-hinges, as shown. Bolt the strap part of the hinge to the outside of the door, and mash the threads so you can back the nuts off. Screw the butt part of the hinge to the door jamb, so that the screw heads will be hidden when the door is closed. These precautions will help to thwart intruders.*

Install aluminum vents in the front and back walls, near the peak of the roof, and in the side wall, near the bottom. This will keep air moving through the barn and prevent moisture from condensing on the contents. To install a vent, cut an opening in the plywood siding with a saber saw. Be careful to work around the 2 x 4 frame; don't cut through it. Put some caulk around the opening, and press the vent in place. Screw the flanges of the vent to the plywood siding.

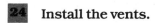 **24** **Install the vents.**

Paint or stain the shed to match your home, or to contrast with it. Use exterior paint or stain to protect the wood from rotting.

25 **Paint the exposed wood surfaces.**

Greenhouse Garden Shed

If you need a storage building *and* you like to do a lot of gardening, here's the perfect structure for you. Half of this building is for storage—lawn mowers, rakes, shovels, barbecue grills, all the things that end up in storage sheds. The other half is a greenhouse. You can use it for starting seedlings early in the spring and for growing flowers and vegetables late into the autumn.

This garden shed sits on solid concrete blocks, so there's no tricky foundation to construct. And the free-standing, gambrel design provides ample head room while angling the glazing to catch the sun. The walls and roof are framed with wood, then one side (the greenhouse side) is covered with corrugated fiberglass glazing, and the other side of this shed (the storage side) is covered with shingles.

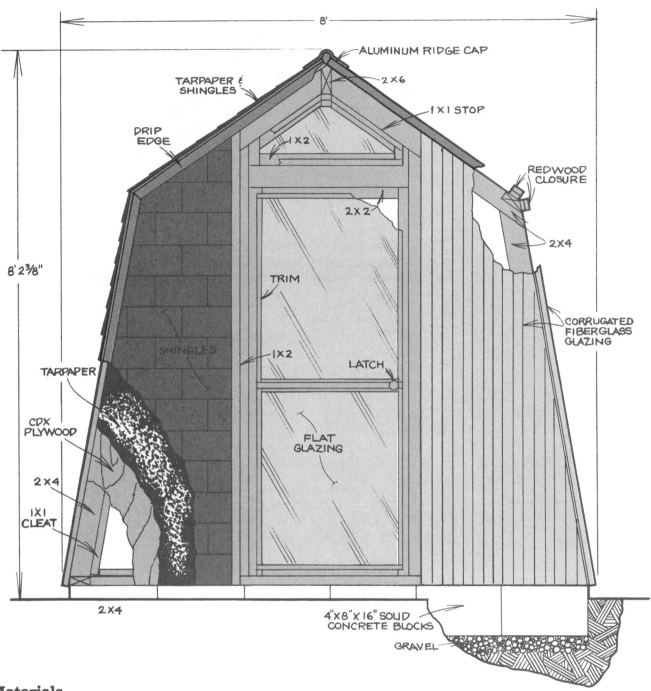

ALUMINUM RIDGE CAP

TARPAPER & SHINGLES

2 X 6

1 X 1 STOP

DRIP EDGE

1 X 2

REDWOOD CLOSURE

2 X 2

2 X 4

TRIM

CORRUGATED FIBERGLASS GLAZING

SHINGLES

1 X 2

LATCH

TARPAPER

FLAT GLAZING

CDX PLYWOOD

2 X 4

1 X 1 CLEAT

8'

8' 2⅜"

2 X 4

4" X 8" X 16" SOLID CONCRETE BLOCKS

GRAVEL

FRONT ELEVATION

Materials

Because of the humidity that will accumulate in this structure, use redwood, cedar, or pressure-treated lumber for all the structural parts. Moisture can rot other woods. If you build from redwood, select kiln-dried heartwood. The sapwood (light colored part) of redwood is not rot-resistant.

Use pressure-treated or redwood 2 x 4's for all the wall and roof framing, unless you have enlarged the size of the garden shed. In that case, check your local building code for the correct lumber size. Purchase one 2 x 6 x 10' for the ridge beam and ½" CDX (exterior) plywood to sheathe the roof and north side of the greenhouse. Use 2' x 12' corrugated fiberglass panels and redwood closure strips to cover the south side.

Purchase 1½″ x 1½″ stock for the door frame and shelf supports and 'one-by' (¾″ thick) stock for moldings, jambs, and window frame. From the one-by material, rip ¼″ x ¾″ stock for the trim.

In addition to these materials, purchase 4″ x 8″ x 16″ concrete blocks, aluminum ridge cap pieces, clear silicone caulk, aluminum window screen, drip edge, tarpaper, shingles, and waterproofing stain. You'll also need ⅝″ roofing nails, aluminum nails with rubber seals, staples, flathead wood screws, galvanized nails, and brads. Use butt hinges to hang the door and window.

Option: Although this garden shed is covered with clear fiberglass panels, you can select acrylic or polycarbonate coverings. Check the chart below for the advantages and disadvantages of each type of covering.

Advantages/Disadvantages of Greenhouse Glazing Materials

Type	Advantages	Disadvantages
Fiberglass	Widely available. Can be made so it is translucent. Very impact resistant. Can be coated to improve weather resistance.	Not all varieties are suitable. Isn't transparent. Weathers poorly, requires scrubbing.
Acrylic	Up to 30 times as impact-resistant as glass. One variety is crystal clear. Remains transparent with age. Widely available.	It's combustible. Chemicals and abrasives can damage it. Reinforcing bars must be added for large pieces.
Polycarbonate	Very impact-resistant. Widely available.	Weathers poorly.

Before You Begin

When selecting a site for your garden shed, keep in mind that plants and vegetables usually require a minimum of four hours of direct sunlight. Also, the more sun your structure gets, the warmer it will be in the early spring and late autumn. So choose a site with an unobstructed southern exposure. Orient the shed so that the side with the fiberglass panels faces south.

Also, take wind-chill into consideration. A greenhouse exposed to the wind will not be as warm as one that isn't. Position the north wall against a hedge, fence, or lee side of your house—or simply shingle the north side as shown in our design.

1 Take advantage of the sun.

As shown in the working drawings, this garden shed is 8′ wide, approximately 8′ high, and stretches 10′ long. However, this may be larger or smaller than you need. If you want a smaller garden shed, simply reduce all three dimensions. The design can be elongated or shortened simply by adding or subtracting rafters and adjusting the length of the ridgeboard. You can raise the roof by using longer rafters. Changing the width is a bit trickier; it requires changing the angle of the gambrel *and* adjusting the length of the short rafter.

Note: If you stretch the width of the garden shed past 8′, use 2 x 6's for the framing members. Check your local building code for the size lumber best suited to your dimensions.

2 Adjust the size of the garden shed.

The type of foundation we use here may or may not be considered permanent under local building codes. Most likely, it won't be, but it's always wise to check. Also check the regulations governing the location of outbuildings. You may have to locate this garden shed several yards back from your property line. If needed, secure a building permit before you start work.

3 Check the building codes.

Laying the Foundation

Use stakes and string to lay out an 8′ x 10′ foundation. The 8′ sides should face east and west, and the 10′ sides north and south. Run a trickle of sand 6″ inside the string to mark the inside location for digging your trenches.

4 Stake out the foundation.

Dig a trench 6″ wide and 12″ deep around the perimeter and fill it with 8″ of gravel for drainage.

5 Dig the trenches.

6 **Cut and assemble the sole plate.**

Cut the sole plate members to size and assemble them using truss plates, as shown in the *Sole Plate Layout* drawing. Make sure the sole plate is square, then temporarily brace it from corner to corner with scrap wood.

7 **Set the concrete blocks.**

Working with a level, lay 4″ x 8″ x 16″ solid concrete blocks on top of the gravel to make a foundation, as shown in the *Foundation Layout* drawing. Put the corner blocks in place first and make sure they're level to each other. Then fill in the sides. Use the assembled sole plate to help you gauge when the blocks are square and level. (See Figure 1.) After the blocks are in place, line the inside of the foundation with 1″ thick rigid foam insulation. Back fill the trenches to hold the insulation in place.

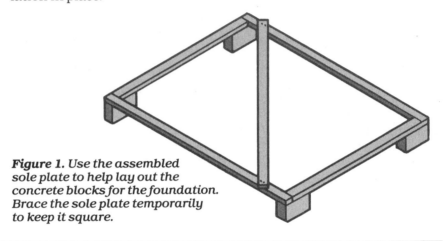

Figure 1. Use the assembled sole plate to help lay out the concrete blocks for the foundation. Brace the sole plate temporarily to keep it square.

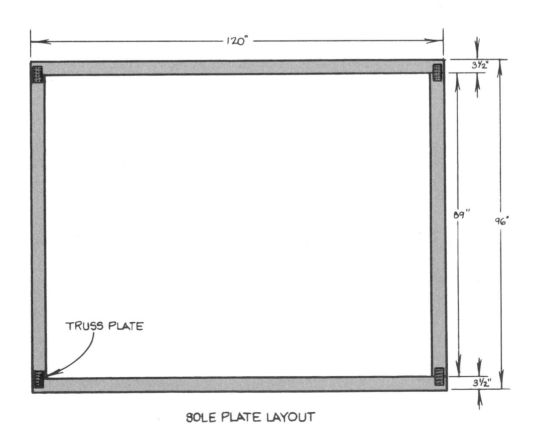

SOLE PLATE LAYOUT

Building the Frame

Line the top of the foundation with a plastic sheet and lay the sole plate on the foundation. If you wish, you can anchor the sole plate to the foundation and the ground with 24″ lengths of #4 rebar. Just drill holes in the sole plate over the joints between the foundation blocks, where shown in the *Foundation Layout* drawing. Then drive the rebar through the holes, between the blocks, and into the ground. After the sole plate is anchored, remove the temporary diagonal brace.

8   **Attach the sole plate to the foundation.**

Cut the long and short rafters as shown in the *Rafter Layout* drawing, and assemble them with 16d nails. Be sure to notch the short rafters for the lookouts.

9   **Cut and assemble the rafters.**

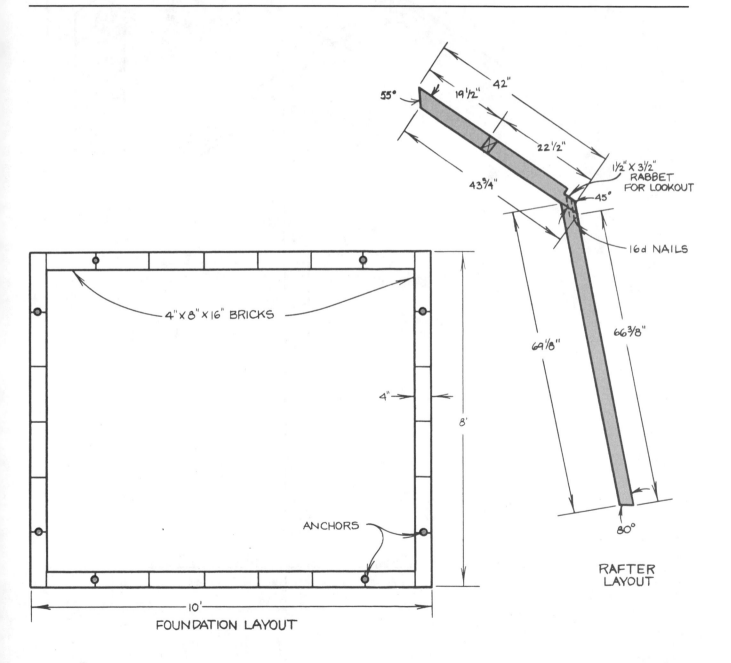

FOUNDATION LAYOUT

RAFTER LAYOUT

10 **Assemble the front and back wall frames.**

Cut the 2 x 4 frame members needed for the front and back walls, as shown in the *Front/Back Wall Frame* drawing. Lay out these pieces and the front and back rafter assemblies on a flat surface, then assemble them with truss plates. To keep all the parts properly aligned until you attach the walls to the sole plate, temporarily brace the parts with scrap wood.

TIP To make the front and back wall frames exactly the same, make the back wall first, then build the front wall on top of the back wall.

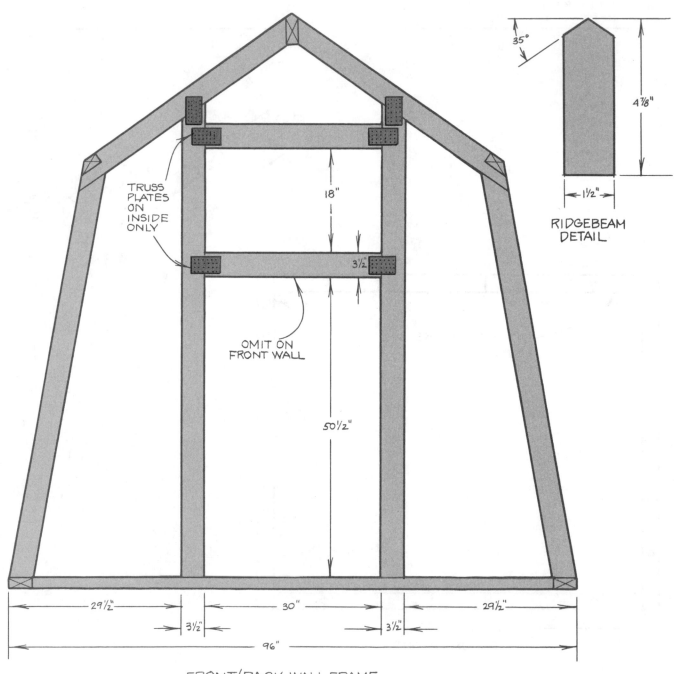

TRUSS
PLATES
ON
INSIDE
ONLY

18"

3½"

OMIT ON
FRONT WALL

50½"

29½"   30"   29½"

3½"   3½"

96"

FRONT/BACK WALL FRAME

35°

4⅞"

1½"

RIDGEBEAM
DETAIL

STORAGE BUILDINGS

With a helper, raise the front and back wall frames in place on the sole plate. Toenail them to the plate with 12d nails, then temporarily brace them upright.

11 **Raise the front and back wall frames.**

Cut the ridge beam, beveling the top edge as shown in the *Ridge Beam Detail* drawing. Then nail the ridge beam to the front and back frames with 16d nails. Mark the location of the rafters, as shown in the *Frame, Side View* drawing. Toenail the rafters to the ridge beam and sole plate, using 12d nails.

12 **Attach the ridge beam and rafters.**

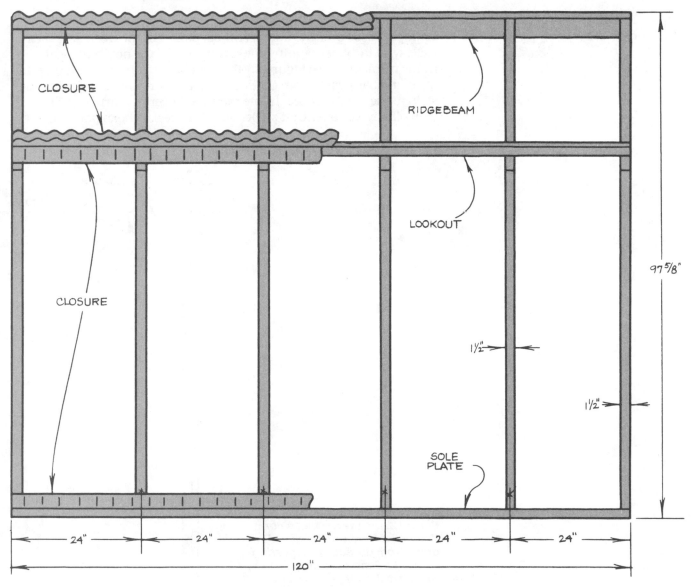

FRAME - SIDE VIEW

13 **Attach lookout members and closure strips.**

Nail the lookout members to the rafters with 16d nails. When the lookout members are in place, remove the braces from the front and back wall frames. On the south side, nail redwood closure strips to the ridge beam, lookouts, and along the bottom of the rafters, just above the sole plate.

Covering the Frame

14 **Sheathe the north side.**

Nail ½″ CDX plywood on the north side wall directly on top of the rafters, from the ridge beam to the sole plate. However, to cover the northernmost panels in the front and back walls, make cleat strips from 1 x 1 stock. Attach these cleats to the edges of the rafters and frame members, as shown in the *Cleat Detail* and *Section A* drawings. Then cut plywood panels to fit *inside* the frame members, and nail the panels to the cleats with 4d nails.

15 **Install the tarpaper on the north side.**

Cover the northern side of the garden shed with a double layer of tarpaper. Also cover the northernmost panels of the front and back walls with tarpaper, lapping the tarpaper over the north vertical frame members. (See Figure 2.) To divide the northern portion of the front and back walls from the southern portion, nail a 1 x 2 molding strip to the north vertical frame members, over the tarpaper. Then trim off any tarpaper that sticks out over the southern portion.

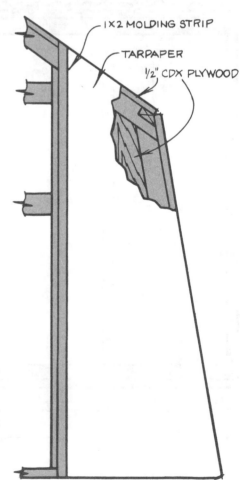

1X2 MOLDING STRIP

TARPAPER

½" CDX PLYWOOD

Figure 2. Set the northern portion of the front and back walls apart from the southern portion with a 1 x 2 molding strip. Nail this strip directly over the tarpaper before you apply the shingles.

Apply shingles to the northern portions of the front and back walls. Then line the northern edges of the side walls with drip edge. Lap the bottom flange of this drip edge over the shingles on the front and back walls, and slide the flange behind the 1 x 2 molding that divide the northern and southern portions of the garden shed. After you've installed the drip edge, apply shingles to the northern side wall, from the ridge to the base. Overhang the shingles ½"-1" over the bottom edge of the sheathing, to protect the wooden frame and sheathing from run-off rainwater. (See Figure 3.)

16 **Install the drip edge and shingles.**

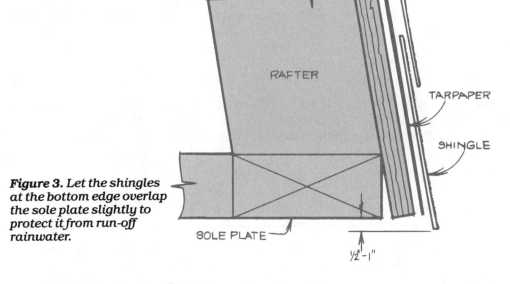

Figure 3. Let the shingles at the bottom edge overlap the sole plate slightly to protect it from run-off rainwater.

TIP Don't use black or dark-colored shingles; they will make the garden shed too hot.

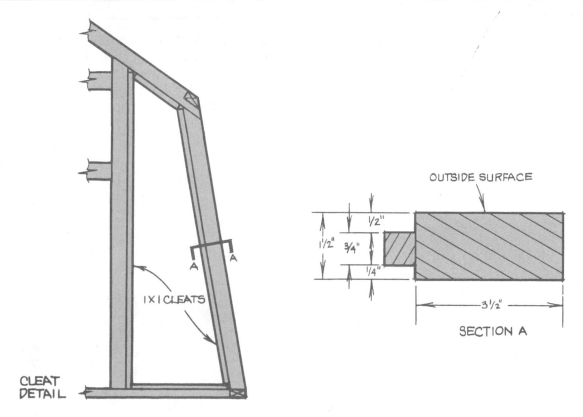

17 Attach the fiberglass panels to the south side.

Cut the fiberglass panels to size with a contractor's saw and a plywood blade. Use some of the leftover sheathing to provide a backup to the blade. Before you cut, reverse the blade so that the teeth are pointing in the wrong direction (against the direction of rotation)—this will give you a much smoother cut in the thin fiberglass material. Attach the fiberglass to the closures on the southern side wall. Begin at the top, using special nails with rubber seals to attach the panels. Seal the joints where the panels overlap with clear silicone caulk. Let the upper panels overhang the minor peak 1″-2″, so that they will cover the lower panels. Let the lower panel extend slightly beyond the sole plate, to protect the wood frame from run-off. After you've installed the fiberglass panels to the south side wall, attach an aluminum ridge cap over the major peak. This cap should lap the glazing on the south side *and* the shingles on the north side.

18 Attach fiberglass panels to the southern portions of the front and back walls.

Cut panels to cover the southern portions of the front and back walls, except for the door, window, and vent openings. Nail these in place directly to the frame—you need redwood closures to attach these panels.

Making the Door, Window, and Vent

19 Cut and assemble the door.

Cut the pieces for the door frame from 2 x 2 stock, and make the dadoes and rabbets in the vertical members where shown in the *Door Layout* drawing. Assemble the door frame with #10 x 1¼″ flathead wood screws. Then cover the door with *flat* glazing to allow more light in the garden shed. Then install trim around the glazing, as shown in the *Section B* drawing.

20 Cut and assemble the window and vents.

Cut the window and vent frames from 1 x 2 stock as shown in the *Window Layout* and *Vent Layout* drawings. Make lap joints in the ends of the frame members. (See Figure 4.) Screw the frame members together with #10 x ¾″ flathead wood screws. Cover the frames with flat glazing, and trim around the edge of the glazing, as you did when you made the door.

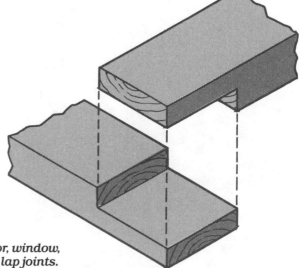

Figure 4. Join the door, window, and vent frames with lap joints.

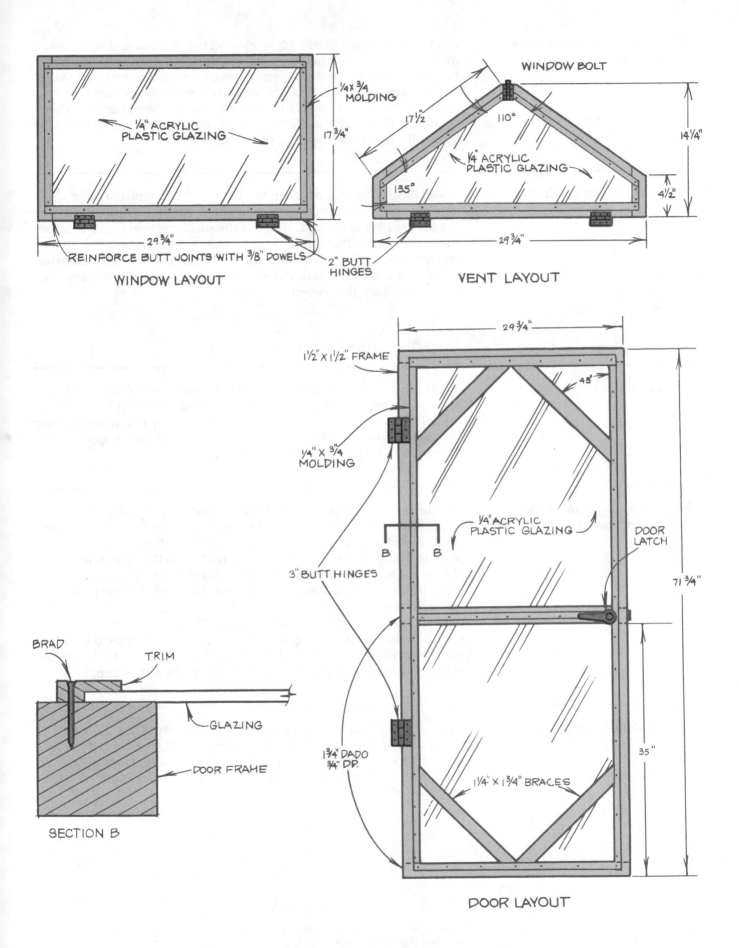

WINDOW LAYOUT

¼ X ¾ MOLDING

¼" ACRYLIC PLASTIC GLAZING

17 ¾"

29 ¾"

REINFORCE BUTT JOINTS WITH ⅜" DOWELS

2" BUTT HINGES

VENT LAYOUT

WINDOW BOLT

17 ½

110°

¼" ACRYLIC PLASTIC GLAZING

135°

14 ¼"

4 ½"

29 ¾"

SECTION B

BRAD

TRIM

GLAZING

DOOR FRAME

DOOR LAYOUT

29 ¾"

1 ½" X 1 ½" FRAME

45°

¼" X ¾ MOLDING

¼" ACRYLIC PLASTIC GLAZING

DOOR LATCH

B   B

3" BUTT HINGES

71 ¾"

1 ¾" DADO ¾" DP.

35"

1 ¼" X 1 ¾" BRACES

GREENHOUSE GARDEN SHED

71

21 Install the door, window, and vent jambs.

Cut the door jambs from 1 x 2 stock, and the window and vent jambs from 1 x 1 stock. Nail the door jambs to the inside surface of the frame members, around the door opening, as shown in the *Vent and Door Jambs Detail* and *Section C* drawings. These jambs overlap the frame members, as shown. Inset the window and vent jambs in their openings. The outside surface of the window and vent jambs must be flush with the outside surface of the frame.

22 Hang the door, window, and vents.

Install hinges, latches, and window bolts to the completed door, window, and vents. Hang the door on its hinges in the frame with the hinges toward the north. Chisel out the south vertical frame member where the latch strikes, and install a striker plate. Hang the window and the vents with the hinges towards the ground. Use a turn-button to hold the window closed, and window bolts to hold the vents closed. Gravity will hold them open.

Installing the Shelves

23 Dig holes for the shelf supports.

To get the maximum use out of the greenhouse side of your garden shed, you'll want to install movable shelves on the south side. To make these shelves, first dig holes 12″ deep and 6″ in diameter in the floor of the building where the ends of the vertical shelf support will be buried. Use a plumb bob to locate these holes.

24 Cut and assemble the shelf supports.

Cut the vertical and horizontal shelf supports for the four middle shelf support assemblies from 2 x 2 stock. Nail the upper ends of the vertical supports to the rafters, then nail the horizontal shelf supports in place between the vertical supports and the rafters, as shown in the *Shelf Supports Detail* drawing. If the wood tends to split, drill pilot holes. Partially fill the holes around the lower ends of the vertical shelf supports with gravel for drainage, then fill the rest of the holes with earth and tamp it down. Check your work with a level to be sure that the vertical supports are plumb. Make the end shelf supports from 1 x 2 stock, and nail these supports to the inside of the garden shed frame, level with the horizontal shelf supports.

25 Install the shelves.

Cut strips of 1 x 2 stock and lay them in place across the shelving supports wherever you want a shelf. Leave a 1½″ space between each strip. This will let the light through the shelves *and* help ventilate all the plants on the shelves. Whenever you want to take a shelf down or add a new shelf, just move the strips.

This project was adapted from a plan originally published by Shopsmith, Inc. Our thanks for letting us borrow some of their materials.

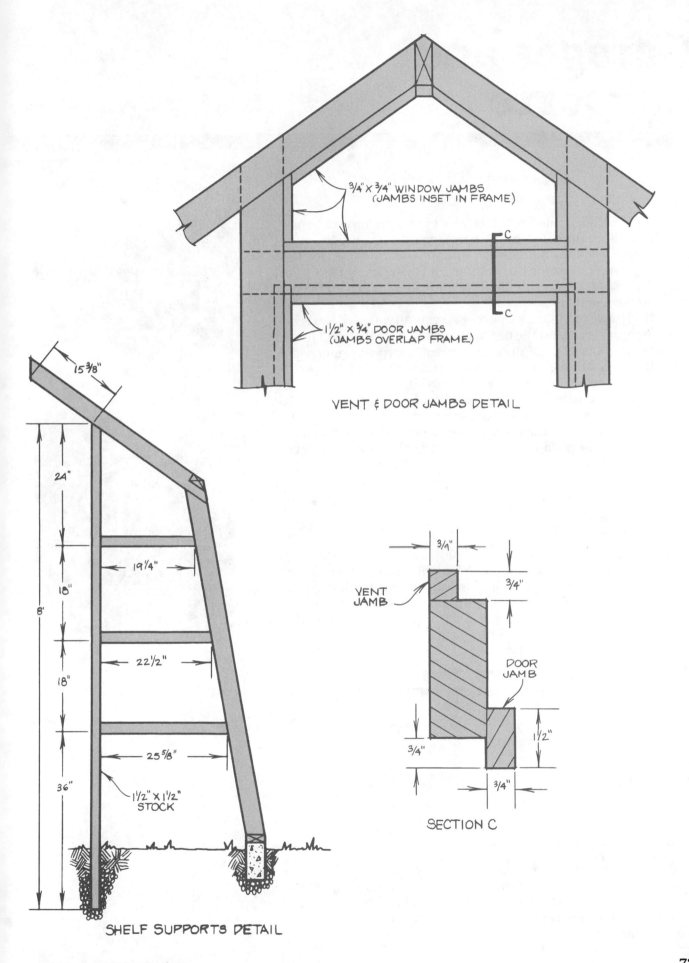

¾" X ¾" WINDOW JAMBS
(JAMBS INSET IN FRAME)

1½" X ¾" DOOR JAMBS
(JAMBS OVERLAP FRAME)

VENT & DOOR JAMBS DETAIL

15⅜"

24"

19¼"

18"

22½"

18"

8'

25⅝"

36"

1½" X 1½"
STOCK

SHELF SUPPORTS DETAIL

¾"

¾"

VENT
JAMB

DOOR
JAMB

1½"

¾"

¾"

SECTION C

Garages and Carports

Providing adequate shelter for your car or truck will not only extend the life of its paint job by keeping it out of the sun, but will save you the hassle of scraping off the windshield on those blustering winter mornings. And, by building your garage or carport yourself, you'll get a structure that's tailor-made to suit your needs as well as your pocketbook.

It's simple to build a carport or garage. All you need is a few basic carpentry tools and the handyman know-how that we'll provide for you in the following chapters. And remember, you can mix and match the foundations, roofs, and wall styles to fit your individual needs. For example, you can frame the walls much as you would for any structure and turn the lean-to carport into an enclosed garage. Where we show a gravel floor with a carport construction, you may install a concrete pad using the steps described in the chapter on enclosed garages. Just flip the pages and you'll find the steps for constructing a shelter for your car or truck that will make it uniquely yours.

Before You Begin

Your first consideration should be the amount of space you have available and its access to the street. This will determine how large your structure will be and whether you will want to attach it to your home or let it stand alone. The following chart giving the dimensions of standard size garages and carports will help you decide which design best meets your space requirements.

Suggested Garage and Carport Sizes

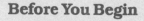

Type of Garage or Carport	Size
1 car	12' wide by 22' long
1½ (with side work area)	18' wide by 22' long
2 car	22' wide by 22' long
2 (with back storage)	22' wide by 24' long
2 (with side and back storage)	24' wide by 24' long
2½ (with side work area, back storage)	28' wide by 24' long

Consider your needs and space restrictions before settling on a building design. How many cars do you have? Do you want to provide an area for storage or for a workshop? An extra half-car area will provide room for a basic workshop, a place to store a small boat, or just a place to tinker.

If you want your structure to serve double-duty—as both recreation space and a home for the car—construct a carport. Use it as a covered patio on warm, sunny afternoons; as a garage when the weather turns lousy. But if keeping your car and outdoor equipment safe from Mother Nature is your top priority, an enclosed garage would be your best bet.

Finally, take a trip to your local building inspections office and find out what the zoning regulations are for your area. This will determine how far your structure will have to set back from the road and from your property lines.

You may also need to obtain a building permit before you begin work. If so, you may have to draw up a complete set of plans to obtain that permit—just showing them the drawings in this book won't do it. If you're not handy with a T-square and a pencil, have an architect draw the plans for you. This will cost you a little money, but it may also save you a big hassle. Most good architects are familiar with exactly what is needed to get that permit, and as part of their fee, they'll help you with the application.

Lean-To Carport

The lean-to carport can be a shelter for your car or truck by night, and a patio by day. Because it's attached to the house or existing garage at one side, it's fast and easy to construct. And, with very little time and trouble, it can be enclosed later to become a full-fledged garage or storage building.

This carport is set on concrete piers, but you could also pour concrete footers or a concrete pad. (If you want to pour footers or a pad, follow the advice in other chapters in this book.) The roof is covered with roofing materials. If your carport will double as a patio, you might want to use fiberglass panels—they let in more light.

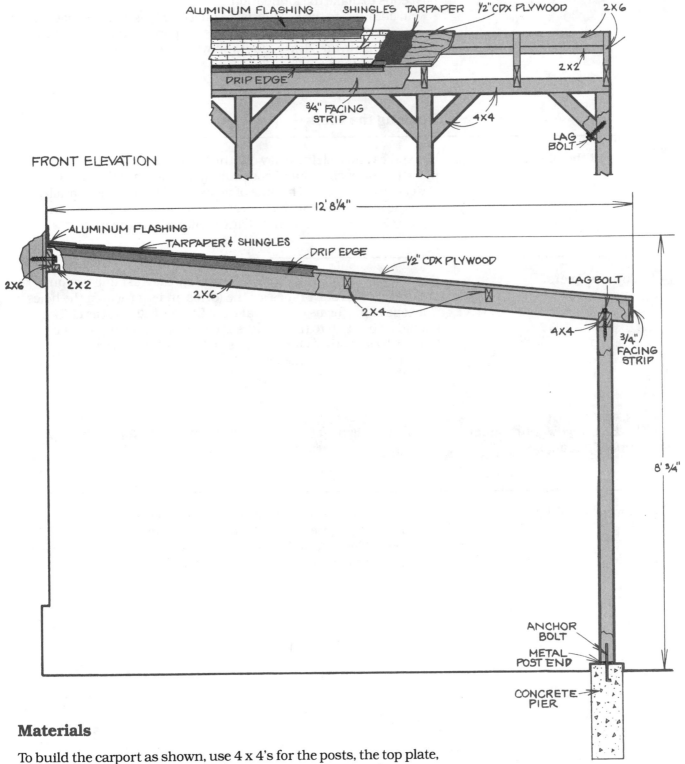

ALUMINUM FLASHING   SHINGLES TARPAPER   ½"CDX PLYWOOD   2×6

DRIP EDGE

¾" FACING STRIP

2×2

4×4

LAG BOLT

FRONT ELEVATION

12' 8¼"

ALUMINUM FLASHING

TARPAPER & SHINGLES

DRIP EDGE

½" CDX PLYWOOD

2×6

2×2

2×6

2×4

LAG BOLT

4×4

¾" FACING STRIP

8' ¾"

ANCHOR BOLT

METAL POST END

CONCRETE PIER

SIDE ELEVATION

Materials

To build the carport as shown, use 4 x 4's for the posts, the top plate, and the braces; and 2 x 6's for the rafters and the header. If your building is a 1½ car carport (16' wide), use 2 x 8's for the rafters and header. The roof is covered with ½" CDX plywood. The ledger strip is a standard 2 x 2, and the facing strip is cut from 1 x 6 stock. You can use pressure-treated lumber for the posts if you want, but it's not really necessary—the posts do not contact the ground. Use untreated lumber for all the other parts.

The posts are set on concrete piers, so you'll need concrete, several lengths of 8" stovepipe to serve as forms, anchor bolts, and metal post ends to make the foundation.

In addition to these materials, you'll also have to purchase galvanized nails, roofing nails, lag screws, tarpaper, shingles, flashing, roofing cement, and drip edge. If you wish to cover the ceiling, you'll also need ¼" tempered hardboard.

Pouring the Piers

1 **Lay out the piers.**

Use stakes and a string to lay out the locations for your posts and piers. Locate each pier and post exactly 4' apart, as shown in the *Post Layout* drawing. This line of posts must be precisely parallel with the side of the house, so measure carefully. Mark the locations with stakes, then remove the string while you dig the holes.

2 **Pour the piers.**

Dig holes for the piers 24"-36" deep, below the frost line for your area. Use 8-inch stove pipes for the forms and set them in the holes. The tops of the forms must be at least 2" above ground level. They should all be flush to the string, so that the tops of the piers will be level with each other. Once you're sure the forms are level, simply mix cement and pour into the form.

TIP Before you pour the concrete, throw 2"-3" of gravel into the bottom of each form. This will help provide drainage.

3 **Set the anchor bolts.**

Before the concrete sets up, position anchor bolts in the center of each pier. (See Figure 1.) Use the string to locate the precise position of the bolts. The tops of the bolts should protrude 6"-8". Wait at least 24 hours for concrete to cure; then remove stovepipe forms.

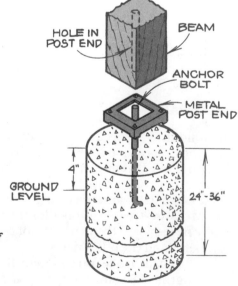

Figure 1. Set the anchor bolt in the pier so that it sticks up 6"-8" into the end of the pier. To keep water from rotting out the base of the posts, use metal post ends to keep them off the surface of the pier.

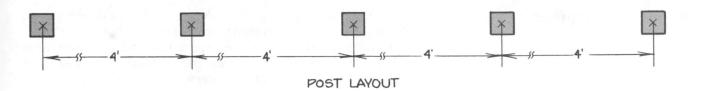

POST LAYOUT

> **TIP** To hold the anchor bolts at the proper height, drive two small stakes on either side of the piers. Wrap a wire around the end of the bolt, then wrap the ends of the wire around the stakes so that the bolt is suspended in the wet concrete.

Building the Freestanding Wall

Drill holes for the anchor bolts in the bottom of the posts, making sure holes are $^1/_{16}''$ larger than the diameter of the bolts. This will give the wood some room to swell during wet weather. Place metal post ends on the piers, then slip the posts over the anchor bolts. The anchor bolts will hold the posts upright.

4 Set the posts in place.

> **TIP** To bore deep holes in the ends of the posts, use 'an aircraft drill', or a 'drill bit extender' and an auger bit.

With a level, straighten the posts so that they are perfectly straight up and down. Hold them in place with stakes and temporary braces. Mark the top of one post 86½" above the ground. Using this mark as a reference, find the tops of the other posts with a string level. (See Figure 2.) Take the posts down and cut them off with a handsaw. Be sure to mark which post belongs on what pier! Put the posts back up and re-attach the temporary stake braces.

Note: Since you poured the piers level to each other, you could simply cut off all the posts at 86½". But the extra work with the string level will compensate for any settling that may have occurred.

5 Cut the tops of the posts level.

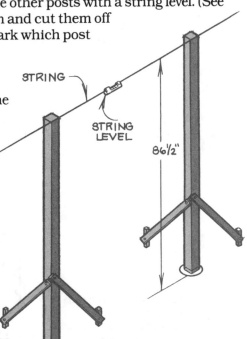

STRING

STRING LEVEL

86½"

Figure 2. Mark the top of one post 86½" above the ground, then use a string level to find the top of the other posts.

6 **Attach the top plate.**

Make a top plate of the proper length. If you can't get a 4 x 4 long enough, you will have to join two shorter boards with a lap joint. Plan this lap joint so that it occurs right over a post. (See Figure 3.) Put the top plate in place, as shown in the *"Freestanding Wall, Front View"* drawing. Attach it to the posts with lag screws, as shown in the *"Top Plate Joinery Detail."*

---

7 **Attach the braces.**

Cut the braces from 4 x 4 stock, mitering the ends at 45°. Attach them to the posts and the top plate with lag screws.

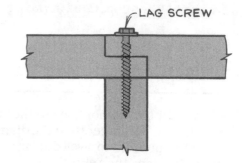

Figure 3. If you can't buy a 4 x 4 long enough for the top plate, join two shorter boards with a lap joint. Plan the joint so that it will be directly over a post.

Making the Roof Frame

---

8 **Measure and mark the house for the header.**

Mark the house where you will attach the header. The header must be parallel to the top plate and above it. When you attach the ledger strip to the header, the top of the ledger strip should be 12″ above the top of the top plate, as shown in the *"Ledger Strip Placement"* drawing. The header must also be properly placed so that the roof frame that you plan to build will be square. To make sure of this, measure diagonally from the proposed ends of the header to the *opposite* ends of the top plate. The two measurements should be the same.

---

9 **Attach the header to the house.**

Cut a header and a ledger strip exactly as long as the top plate. Nail the two pieces together with 12d nails, passing the nails through from both the front and back of the assembly. This will strengthen the joint. The ledger strip should be flush with the bottom of the header, as shown in the *Ledger Strip, Front View* and *Side View* drawings. Attach the completed assembly to the house with lag screws. These screws must bite into the frame studs so that the ledger strip can support the weight of the roof. (See Figure 4.)

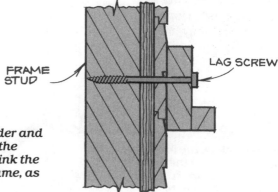

Figure 4. Attach the header and ledger strip assembly to the house with lag screws. Sink the screws into the house frame, as shown.

Cut the rafters from 2 x 6 stock, as shown in the *Rafter Layout* drawing. Miter each end at 84½°, and notch the rafters so they fit over the ledger strip and the top plate. These notches must be slightly angled so that they are square to the mitered ends.

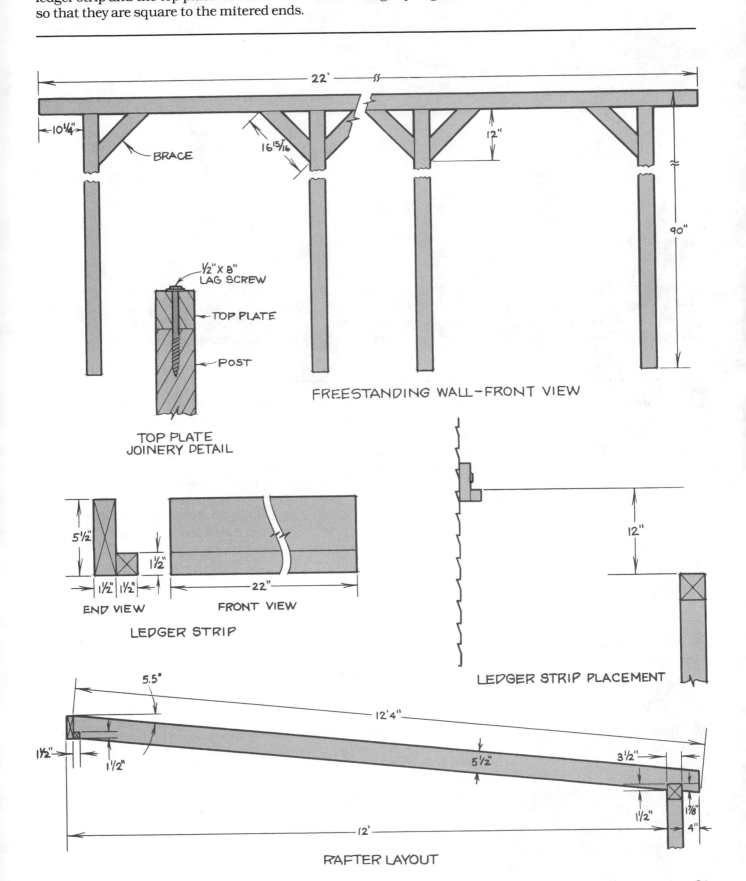

TOP PLATE
JOINERY DETAIL

FREESTANDING WALL—FRONT VIEW

LEDGER STRIP

LEDGER STRIP PLACEMENT

RAFTER LAYOUT

11 Build the roof frame.

Carefully measure along the ledge strip and the top plate, marking where the rafters will go. They should be spaced every 2 feet, as shown in the *Roof Frame, Top View* layout. Nail the rafters in place with 16d nails. Cut a facing strip for the lower ends of the rafters, beveling the top edge at 84½°. (See Figure 5.) Attach this strip to the ends of the rafters with 8d nails.

Note: If you're going to install the optional ceiling, rip the facing strip ¼″ wider than the ends of the rafters. This extra width will hide the edge of the ceiling material.

Figure 5. *Bevel the top edge of the facing strip to match the mitered ends of the rafters.*

5.5°

TIP Just tack the roof frame pieces in place at first; don't drive the nails all the way into the wood. Build the entire frame and check it for squareness. If a rafter or two has to be realigned, it will be much easier to remove the nails. After you're sure the frame is square, hammer the nails home.

---

12 Install crossbraces.

To add strength to your roof, you may want to install crossbraces between the rafters. These crossbraces help evenly distribute the weight of the roof over all the rafters. This is especially important in areas of heavy snowfall, where a lot of weight could pile up in one spot on the roof. If you're going to finish the underside of your roof with a hardboard ceiling, use metal crossbraces—these save a lot of time. If you're not going to install a ceiling, you'll want to do something slightly more decorative. Cut 2 x 4 'spacers' and nail them in between the rafters with 16d nails. Stagger the positions of the spacers from rafter to rafter, as shown in the working drawings.

Finishing the Roof

---

13 Cover the roof frame with plywood.

Attach sheets of ½″ CDX (exterior) plywood sheathing to the roof frame with 8d nails. If you're not going to install a ceiling, be sure to turn the good side down.

---

14 Install a ceiling, if you want one.

Just as you covered the top of the roof frame with plywood, cover the bottom with ¼″ tempered hardboard or chipboard. Use short drywall screws to attach this 'ceiling'. Butt the ceiling material up against the house, the facing strip, and both sides of the top plate. When the ceiling is in place, cut and install side facing strips, as shown in the *Optional Hardboard Ceiling* drawing, to hide the edges of the ceiling material.

TIP If you're going to put a light fixture in your carport, run the electrical lines through the rafters *before* you install the ceiling.

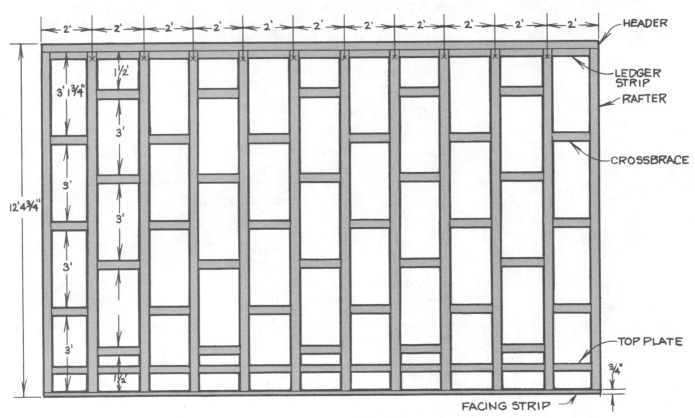

ROOF FRAME - TOP VIEW

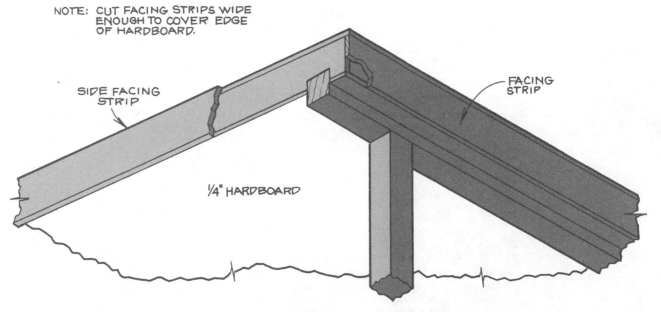

NOTE: CUT FACING STRIPS WIDE
ENOUGH TO COVER EDGE
OF HARDBOARD.

OPTIONAL HARDBOARD CEILING

LEAN-TO CARPORT

15 Install the roofing materials.

Run drip edge around all three sides of the carport. To make the drip edge turn the corners, cut it in two, then snip the bottom flanges where shown in Figure 6. Remove a little piece of one flange, and bend the other back at 90°. Then fit the two pieces together at 90°, as shown in Figure 7. When you attach the drip edge to the roof, the bent flange should be *behind* the other piece. Once the drip edge is in place, cover the entire roof with tarpaper, and then shingles or roofing felt, as shown in the *Roof Cutaway* drawing. Finally, install flashing at the top edge of the roof, to seal the seam between the house and the carport. Use lots of roofing cement under the flashing, to be sure the seam doesn't leak.

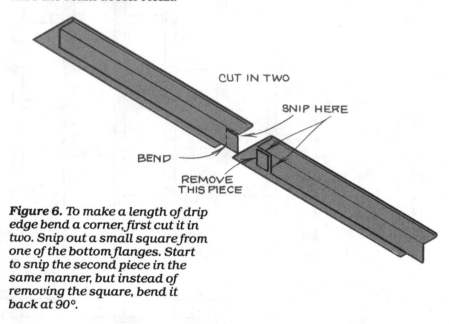

Figure 6. To make a length of drip edge bend a corner, first cut it in two. Snip out a small square from one of the bottom flanges. Start to snip the second piece in the same manner, but instead of removing the square, bend it back at 90°.

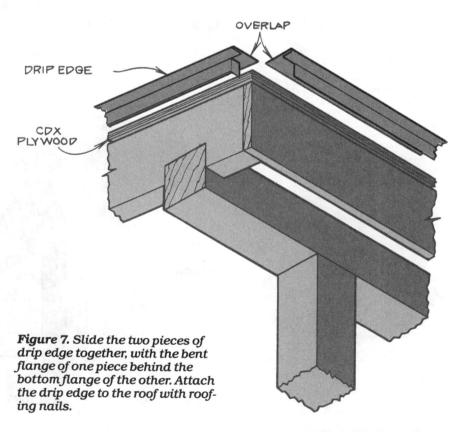

Figure 7. Slide the two pieces of drip edge together, with the bent flange of one piece behind the bottom flange of the other. Attach the drip edge to the roof with roofing nails.

Finishing Up

Paint or stain the carport to match your home. Paint the underside of the roof or the ceiling (if you've installed one) white. This will make the area under the carport seem brighter and cheerier.

16 Paint or stain all exposed wood surfaces.

If the space under the carport isn't paved, you'll want to put down a hard bed of some sort. Grade and level the area as you would in preparation for pouring a concrete pad. Then lay a base of crushed rock or stone. As you spread the stone, keep the base level by checking it with a string level. Finally, spread small-size 'pea' gravel over the base, and tamp it down until the surface is well compacted.

17 Install a gravel floor, if needed.

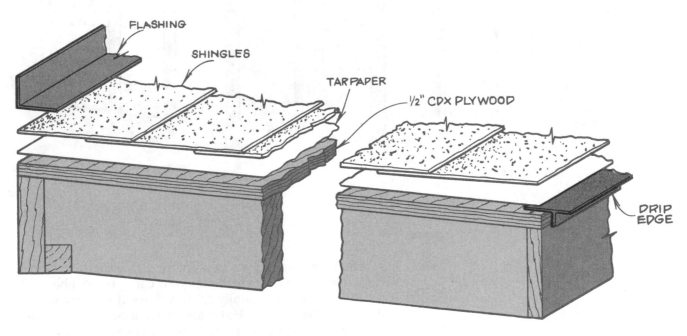

FLASHING

SHINGLES

TARPAPER

½" CDX PLYWOOD

DRIP EDGE

ROOF CUTAWAY

Stand-Alone Carport

This stand-alone carport has many of the virtues of the attached variety. It will serve as a shelter for your car or truck, or as a covered patio for those family get-togethers. It's relatively simple and inexpensive to build. But it also has one additional virtue: You can choose where you want to put it; you don't have to set it right next to your home.

As shown, the carport uses pole construction and a gabled roof. The roof is covered with shingles and the gable ends are covered with plywood siding. You can match these materials with your house to make your carport blend in with your property. Or, if you want your carport to serve double-duty as a patio, you might want to use fiberglass panels instead of solid roofing materials and siding. This will let in more light.

GARAGES AND CARPORTS

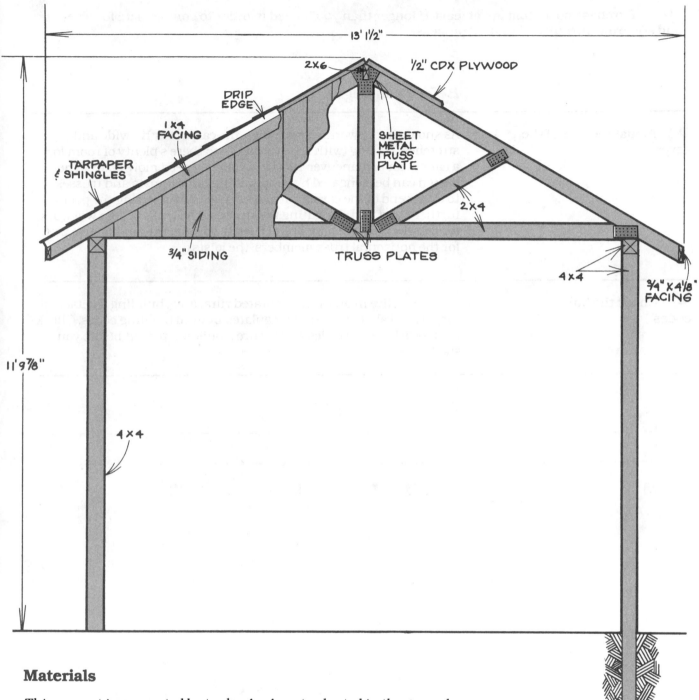

13' 1½"

2 X 6   ½" CDX PLYWOOD

DRIP EDGE

1 X 4 FACING

TARPAPER & SHINGLES

SHEET METAL TRUSS PLATE

2 X 4

¾" SIDING

TRUSS PLATES

4 X 4

¾" X 4⅛" FACING

11'9⅞"

4 X 4

FRONT ELEVATION

Materials

This carport is supported by twelve 4 x 4 posts planted in the ground, so you'll need large rocks and gravel to serve as a drainage base. Select posts that have been pressure-treated or are made of naturally rot-resistant lumber such as redwood so that the ground moisture won't rot them.

Also use 4 x 4 posts for the top plates, a 2 x 6 for the ridgeboard, and 2 x 4's for the trusses. Make the truss plates from ½" thick scrap plywood or purchase metal truss plates. The gable ends are covered with ¾" plywood siding and the facings are cut from 1 x 6 stock.

In addition to these materials, you'll have to purchase ½" CDX (exterior) plywood roof sheathing, metal drip edge, tarpaper, shingles, roofing cement, galvanized nails, roofing nails, lag screws, and gravel for your floor.

Purchase posts that are at least 1′ longer than you'll need in order to compensate for leveling and squaring during the building process.

Before You Begin

1 Adjust the size of the carport.

As shown in the working drawings, the carport is 12′ wide and stretches 21′ long (with the overhang), so there's plenty of room for a car or van to maneuver. However, you may need more space. The design can be elongated by simply adding more posts and trusses to either end of the carport. It can be widened by setting the posts further apart and elongating the trusses. The truss design will work for trusses up to 16′ long. If you need to go longer than that, use 2 x 6's for the horizontal truss members, the joists.

2 Check the building codes.

Unless you live in an unincorporated rural area, building a detached carport will almost surely be regulated by local building codes. Check your local building codes and secure a building permit before you start work.

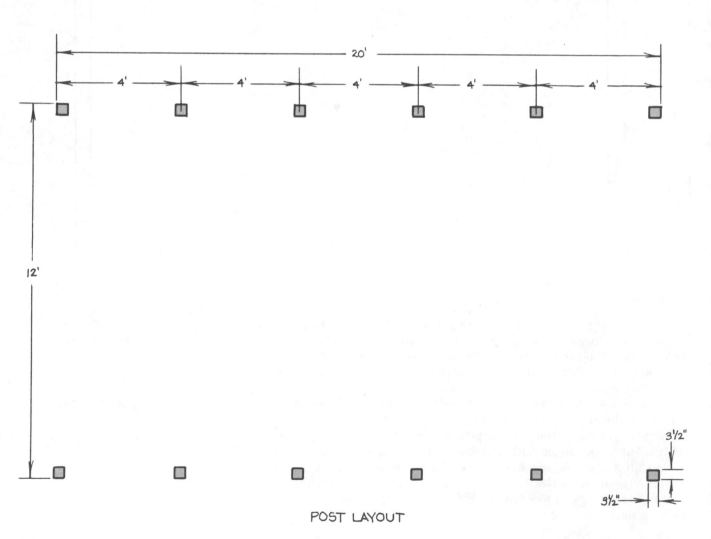

POST LAYOUT

Setting the Posts

Use stakes and string to mark the locations of your posts. For this carport, you'll need to plant 6 posts on each side. (See Figure 1.) Follow the *Post Layout* drawing to determine the exact location of your posts. If you have changed the dimensions of the carport, remember to space your posts every 4' to give the structure adequate support. Mark the location of the posts with stakes, then dig the holes 24"-36" deep. The holes must be deeper than the frost line for your area, and at least twice as wide as your posts. This will allow space for packing gravel and dirt around the posts.

3 **Lay out the posts.**

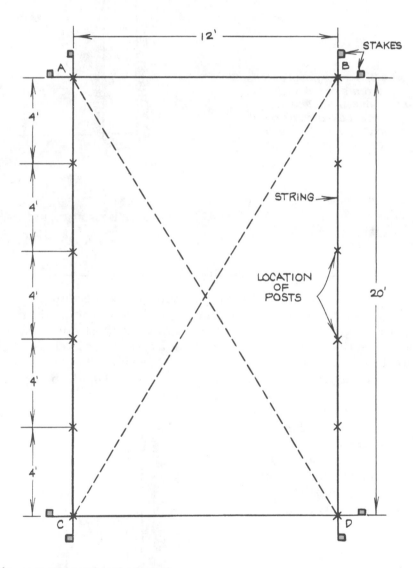

Figure 1. Determine the location of your posts by running string between eight stakes and then measuring along the string to find the position of each post. Mark the position of each post with a stake. You can check to be sure your layout is square by measuring diagonally from corner to corner. Line AD must equal BC.

4 Set the posts in the ground.

Place a large rock in the base of each hole to keep the posts from settling. These rocks should be twice the diameter of the posts—at least 8″ in diameter. (See Figure 2.) With the help of a friend, raise the posts in place, then shovel gravel into the hole to a depth of 12″. This will help drain the ground water away from the posts. Finally, fill the rest of the hole with dirt and tamp lightly around the post.

Figure 2. Set the posts at least 2′ in the ground or below the frost line in your area. Rest the posts on a large rock and surround them with gravel to help drain water away from the posts.

24″-36″

TIP Don't tamp the earth completely until after you have aligned the posts with a level and braced them so that they are perfectly straight up and down.

5 Cut the posts to the proper height.

Once the posts are in the ground, you should brace them upright by driving stakes about 3′ away and nailing scrap lumber from the post to the stakes. You must use at least two braces per post, and these must be at right angles to each other. Use a carpenter's level to make sure they are plumb. When the posts have been braced, cut them to the proper height. To do this, measure one post and mark the top 92½″ above the ground. Using this mark as a reference, find the tops of the other posts with a string level. (See Figure 3.) Remove the string and cut the posts off with a handsaw. Check again that the posts are still plumb, and tamp the dirt down as tight as you can.

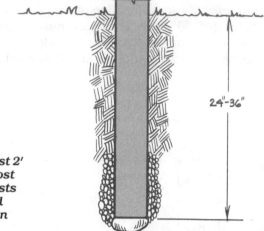

STRING

STRING LEVEL

86½″

Figure 3. Brace the posts to hold them plumb—each post needs at least two braces, at right angles to each other. Use a string and string level to mark the tops of the posts. After you've marked the posts, cut them off at the proper height with a handsaw.

90

Alternate method: If you're uncomfortable up on a ladder cutting the posts off, try this instead. Don't pack any gravel or dirt around the posts before you mark them, just brace them upright. Find the tops with a string level, then remove the braces. Take the posts out of their holes and lay them across two sawhorses to cut them off. *Be sure to mark which post goes in what hole.* After cutting them, put them back in their holes, pack the holes with gravel and dirt, and brace the posts upright. When you're satisfied they're all plumb, tamp the dirt down.

Building the Roof Frame

Cut the top plates from 4 x 4 stock. The top plates run parallel to each other, along the top of each row of posts. There are no top plates running between the rows—the open walls are tied together by the trusses. If you can't get a 4 x 4 long enough, you will have to join two shorter boards with a lap joint. Plan this lap joint so that it occurs right over a post. (See Figure 4.) Put the top plate in place and attach it to the posts with lag screws, as shown in the *Top Plate Joinery Detail* drawing. Cut the braces from 4 x 4 stock, mitering the ends at 45°, as shown in the *Wall Frame, Side View* drawing. Attach them to the posts and the top plates with lag screws.

6 Attach the top plates and braces.

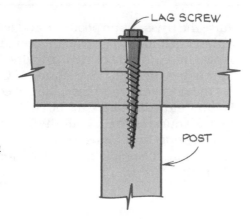

Figure 4. *If you can't buy a 4 x 4 long enough for the top plates, join two shorter boards with a lap joint. Plan the joint so that it will be directly over a post.*

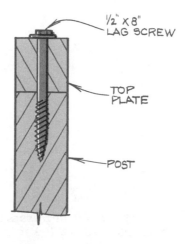

TOP PLATE JOINERY DETAIL

7 **Build the trusses.**

Cut the trusses from 2 x 4 stock and miter the ends as shown in the *Truss Layout* drawing. Attach the truss parts—joists, rafters, and braces—with 4d nails and truss plates. You can buy most of these truss plates, but you'll have to make the plate at the peak. Note that there is a 1½" gap at the peak between the two rafters. The ridgeboard fits in this gap. The truss plate at the peak must be 'Y' shaped, with a corresponding gap (as shown in Figure 5), to tie the 2 x 4 parts together properly. Make this special truss plate from plywood scraps or galvanized sheet metal.

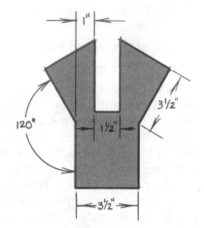

Figure 5. Cut the peak truss plates as shown from plywood or galvanized sheet metal. The ridgeboard rests in the notch of the 'Y'.

---

8 **Attach the trusses to the top plate.**

Raise the trusses into position, and temporarily brace them upright with scrap lumber. The trusses should be spaced 24" on center, as shown in the *Truss Location* drawing. Toenail the trusses to the top plate using 16d nails. (See Figure 6.) Once the trusses are in place, cut the ridgeboard and bevel the top edge, as shown in the *Ridge Board Detail* drawing. Cut the ridgeboard 10½" longer than the span of your roof to allow for the 5¼" overhang of the roof. (For a 20' roof, the ridgeboard must be 20' 10½" long.) Once the ridgeboard is cut, set it in the notch in the top of the trusses and secure it using 16d nails.

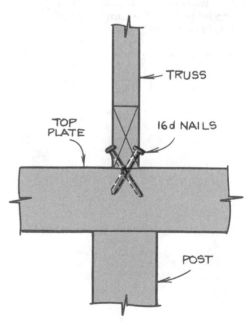

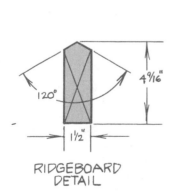

RIDGEBOARD
DETAIL

Figure 6. Toenail the trusses to the top plates using 16d nails.

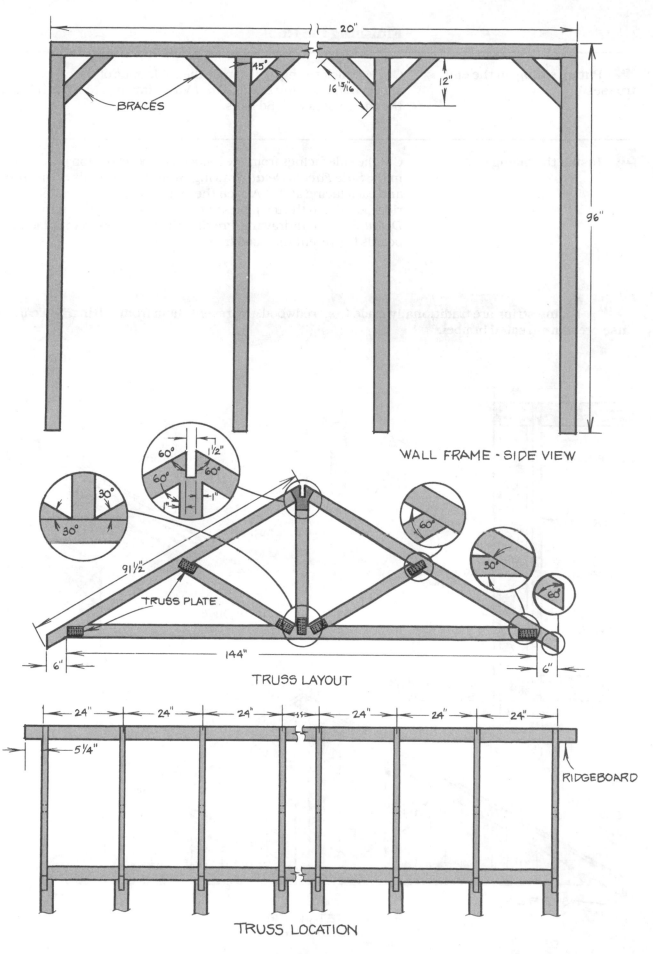

20"

45°

BRACES

16 15/16"

12"

96"

WALL FRAME · SIDE VIEW

60° 1½"

60° 60°

1"

1"

30°

30°

60°

30°

60°

91½"

TRUSS PLATE

144"

6"

6"

TRUSS LAYOUT

24" 24" 24" 24" 24" 24"

5¼"

RIDGEBOARD

TRUSS LOCATION

9 **Put up siding on the end trusses.**

Cut ¾″ thick siding to fit across the gable ends of the carport, as shown in the *Facing Detail, Front View* drawing. Nail the siding to the end trusses with 6d nails.

10 **Install the facing.**

Cut the side facings from 1 x 6 stock and bevel the top edge as shown in the *Side Facing Detail* drawing. Miter the top ends of the front and back facing at 30°. Attach the facing strips to the rafters and ridgeboard and the top plates with 6d nails, as shown in the *Facing Detail, Side View* drawing. Pre-drill the nail holes in your facing boards to prevent the wood from splitting.

TIP Facing strips are traditionally made from redwood, to prevent them from rotting. You can also use pressure-treated lumber.

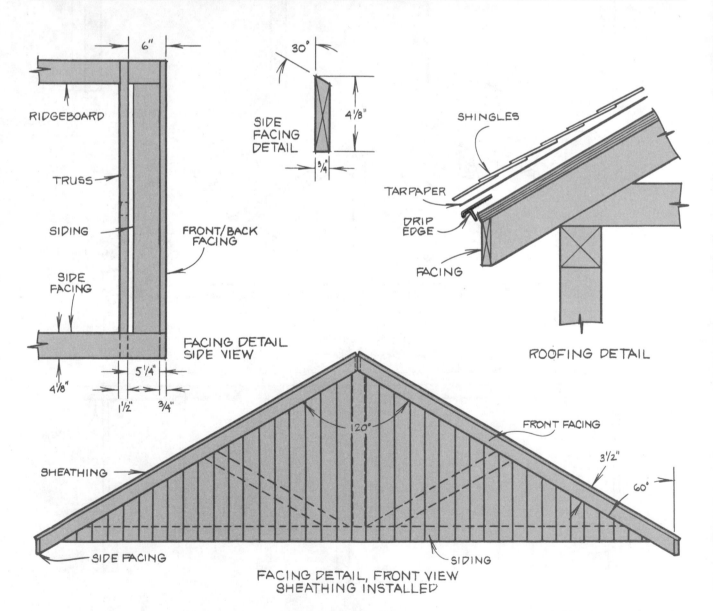

Attach sheets of ½″ CDX (exterior) plywood sheathing to the roof frame with 6d nails. If you're not going to install a ceiling, be sure to turn the good side down.

11 Install the roof sheathing.

Run drip edge around all four sides of the carport. To fit the drip edge to the peaks, snip the bottom flanges where shown in Figure 7. Bend the drip edge to the proper angle and nail it in place with roofing nails. Once the drip edge is in place, cover the entire roof with tar-paper, then shingles, as shown in the *Roofing Detail* drawing.

12 Install the roofing materials.

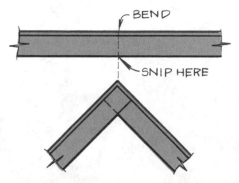

Figure 7. *To make a length of drip edge bend at the peak, snip the bottom flange, then bend it to fit the peak.*

TIP Lengthen the life of your roof by double-wrapping shingles around the highest part of the roof to form a ridge cap.

Finishing Touches

Paint or stain all exposed wood surfaces to prevent them from becoming damaged by the weather. If you've built your carport entirely from pressure-treated lumber, redwood, or cedar, you need not paint or stain it.

13 Paint the wood surfaces.

Level the area, digging out the high spots and filling in low areas. Then spread a base of rock gravel or crushed limestone and roll it until it is smooth and level. Spread pea-size gravel over this base and roll it until it is also smooth and level. Tamp down both layers to compact the gravel.

14 Install a gravel pad.

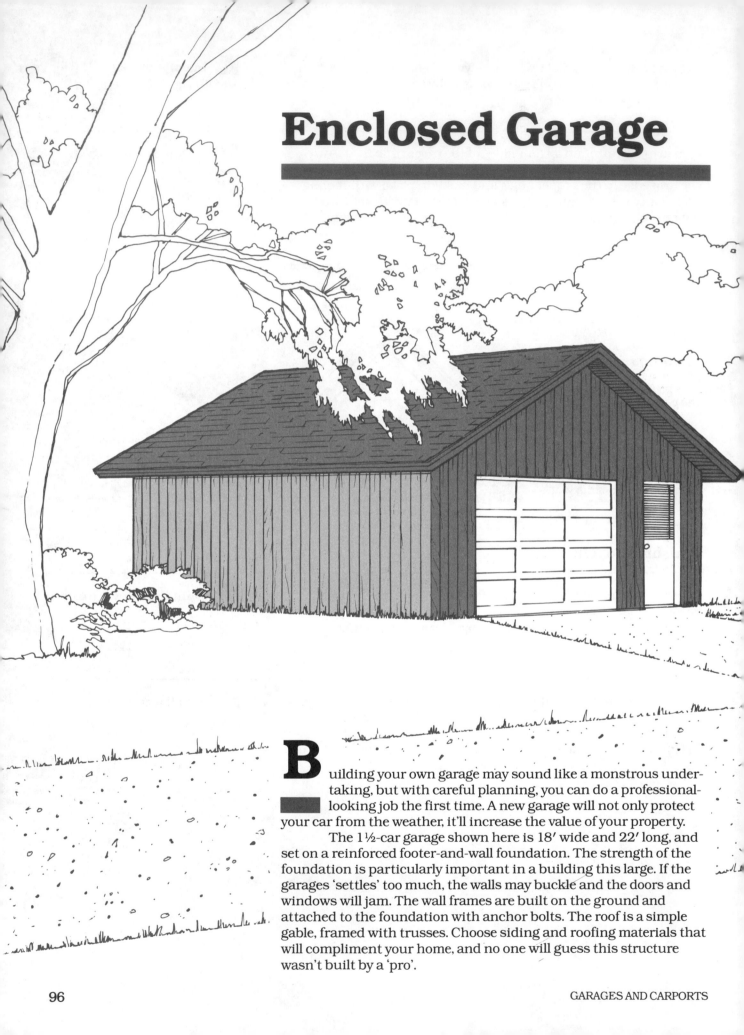

Enclosed Garage

Building your own garage may sound like a monstrous undertaking, but with careful planning, you can do a professional-looking job the first time. A new garage will not only protect your car from the weather, it'll increase the value of your property.

The 1½-car garage shown here is 18' wide and 22' long, and set on a reinforced footer-and-wall foundation. The strength of the foundation is particularly important in a building this large. If the garages 'settles' too much, the walls may buckle and the doors and windows will jam. The wall frames are built on the ground and attached to the foundation with anchor bolts. The roof is a simple gable, framed with trusses. Choose siding and roofing materials that will compliment your home, and no one will guess this structure wasn't built by a 'pro'.

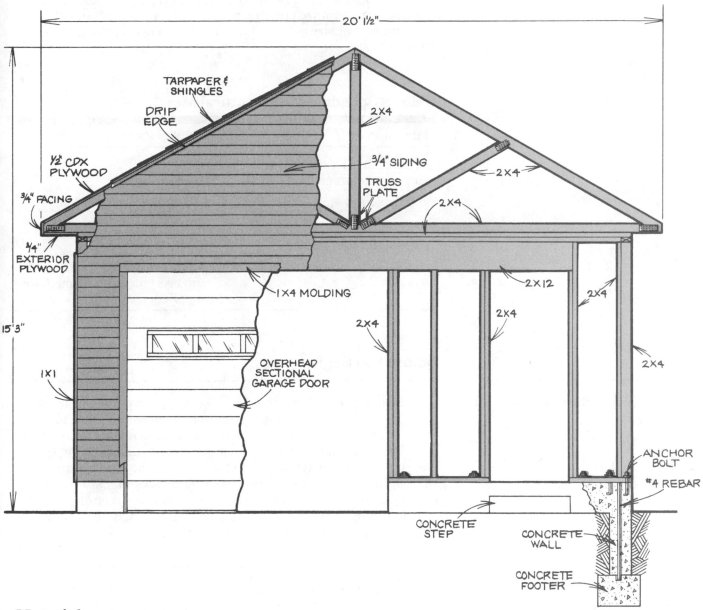

20' 1½"

TARPAPER &
SHINGLES

DRIP
EDGE

2X4

¾" SIDING

TRUSS
PLATE

2X4

2X4

½" CDX
PLYWOOD

¾" FACING

¾"
EXTERIOR
PLYWOOD

2X12

2X4

1X4 MOLDING

2X4

15'3"

2X4

1X1

OVERHEAD
SECTIONAL
GARAGE DOOR

ANCHOR
BOLT

#4 REBAR

CONCRETE
STEP

CONCRETE
WALL

CONCRETE
FOOTER

Materials

To build the garage as shown, you'll need 2 x 4's, 2 x 8's, and 2 x 12's to complete the frame. Most of the wall frames and roof trusses are built from 2 x 4's. Make the sole plates from pressure-treated or rot-resistant lumber, since these will sit on the concrete foundation. The roof joists are 2 x 8's and the headers, above the doors and windows, are 2 x 12's. You'll also need ½" thick CDX plywood to sheath the roof, and some 'one-by' (¾" thick) stock to make the facing strips and other trim.

To make the foundation, purchase #4 reinforcing bars, stakes, 1 x 2 spreaders, wire ties, anchor bolts, and (of course) concrete. You'll also need ¾" plywood and 2 x 4 scrap lumber to build the forms. To finish the garage roof, buy drip edge, tarpaper, and shingles. To finish the walls, purchase siding that matches or compliments your home.

FRONT VIEW

Depending on the type of siding you buy, you may also need sub-siding.

Purchase the windows, entrance door, and overhead garage door with ready-made casings and trim from your local lumberyard or building supplier. In addition to these materials, you'll also need truss plates, galvanized roofing nails, and lots of common nails. To determine what size nails and how many you need, refer to the following chart:

Nailing Schedule for Structural Members

Material	Number and type of nail
Plate to stud—end nail	2—16d
Stud to plate—toenail	4—8d or 3—16d
Doubled studs—face nail	16d, spaced every 12″
Doubled top plates—face nail	16d, spaced every 12″
Trusses to plate—toenail	2—16d
½″ plywood roof sheathing	8d spaced every 6″ at the edges and every 12″ at intermediate supports

Before You Begin

1 Adjust the dimensions of the garage.

As shown in the working drawings, the garage is 18′ wide and 22′ long. This will make a roomy 1 ½-car garage. However, if this is too small or too large for your needs, adjust the dimensions to suit yourself. There is a chart in the beginning of this section showing the standard sizes of all types of garages, from a basic 1-car garage, all the way up to a 2-car garage with some extra space for storage. You can expand or shrink the plans for this garage in either direction by simply adding or subtracting wall studs, and stretching or shortening the trusses. You can raise the roof by using longer wall studs.

2 Plan the door and window openings.

After you've decided on the overall dimensions of your garage, carefully plan the wall frames, positioning the door and window openings. Read the directions supplied by the manufacturers of these ready-made door and windows to see just how big the openings in the frames should be, and where they should be placed.

3 Check the building codes.

Unless you live in a rural area or unincorporated district, this garage will likely be affected by building codes. Check your local codes and pay particular attention to regulations governing the location of outbuildings. You'll probably find that, unless you apply for a variance, you'll have to locate this garage several yards back from your property line. If needed, secure a building permit before you start work.

Pouring the Foundation

Use stakes and string to lay out the foundation. Locate the first line by measuring carefully from the property line or from an existing structure. Drive the stakes outside the foundation lines—the intersection of the strings will mark the corners. (See Figure 1.) Make sure the corners are square by measuring diagonally from corner to corner. Mark the garage door opening with stakes, as shown in the *Foundation Wall Layout* drawing. Mark the borders of the trench footing on the ground with a trickle of sand and drive the stakes clear of the digging area to fix the marks.

4 Stake out the foundation.

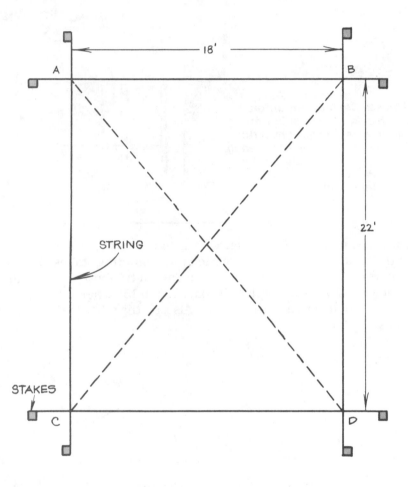

Figure 1. Use stakes and strings to lay out the foundation of your garage. Measure diagonally from corner to corner to be sure the foundation is square. AD must equal BC.

TIP Pouring a deep foundation is sometimes best left to professionals. It requires an enormous amount of heavy equipment and materials to move the earth, put up the proper forms, pour the concrete, and backfill the trench after the forms are removed. We'll show you how to do it so you know what's involved, but you can save yourself a big headache by having the foundation done by someone who specializes in concrete work.

5 Dig the trench.

Dig a trench that is 24″ to 36″ deep, so that it extends below the frost line for your area. Make the trench wide enough on both sides to allow you to stand and stoop in it. When the trench is completed, dig another trench that is 12″ deep and 16″ wide for the footer. (See Figure 2.) Along both sides of the footer trench, about 4″ from the sides and at 3′ intervals, drive grading stakes made from 18″ lengths of 'rebar' (reinforcing rod). Use a string level to make sure that the tops of these stakes are level with each other, all around the foundation.

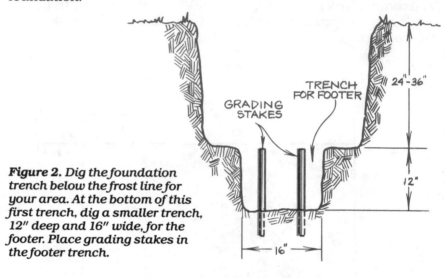

Figure 2. Dig the foundation trench below the frost line for your area. At the bottom of this first trench, dig a smaller trench, 12″ deep and 16″ wide, for the footer. Place grading stakes in the footer trench.

6 Set rebar in the footer trench.

Your footer must be strengthened with #4 reinforcing rod or 'rebar'. Set the rebar along the inside of each row of grading stakes. Support the bars on bricks or stones so they are about 3″ above the ground, as shown in Figure 3. Overlap the bars 12″ to 15″ where they meet and tie them together with wire ties. Also, tie the rebar to the grading stakes.

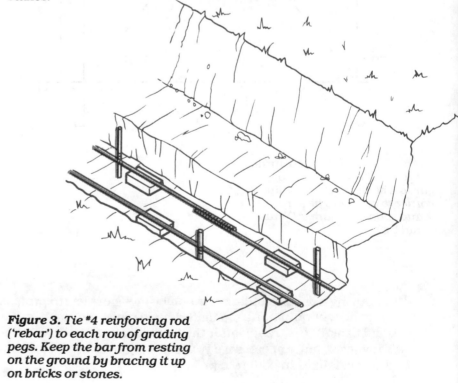

Figure 3. Tie #4 reinforcing rod ('rebar') to each row of grading pegs. Keep the bar from resting on the ground by bracing it up on bricks or stones.

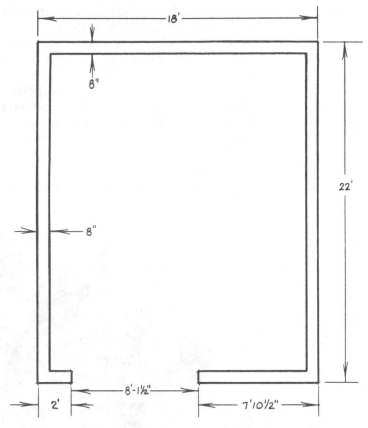

FOUNDATION WALL LAYOUT

With the help of a friend, pour the concrete into the footer trench and spread it with square-tipped shovels. Dig into the concrete with the shovels or a yardstick to break up any air pockets. Make a 'float' from 2 x 4's nailed together in a 'T' to form a handle and a flat working surface. (See Figure 4.) Using this float, level the concrete as you pour. When the footer is filled, pat down the concrete, then zigzag the float horizontally across the surface until the concrete is fairly smooth. Finally, sweep the surface with the trailing edge of the float working diagonally toward you. The tops of the grade pegs should be barely visible above the surface of the concrete. Make a notch or key in the footer by pressing a 2 x 2 into the center of the wet concrete. Allow the concrete to cure at least 24 hours, then remove the 2 x 2.

7 **Pour the concrete for the footer.**

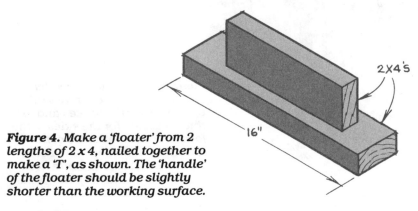

Figure 4. *Make a 'floater' from 2 lengths of 2 x 4, nailed together to make a 'T', as shown. The 'handle' of the floater should be slightly shorter than the working surface.*

8 Build the forms for the foundation walls.

Use 2 x 4 studs and ¾″ plywood sheathing to build the forms, as shown in Figure 5. Make sure the wooden frame extends far enough above the ground to allow you to pour the concrete wall 12″ above the ground. Attach 2 x 4 cross bracing to the forms to tie them together above the plywood sheathing. In addition to the braces across the top of the forms, use wire ties and spreader blocks to keep the sides of the forms from shifting. The wire ties keep the form sides from bowing out while you're pouring the concrete. To put them in place in the forms, drill small holes in the sides and pass the wire through the holes. Place a 1 x 2 spreader near each tie so you won't pull the form sides too close together when you tighten the wire. Then twist the wire with a screwdriver or small block of wood until it is tight.

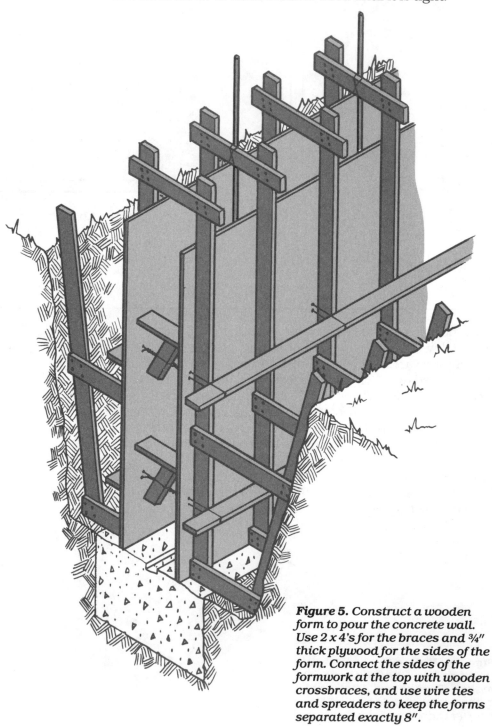

Figure 5. Construct a wooden form to pour the concrete wall. Use 2 x 4's for the braces and ¾″ thick plywood for the sides of the form. Connect the sides of the formwork at the top with wooden crossbraces, and use wire ties and spreaders to keep the forms separated exactly 8″.

Cut #4 reinforcing bars so they will extend about 1' above the forms. Place the rebars vertically inside the forms and attach them to the wire ties with a short length of wire or string. Space the rebar about 24" apart throughout the foundation wall.

9 Set the rebar vertically in the forms.

With the help of a friend, pour the concrete into the forms. You will probably want to order a ready-mix delivery, rather than go through the hassle of mixing your own. If the concrete truck cannot back up to the site, arrange for extra wheelbarrows or a pump truck. As the concrete is being poured, remove the spreaders. Leave the wire ties in place—you can cut them flush with the sides of the concrete after it has cured.

10 Pour the concrete wall.

> **TIP** After the concrete has been poured, use a concrete vibrator to settle the concrete and remove any air pockets. You can rent these vibrators at most tool rental companies.

While the concrete is still wet, set the anchor bolts near the corners and every 24" on center around the walls. Be sure that the position of the bolts doesn't coincide with any of the wall studs. The bolts should extend 2½" from the top of the foundation, as shown in Figure 6. Let the concrete cure at least 24 hours before you continue construction. Then remove the wooden forms, cut the wire ties flush with the sides of the wall, and saw the rebar flush with the tops of the walls. Spray the walls with a sealant or run a plastic sheet along the sides of the wall to keep ground moisture from soaking into and weakening the concrete. When this vapor barrier is in place, throw gravel into the foundation trench and fill it with earth, tamping as you go.

11 Set the anchor bolts.

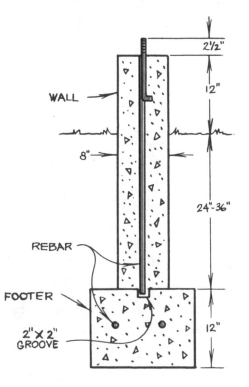

Figure 6. The anatomy of a solid footer-and-wall foundation includes a reinforced footer with a 2 x 2 groove to help hold the wall in place. The wall should also be reinforced, and should extend at least 12" above the ground. Set the anchor bolts in the top of the wall so that they protrude 2½".

12 **Cut and drill the sole plates.**

Cut the sole plates as shown in the *Back Wall Layout*, *Side Wall Layout*, and *Front Wall Layout* drawings. Lay these in place on the foundation wall, next to the anchor bolts. With a carpenter's square, mark the position of the anchor bolts. (See Figure 7.) Then drill the holes about ¼" larger than the diameter of the bolts so the wood will fit easily over the bolt but still be snug enough to prevent shifting. For example, drill a ¾" hole to fit over a ½" bolt. Lay the sole plate in place to check that the holes fit over the bolts.

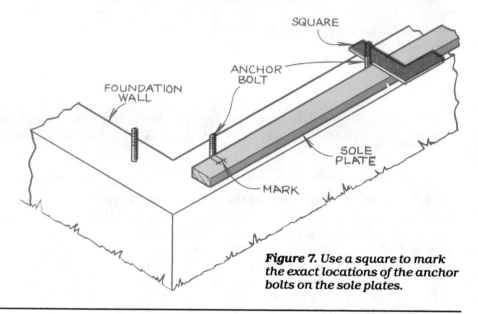

Figure 7. *Use a square to mark the exact locations of the anchor bolts on the sole plates.*

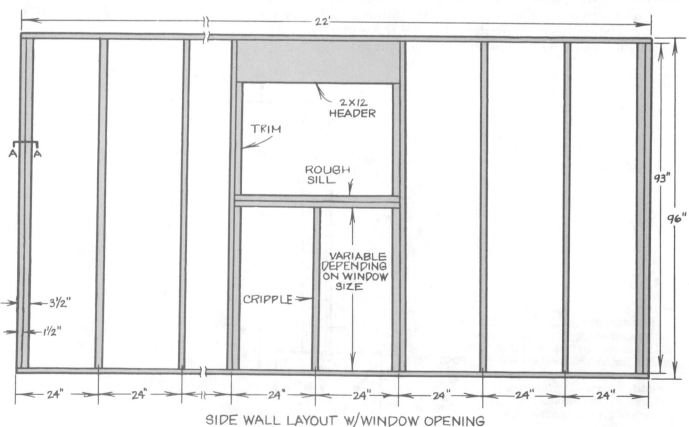

SIDE WALL LAYOUT W/WINDOW OPENING

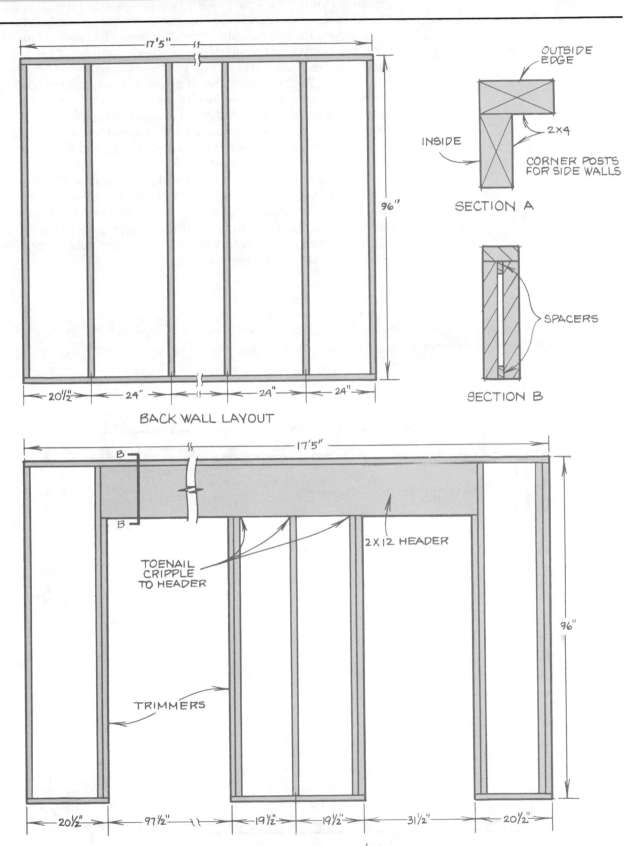

17'5"

96"

20½"   24"   24"   24"

BACK WALL LAYOUT

OUTSIDE EDGE

INSIDE

2×4

CORNER POSTS FOR SIDE WALLS

SECTION A

SPACERS

SECTION B

B

17'5"

2×12 HEADER

TOENAIL CRIPPLE TO HEADER

96"

TRIMMERS

20½"   97½"   19½"   19½"   31½"   20½"

FRONT WALL LAYOUT W/DOORS

ENCLOSED GARAGE   105

13 Build the wall frames on the ground.

Lay the sole plates top side up on the ground. Cut the top plates and lay them, bottom side up, next to the sole plates. Using a framing square, mark the position of the studs according to the *Stud Layout* drawing. Cut the 2 x 4 parts—studs, cripples, trimmers, and sills as shown in the drawings. Build the frames on the ground and nail the top plate, studs, and sole plate together using 16d nails. Notice that the corner posts for the side walls are made from two 2 x 4's, as shown in the *Section A* drawing. To frame the windows and doors, use two 2 x 12's for headers. Set them on edge, separated with ½" blocks, as shown in the *Section B* drawing. (The spacers bring the total width of the header to 3½", to match the top plate.) Nail the headers together. Place the assembled headers in the frames and nail them to the frame parts with 16d nails. Finally, nail the cripples, sills, and trimmers in place around the door and window openings.

Note: Make sure your cripples and trimmers meet local codes. Some areas may require larger stock than what we show, particularly around the main door. Also, check garage door manufacturer's instructions for minimum clearance required for garage door hardware and automatic door opener before you construct the door frame. You may need to adjust the dimensions.

TIP If you are using wall sheathing and siding, it's easier—and faster—to install the sheathing while the wall is on the ground. Just be certain the walls are square before nailing the sheathing in place.

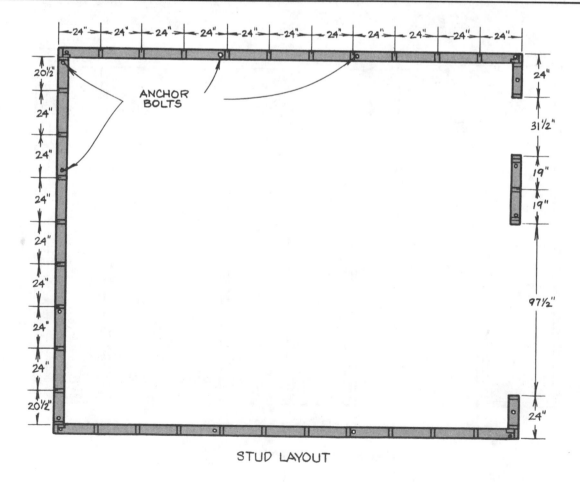

STUD LAYOUT

Put a plastic strip over the top of the foundation wall to serve as a vapor barrier between the sole plate and the concrete. Cut this barrier so that the anchor bolts stick up through it. With the help of a friend, gently lift the side wall frames in place and fit the sole plates over the anchor bolts. Fasten the sole plates to the foundation with washers and nuts. Then raise the front and back walls and attach them to the side corner posts using 16d nails placed 24" on center. (See Figure 8.) Use a carpenter's level to make sure all the studs are plumb vertically and the top plate is level horizontally. Then temporarily brace the frame to keep it square with some long 1 x 4's, nailed to the wall frames at a diagonal.

14 **Raise the wall frames and temporarily brace them.**

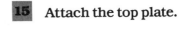

Figure 8. When the walls are all nailed together, the arrangement of studs at the corners should look like this. Not only does this arrangement strengthen the corners, it provides a nailing surface for drywall or paneling, should you want to finish the inside of the building.

TIP To level a wall, drive a wooden wedge under the sole plate until the top is level. Then place washers and nuts on the anchor bolts and tighten. Later, push cement grout under the bottom plate to give the wall added support.

To tie the wall frames together, nail a second top (or 'cap') plate down onto the top plate using two 16d nails placed 16" on center. Be sure that the ends of the cap plates lap the joints between the walls. (See Figure 9.)

15 **Attach the top plate.**

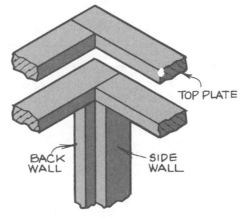

Figure 9. Nail a second top (cap) plate over the first to tie the walls together. The cap plates must overlap the joints between the wall frames.

16 **Build the trusses.**

Cut the joists, braces, and rafters for the roof trusses from 2 x 4 stock. Miter the ends of the parts at either 60° or 30°, as shown in the *Truss Layout* drawing. Lay out a single truss on the ground. (As you assemble this truss, remember that the roof overhangs the wall frame 1' on both sides.) Nail the parts together with metal truss plates at each joint. Turn the truss over and put truss plates on the other side. Use this first truss as a 'template' for the rest of the trusses —build them on top of the first. That way, all the trusses will be exactly the same.

TIP Use pre-engineered and pre-assembled trusses to save time. These are available at most building supply companies. If you can't find them to fit the span you need, you can have them cut and assembled for you. Check local building codes to be sure your trusses meet local requirements.

17 **Toenail the trusses to the top plate.**

Place the trusses every 24" on center as shown in the *Truss Spacing* drawing. Important: No matter what the length of the garage, don't space the trusses any further apart than 24". Any further apart, and the roof won't have the proper support. Toenail the trusses to the top plate using 16d nails. (See Figure 10.) Or use metal framing anchors to attach the trusses to the top plate. These eliminate the need for toenailing and provide a stronger connection. Temporarily brace the trusses to hold them upright.

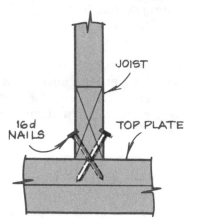

Figure 10. Toenail the trusses to the top plate using 16d nails.

Finishing the Roof

18 **Install the soffit.**

Cut the soffit boards from ¾" thick exterior plywood and nail them to the bottom of the overhang with 6d nails. The edge of the soffit must be flush with the ends of the joists.

Although the siding is traditionally installed after the roof is finished, it's easier to put the siding on just the front and back walls at this point. The reason is that the roofing materials overlap the siding on the gable ends, and therefore the siding has to be installed before the roof. First, cover the gable ends with sheathing (if you're using sheathing). Then nail siding to the front and back walls using 6d galvanized box nails. Make sure the siding goes down to the bottom of the sole plate, but doesn't touch the concrete. Moisture from the concrete can cause the siding to rot.

19 **Install the siding on the front and back walls.**

TIP Save time and money by using ready-to-finish structural siding such as square edge rough sawn pine that will act as a wall sheathing and siding. It is available in 4′ x 8′ sheets and will save you the time and expense of installing both sheathing and siding.

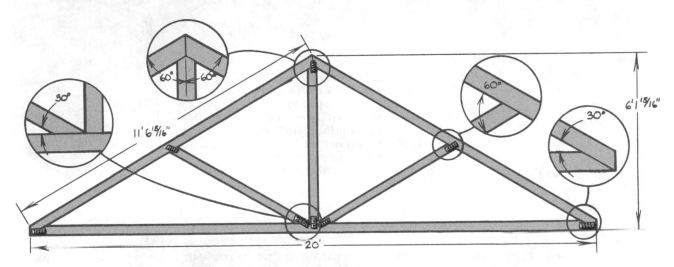

TRUSS LAYOUT

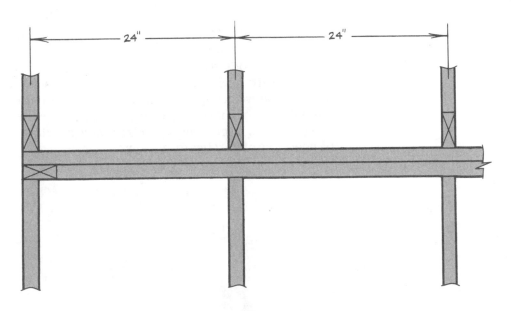

TRUSS SPACING

20 **Attach the facing strips to the truss ends.**

Rip the side facings from 'one-by' (¾″ thick) stock, beveling the upper edge at 30°, to match the pitch of the roof. The facings should lap the soffit, as shown in Figure 11. Attach the facing to the truss ends with 6d nails.

SIDING

SHEATHING

FACING STRIP

SOFFIT

CORNER BLOCK

Figure 11. Install soffit first, then the front and back wall siding. Attach the facing strip to the truss ends so that it laps the edge of the soffit. Finally, install the roof sheathing so that it's flush with the top of the siding at the front and back, and the facing strip at the sides.

TIP The facings are exposed to a lot of moisture from rainwater run-off. To help prevent rot, make them from redwood or pressure-treated lumber.

---

21 **Cover the trusses with sheathing.**

Attach sheets of ½″ CDX (exterior) plywood sheathing to the roof frame so that it is flush with the edge of the facing at the sides and flush with the siding on the front and back, as shown in Figure 11. Lay the sheathing on the roof so that the panels are 'staggered'—no two joints between adjacent sheets should fall on the same rafter. This will give the roof extra strength. Nail the sheathing to the trusses using 6d nails. If you're not going to install a ceiling, be sure to turn the good side of the sheathing down.

Run drip edge around all four sides of the roof. Once the drip edge is in place, cover the entire roof with a double layer of tarpaper or roofing felt, lapping this underlayment at least 2″ at all horizontal joints and 4″ at all vertical (end) joints. Install the shingles according to the manufacturer's instructions. (See Figure 12.) Double-wrap the shingles along the roof ridge to make a ridge cap.

22 **Install the roofing materials.**

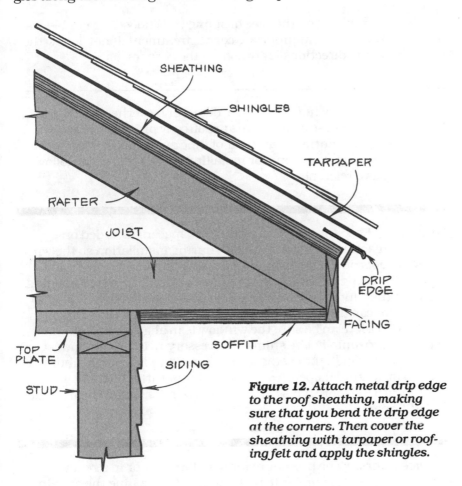

Figure 12. *Attach metal drip edge to the roof sheathing, making sure that you bend the drip edge at the corners. Then cover the sheathing with tarpaper or roofing felt and apply the shingles.*

Finishing Up

Nail sheathing and siding to the side walls using 6d galvanized box nails. As before, make sure the siding goes down to the base of the sole plate but doesn't touch the concrete or the ground. Fit the siding flush with soffit at the top of the wall. At the corners, *do not* install the side wall siding over the front or back wall siding. Instead, install the side wall siding so that the ends are flush with the side wall frames, as shown in the *Corner Block Detail* drawing. Once all the siding is installed, you can remove all the temporary bracing from the wall frames.

23 **Install the siding on the side walls.**

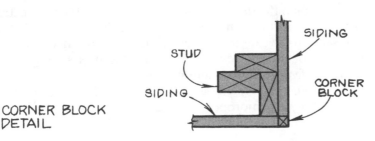

CORNER BLOCK DETAIL

24 **Install the corner blocks.**

To finish the siding, run 1 x 1 strips down the corner edges, hiding the ends of the siding. First, apply a generous bead of caulk down the space between the wall sidings to help keep moisture from seeping through the seams. Then press the corner blocks into the space and nail them in place with 6d nails, placed 6" on center. You may want to pre-drill the nail holes in the corner blocks to keep them from splitting.

Note: Depending on the type of siding you choose, corner blocks may or may not be an appropriate corner treatment. If not, follow the manufacturer's directions for finishing the corners.

---

25 **Install the windows, doors, and garage door.**

Use manufactured windows, service doors, and overhead garage door for your garage. These components are already pre-engineered, and they save you an enormous amount of time and expense. Follow the manufacturer's instructions for installing them and use the following guides as a reference.

Windows. Keep the temporary braces or spacers included on your manufactured windows in place during installation so the windows will not come out of alignment. To install, cut 8" to 10" wide strips of building paper and tack in place around the rough opening. Then set the window unit in the rough opening from the outside so the exterior casing or the installation flange overlaps the siding. Check with a level to be sure the window is in alignment both vertically and horizontally. Use shims, if necessary, to level the window. Drive nails through the casing at the corners and check the movable sash to make sure it operates smoothly. Finish attaching the windows by driving more nails through the casing, spacing them about 10"-12" apart.

Service Doors. The procedure for installing a door in a ready-made casing is very similar to installing a window. Line the opening with building paper, then set the unit in place. Check that the door is square. If it isn't, shim the casing so that it is. Put a few nails through the casing to hold it in place, then check the action of the door. If it opens and closes smoothly, finish nailing the casing in place.

Overhead Door. To install a sectional door, first frame in the opening according to the manufacturer's directions. You will probably have to make your own casing from 'one-by' stock. Place the first section in position and level it. Cut the bottom edge to conform to the garage floor. Apply weather stripping to the bottom of the first section to help seal the gap between door and floor, and to act as a cushion when the door closes. Next, install the hardware, track, and springs according to the manufacturer's directions and attach the door sections to the track. After the door is attached, install an electric door opener according to the manufacturer's directions.

Note: If you choose a door with torsion-springs you may want to have it professionally installed. Torsion-springs can cause serious injury if they are improperly handled.

Cut 1 x 3 stock to trim the windows and doors. (In some cases, the trim will be supplied with the ready-made casings.) Miter the ends so the trim pieces fit together smoothly at the corners. Attach the trim to the edge of the casing or jamb using 6d finishing nails placed about ⅜″ from the inside edge. Nail along the outside of the trim using 6d or 8d casing nails.

26 Install the door and window trim.

Paint all exposed wood surfaces with exterior paint. If you choose to stain your garage instead, make sure you apply at least two coats of stain to adequately protect your wood from the weather.

27 Paint all exposed wood surfaces.

Attach metal gutters around the sides of the roof according to the manufacturer's instructions. Be sure to plan the location of your downspouts so they are as inconspicuous as possible. In some cases, the manufacturer will suggest you install the gutters after the roof sheathing is in place, but *before* the finish roofing goes on. If that's the case, be sure to paint the facing before you install the gutters, and wait until you've installed the siding before you put up the downspouts.

28 Install gutters, if needed.

Grade and level the area, then lay a base of crushed rock or limestone. As you spread the stone, keep the base level by checking it with a string level. Finally, spread small-size (pea) gravel over the base, and tamp it down until the surface is well compacted.

29 Install a gravel pad, if needed.

Install a step easily by digging out the area about 3″ deep and setting pre-cast concrete blocks in place. Or, build a wooden form from scrap lumber, set it in place, and pour concrete in it. Wait at least 24 hours for the concrete to cure before you use the step.

30 Install or pour a block for the service door step.

Gazebos and Summer Spaces

When summer arrives, there's no better place to enjoy it than your own backyard. To make your summer all the more enjoyable, why not add a gazebo, a sunspace, or a summer house to expand your warm-weather living space? Gazebos and summer houses are enjoying something of a revival. These provide private spaces, away from the house, where you and your family can get away and enjoy the sunshine—sort of an instant vacation. If you have a green thumb, perhaps you'd rather have a sunspace. These are really 'attached' greenhouses where you can grow things for most of the year.

There is nothing unusual or difficult to build these projects. Like the rest of the projects in this book, they require simple carpentry skills, for the most part. The gazebos are perhaps the most difficult to build because they have eight sides. Most carpentry tools and materials are oriented toward building four-sided structures. We're accustomed to making 90° corners; 45° corners are a new experience. But as long as you take your time to carefully plan each step and measure each board, your eight-sided gazebo will go up as easily as a four-sided storage barn or shed.

Remember that you can alter all of these projects, borrowing components and features from other projects in this section, or elsewhere in the book. Just because we show a sunshade gazebo on a pole foundation doesn't mean you can't put it on a concrete pad. Just flip the pages to find the instructions for making a gazebo or summer space with the features you want.

Before You Begin

First, select a site. Since the primary purpose of a summer space is to help you enjoy the outdoors, put your porch or gazebo where you'll get the best view. If you're using the screen house as a fishing lodge, place it where you'll get a bird's-eye view of the lake. Also consider whether you'll want to have electricity or plumbing in your structure. In either case, this will affect your site selection because it's less costly —and time consuming—to locate the project near a power or water source. Finally, think about the traffic flow to and from your shelter— you don't want the kids leaping across the flower beds or wearing a path through the tomatoes to get to your gazebo or summer house.

After you've chosen a good site for this project, adjust the design of your shelter so that it blends in with your home and its surroundings. For example, if you already have a patio or deck, attach a gazebo to one corner. Also, alter the dimensions of your shelter to fit your individual needs and your site. If you are building on a small site

114

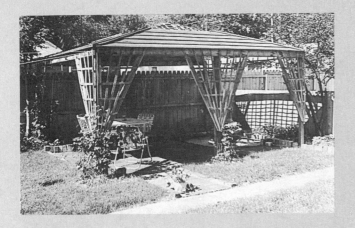

with lots of trees, keep the gazebo or summer house small and cozy so it doesn't overwhelm the scenery.

Once you've chosen a location and a design, take a trip to your local building inspector's office and brush up on local building code requirements. You may need to obtain a permit before you begin building. This is also the best place to find out what foundations and roofing materials work best in your area.

Gazebo Shelter

If you're tired of looking for a quiet place to read or to entertain friends, away from the house but out of the sun and the rain, then this gazebo is just for you. The straight-peak roof keeps out the sun's rays—and rain showers—but still lets you soak up the scenery. Open to the fresh air, yet shaded from the sun, this inviting gazebo will be a family favorite for a long time to come.

This eight-sided gazebo has a wood deck that sits on a pier-and-beam foundation. The peaked roof is sheathed and covered with ordinary roofing materials—shingles or shakes. The shaped posts and decorative handrail add a professional touch—but there's no lathe work to do! All these parts are store-bought.

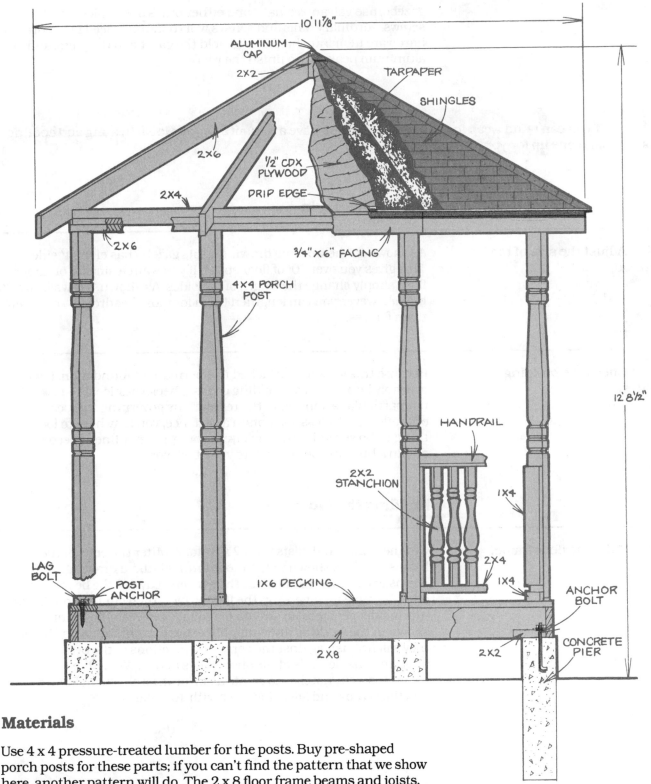

10' 11⅞"

ALUMINUM CAP

2X2

TARPAPER

SHINGLES

2X6

½" CDX PLYWOOD

2X4

DRIP EDGE

2X6

¾"X6" FACING

4X4 PORCH POST

HANDRAIL

2X2 STANCHION

1X4

12' 8½"

LAG BOLT

POST ANCHOR

1X6 DECKING

2X4

1X4

ANCHOR BOLT

2X8

2X2

CONCRETE PIER

SIDE ELEVATION

Materials

Use 4 x 4 pressure-treated lumber for the posts. Buy pre-shaped porch posts for these parts; if you can't find the pattern that we show here, another pattern will do. The 2 x 8 floor frame beams and joists, 1 x 6 decking, and 2 x 2 cleats should also be pressure-treated to protect them from water or insect damage. Purchase 2 x 6's for the rafters, top plate, and cap plate; a 2 x 2 for the key; and handrail stock, pre-shaped spindles, and 2 x 4's for the handrails. Use 'one-by' (¾" thick) stock for the facing strips and the handrail spacers. Sheath the roof with ½" CDX (exterior) plywood.

You'll need some ready-mix concrete for the piers, and drip edge, tarpaper, and shingles to cover the roof. When you assemble this gazebo, use galvanized nails and either brass or stainless steel screws—ordinary nails and screws will rust. You'll need two more hardware items: anchor bolts to hold the gazebo to the piers, and an aluminum peak cap to finish the roof.

TIP If you can't find a cap for an eight-sided roof, have a tinsmith at your local heating and cooling shop make one up for you.

Before You Begin

1 **Adjust the size of the gazebo.**

As shown in the working drawings, this gazebo has eight 4' sides. This gives you over 30' of floor space. If you want a smaller or larger floor, simply change the length of the sides. As shown, the walls are 8' high; however, you can lengthen them for more headroom—or lower them for less.

2 **Check the building codes.**

Because this gazebo is attached to a permanent foundation, it will probably be affected by building codes. Check your local codes and pay particular attention to the regulations governing the location of outbuildings. Unless you obtain a variance, you may have to locate this gazebo several yards back from your property line. If needed, obtain a building permit before you start work.

Building the Floor

3 **Build the floor frame.**

Cut the beams and joists from 2 x 8 stock. Miter the ends of the beams 67½°, as shown in the *Floor Frame Layout* drawing, so the beams will fit together flush at the corners. Lay out the beams (the outside frame members) on the flat surface, then cut the floor joists (the inside members) to stretch diagonally between the beams, as shown in the working drawings. Miter the ends of the joists 135° so they will fit flush against the corner connections of the beams. Also miter the 'inside' ends of the short joists at 90°. Where the two 'long' joists cross, cut notches to create a lap joint, as shown in Figure 1. Nail the beams and joists together with 16d nails.

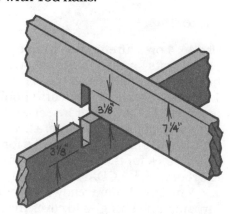

Figure 1. Cut notches in the long floor joists where they overlap to create a lap joint.

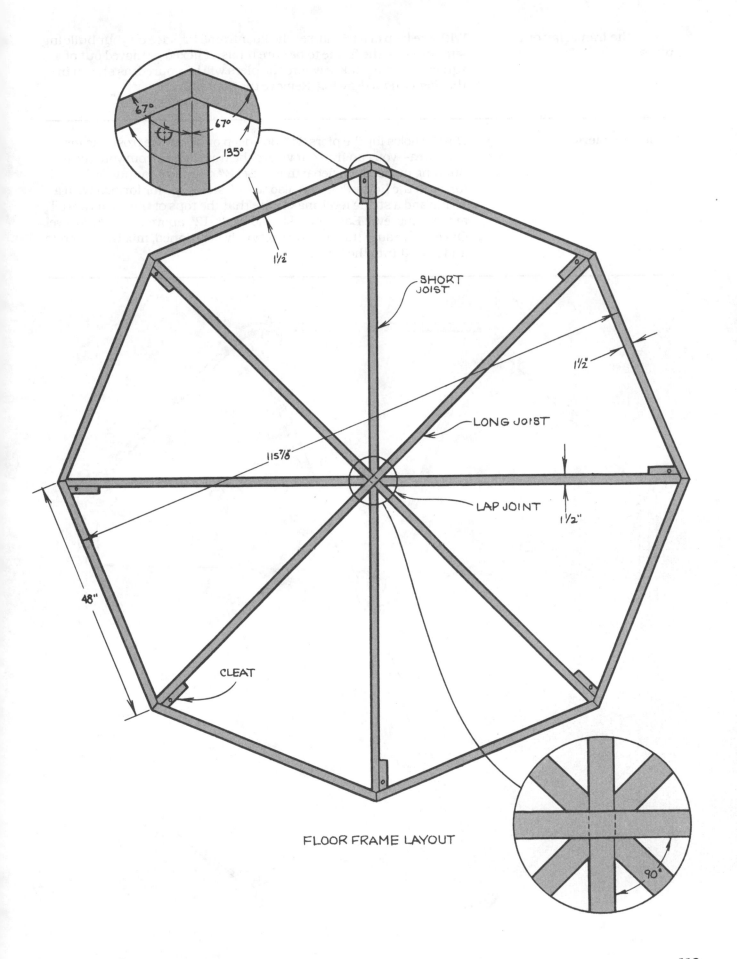

67°

67°

135°

1½"

SHORT
JOIST

1½"

LONG JOIST

115⅞"

LAP JOINT

1½"

48"

CLEAT

90°

FLOOR FRAME LAYOUT

4 **With the frame, lay out the piers.**

With the help of a friend, set the floor frame in place on your building site. Measure the frame to be sure it hasn't flexed or moved out of square. Drive in stakes where the piers will be poured, as shown in the *Pier Layout* drawing. Remove the floor frame.

5 **Pour the piers.**

Dig the holes for the piers 24"-36" deep or below the frost line for your area—you can find out where the frost line is from your local building inspections department. Set 8" diameter round cardboard forms in the holes. (You can also use 8" stovepipe for forms.) With a string and a string level, make sure that the tops of the forms are all at the same level. Each form should be 8"-12" above the ground level. Once you're sure the forms are properly positioned, mix the concrete and pour it into the form.

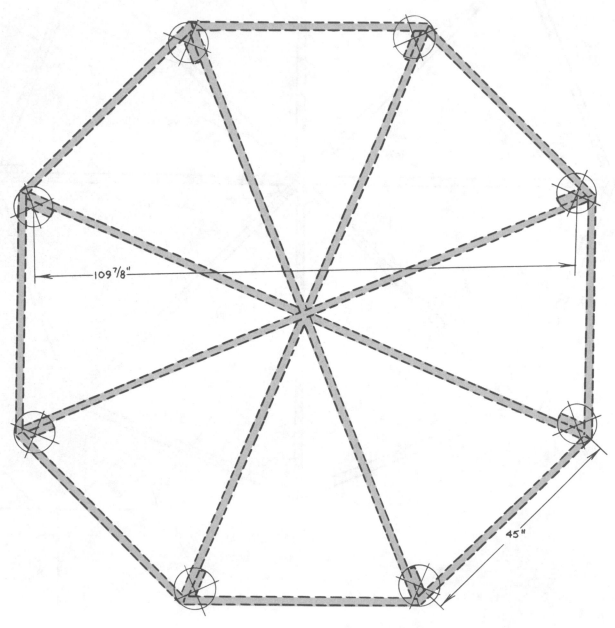

PIER LAYOUT

6   **Set the anchor bolts.**

Before the concrete cures, position anchor bolts in the center of each pier. (See Figure 2.) The tops of the bolts should protrude 2½". Wait at least 24 hours for the concrete to cure, then remove the cardboard forms.

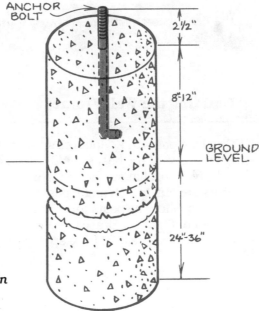

Figure 2. Set the anchor bolts in the piers so that they protrude 2½" from the pier.

7   **Attach the floor frame to the piers with cleats.**

Cut 2 x 2's for cleats and drill slightly oversized holes in the center of each cleat so they will fit over the anchor bolts with some slop. This slop will let you adjust the cleat position slightly. Put the cleats in place on the piers. Then set the floor frame in place around the cleats. Nail the cleats to the frame with 16d nails. Check the frame to be sure it is level and the corners are in line. Secure the cleats to the piers with fender washers and nuts.

8 Install the deck floor.

Cut 1 x 6 decking and miter the ends 67½°, as shown in the *Floor Installation Detail* drawing. The edges of the decking must fit together flush at the joists. After the decking has been cut, lay the first piece flush with the outside edge of the floor beams and nail it to the joists with 12d nails. Lay the next decking piece in place, leaving a ½″ space between the boards to allow the wood to swell. Continue installing decking, leaving ½″ spaces between the boards, using two nails in each end of each floor board.

TIP Reduce the risk of splitting the deck wood by using serrated square-shanked nails. These nails also hold better than common nails.

Building the Wall Frame

9 Set the posts in place.

Cut 4 x 4 posts 91½″ long, as shown in the *Post Layout* drawing. The bottom 36″ of the post should be square—not shaped or turned. This will give you a flat surface on which to attach the railing. After the posts are cut, attach metal post anchors to the floor frame at the corners. Slide the posts into the anchors and nail them in place, as shown in Figure 3. Brace the posts upright with temporary braces. Use a carpenter's level to be sure each post is plumb and level.

Figure 3. Attach a metal post anchor to the floor frame, then slip the post into the anchor. Secure the post in the anchor with screws or nails.

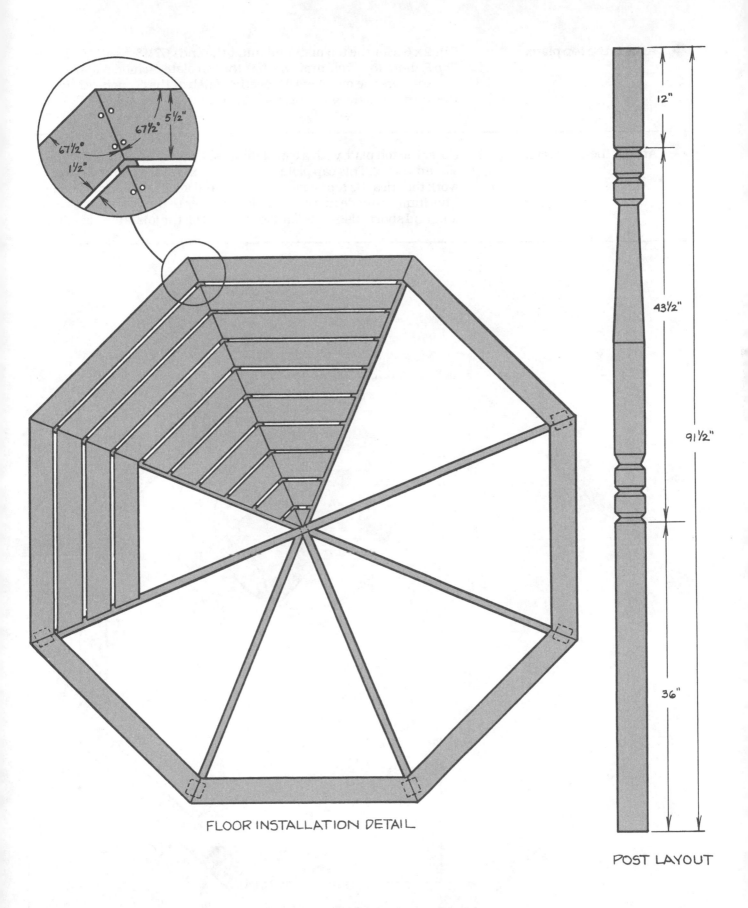

67½° 5½"

67½°

1½"

FLOOR INSTALLATION DETAIL

12"

43½"

91½"

36"

POST LAYOUT

10 Attach the top plate.

Cut 2 x 6's for the top plate and miter the ends 67½°, as shown in the *Top Frame, Top View* drawing. Lift the top plate members in place, making sure the members fit together flush at the corners. Attach the top plate to the posts using 16d nails.

11 Attach the cap plate.

Cover the top plate with a cap plate, as shown in the *Cap Plate, Top View* drawing. This cap plate consists of 2 x 6 spacers and a framework that ties the top corners together and keeps them from flexing. This frame is made in a very similar manner to the floor frame, with long and short joists. The 'outside' ends of all the joists are cut square.

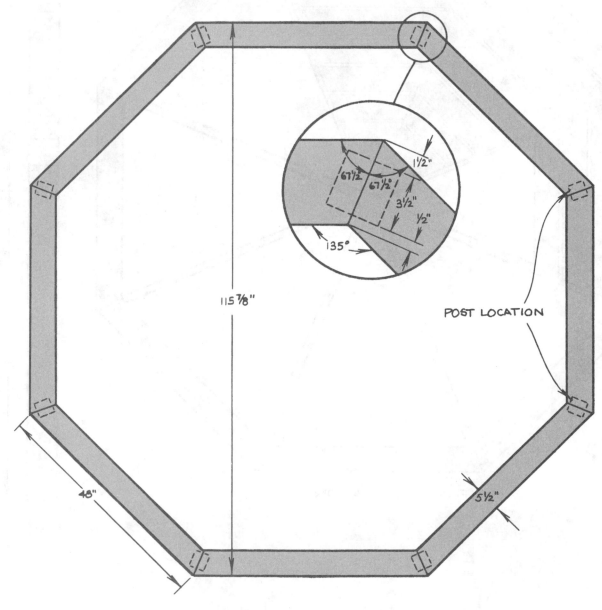

TOP FRAME, TOP VIEW

(See Figure 4.) The 'inside' ends of the short joists are mitered, and the long joists are lapped in the middle. With a friend, raise the joists and put the spacers in place, making sure all the parts fit properly. When you're sure, attach the cap plate parts to the top plate using 16d nails.

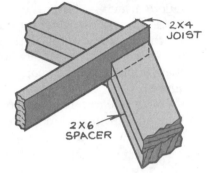

Figure 4. *The cap plate joists sit on the top plate. These joists are kept from shifting by 2 x 6 spacers.*

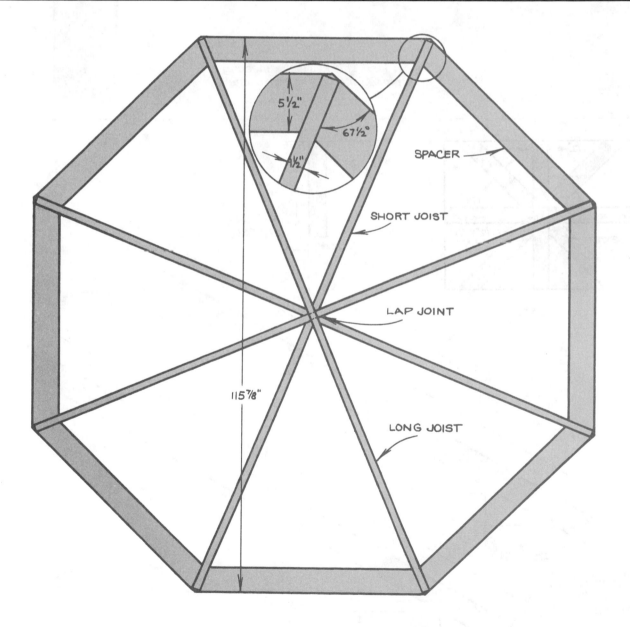

CAP PLATE, TOP VIEW

12 Cut the rafters and the key.

Cut eight rafters from 2 x 6 stock, making four of the rafters slightly longer than the other four, as shown in the *Long* and *Short Rafter Layout* drawings. Miter the ends of the rafters at 60°. Then bevel the 'inside' ends of the short rafters at 90°, as shown in the *Key Detail*, and the outside ends of *all* the rafters at 135°, as shown in the *Rafter End Detail* drawing. Finally, cut a 'bird's mouth' notch in each rafter where it will fit over the cap plate. Also, cut the 2 x 2 key to size. (See Figure 5.)

Note: The rafters should overhang the top plate by 8″. If you have changed the dimensions of the gazebo, adjust the length of your rafters to allow for the overhang.

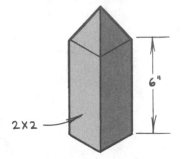

Figure 5. *Cut 2 x 2 stock to form a 'key'. This key is the 'meeting place' for all the rafters at the peak.*

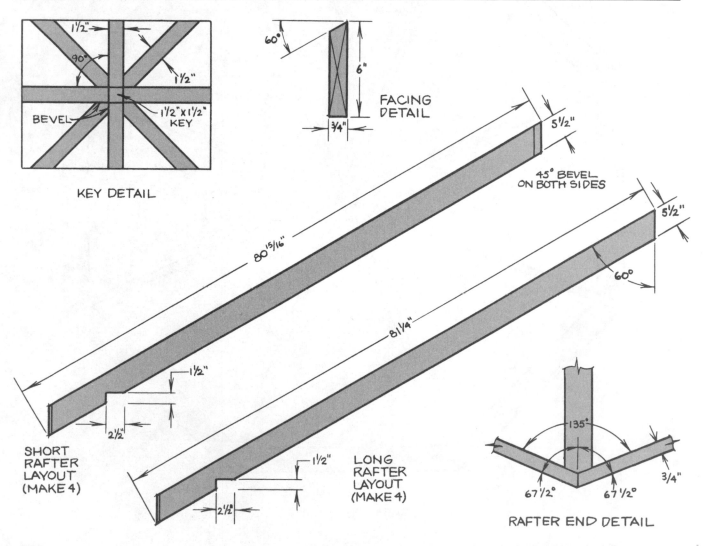

KEY DETAIL

FACING DETAIL

SHORT RAFTER LAYOUT (MAKE 4)

LONG RAFTER LAYOUT (MAKE 4)

RAFTER END DETAIL

With a friend, lift the first two rafters in place and brace them with scrap lumber. Put the key in place between the rafters, as shown in the *Roof Frame, Cutaway View* drawing. Attach the rafters to the key and the top plate with 16d nails. Put the remaining long rafters in place, then the short rafters. Nail then to the key with 16d nails, as shown in the *Key Detail* drawing.

13 Erect the rafters.

Cut facing strips from 'one-by' stock to fit across the rafters at the overhang of the roof. Bevel one edge of the facing strips 30°, as shown in the *Facing Detail* drawing, to match the slope of the roof. Miter the ends at 67½°, so they butt together where they are attached to the rafters. Nail the strip to the rafters with 6d nails.

14 Attach the facing strips.

Cut ½″ CDX (exterior) plywood roof sheathing to fit over each roof section. Cut one piece, put it in place to be sure it fits, then use it as a guide for cutting the other sections. Put the sheathing in place and nail it to the rafters and facing strips using 6d nails spaced every 6″ along the edges.

15 Install the roof sheathing.

Run metal drip edge around the bottom edges of the roof. This drip edge will keep the rain water from collecting under the shingles at the edge of the roof and possibly rotting out the sheathing. Once the drip edge is in place, cover the entire roof with a double-layer of tarpaper, then install the shingles. (See Figure 6.)

16 Install the drip edge, tarpaper, and shingles.

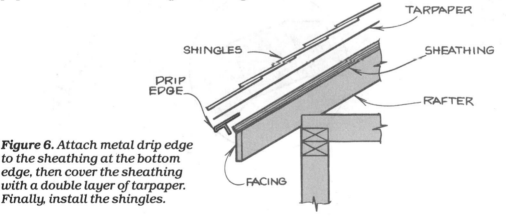

Figure 6. Attach metal drip edge to the sheathing at the bottom edge, then cover the sheathing with a double layer of tarpaper. Finally, install the shingles.

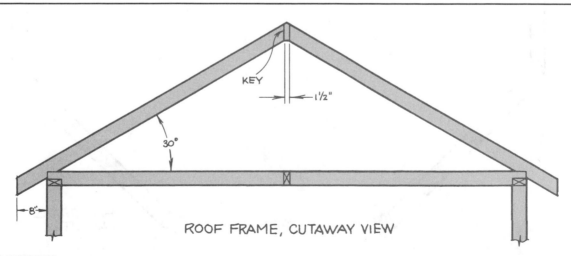

ROOF FRAME, CUTAWAY VIEW

17   Install the peak cap and ridge caps.

Create a ridge cap at each 'hip' by 'double-wrapping' the shingles, just as you would on the ridge of an ordinary roof. (See Figure 7.) Finish the roof by installing an aluminum cap over the peak.

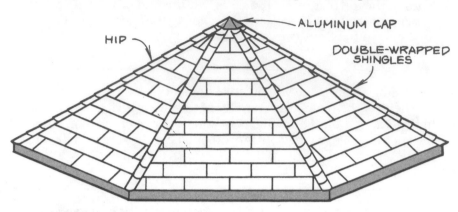

Figure 7. Double-wrap shingles over the hips to form a cap. Finish the roof by installing an aluminum cap at the peak.

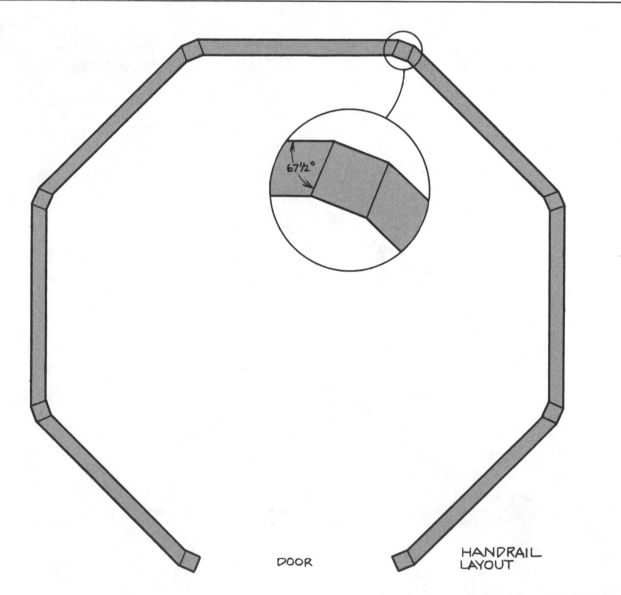

67 1/2°

DOOR

HANDRAIL LAYOUT

Finishing Up

Make the handrail from 3½" wide handrail stock (available at most building suppliers), pre-shaped spindles, and 2 x 4's. Cut and miter the handrail stock (the top members) and the 2 x 4's (the bottom members) at 67½° to fit between the posts, as shown in the *Handrail Layout* drawing. Also, cut the spindles to length for the 'stanchions'. Lay the handrail top members and 2 x 4 bottom members side by side. Using a carpenter's square, mark the location of each stanchion on the rail and plate. Lay the top members and bottom members on edge and nail the stanchions in place using 12d nails. To attach the handrail assemblies to the posts, first cut 1 x 4 ledger blocks, as shown in Figure 8. Attach the blocks to the posts with 6 nails. Slide the handrail assemblies in place atop the blocks, and toenail them to the posts using 12d nails.

18 Install the handrail.

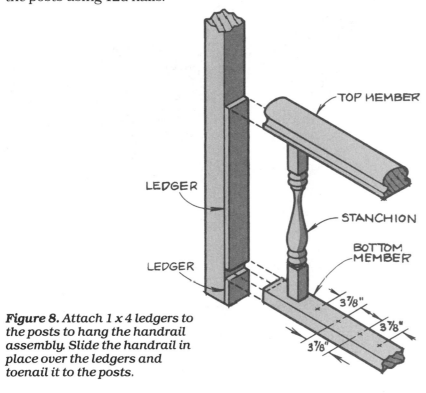

TOP MEMBER

LEDGER

STANCHION

LEDGER

BOTTOM MEMBER

3⅞"

3⅞"

3⅞"

Figure 8. Attach 1 x 4 ledgers to the posts to hang the handrail assembly. Slide the handrail in place over the ledgers and toenail it to the posts.

TIP If you don't want to go to the expense of buying pre-shaped spindles, you can also use ordinary 2 x 2's for the stanchions.

Use exterior paint or waterproofing stain to protect the wood from moisture or insect damage. Paint or stain all exposed wood surfaces.

19 Paint or stain all exposed wood surfaces.

Shaped-Roof Gazebo

This delicate and gracious gazebo is a modern derivative of the Victorian gazebo with its open "windows" and graceful shaped-peak roof. Over 9′ wide, with more than 30 square feet of space, this garden shelter complements the grounds of an older home—or the grounds of a new home with traditional architecture.

This octagonal gazebo sits on a slab foundation, so there are no tricky floor frames to construct. The tricks come when you get to the roof: The rafters are built up and then cut to form a curve. When the roof is covered, these simple curves become compound curves. The effect is striking; but it is, in fact, quite simple to create. We'll show you the tricks as we go along.

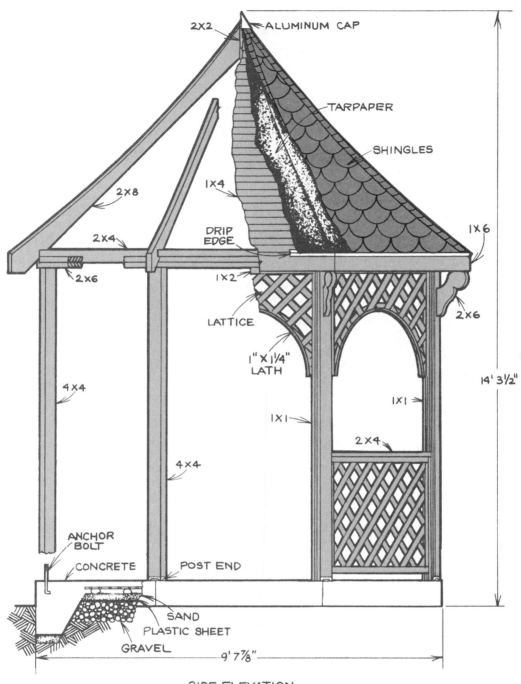

2X2 — ALUMINUM CAP

TARPAPER

SHINGLES

2×8

2×4

1X4

DRIP EDGE

1X6

1X2

LATTICE

1" X 1¼" LATH

2×6

2×6

1X1

1X1

4×4

2×4

4×4

ANCHOR BOLT

CONCRETE

POST END

14' 3½"

SAND

PLASTIC SHEET

GRAVEL

9' 7⅞"

SIDE ELEVATION

Materials

To build the gazebo as shown, use pressure-treated 4 x 4's for the posts. Purchase 2 x 6's for the top and cap plates, and 2 x 4's for the handrails. The rafters are cut from 2 x 8's, or you can save expense by joining two 2 x 4's. The rafters are joined at the top with a 2 x 2 key. The facing strips, lattice frame, and cleats are all ripped from 'one-by' (¾" thick) stock. The lattice is available ready-made from most lumberyards in large sheets.

For the concrete slab, you'll need 2 x 8's for the forms, pea-size gravel, plastic sheeting (for vapor barrier), sand, 6" x 6" reinforcing mesh, anchor bolts, and—of course—concrete. In addition to these materials, you'll also have to purchase galvanized nails, tarpaper, metal post ends, shingles, and drip edge.

Before You Begin

1 **Adjust the size of the gazebo.**

As shown in the working drawings, this gazebo has eight 4' sides. If you want a smaller or larger floor space, simply change the length of the sides. Pour a larger or smaller slab by using longer or shorter members when building the form. The posts, as shown, are 8' high; however, you can lengthen them for more headroom—or lower them for less.

2 **Check the building codes.**

Because this gazebo is attached to a permanent slab foundation, it will probably be affected by building codes. Check your local codes and pay particular attention to the regulations governing the location of outbuildings. Unless you obtain a variance, you may have to locate this gazebo several yards back from your property line. If needed, obtain a building permit before you start work.

Note: Choose the site for this gazebo carefully. Once you've poured the concrete it'll be next to impossible to move it. Give some thought not only to the location of the gazebo, but the orientation of the entrance. Make the opening conform to existing traffic patterns. You don't want the kids jumping over the handrail all the time.

Pouring the Pad

3 **Build the form.**

Cut the 2 x 8 form members 49¼" long, as shown in the *Form Layout* drawing. Miter the ends of the members at 67½° so they will fit flush together at the corners. Join the members with steel straps, connecting them at the corners, as shown in Figure 1. Make sure each corner measures 135° on the inside—otherwise the sides of your octagonal gazebo won't fit together properly.

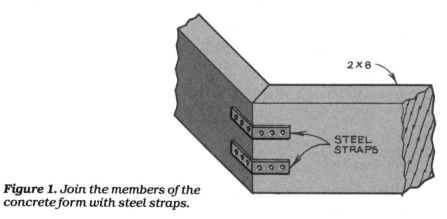

Figure 1. *Join the members of the concrete form with steel straps.*

4 **Lay out the pad with the form.**

With the help of a friend, set the form in place on your building site. Measure the form again to be sure it hasn't been moved out of alignment. Level the form, and drive stakes in the ground along the outside of the form to hold it in position. Then brace the form by driving in another row of stakes about 2' from the first stakes and nailing scrap lumber between the stakes.

Dig a trench inside the form 8″ wide and 6″-12″ deep. Level the bed inside the trench with a flat shovel; then tamp the bed down. Next, peel up the sod and remove any large rocks, debris, or plants from the area where you will pour the slab. It's important that the top of the bed be 10″-12″ below the tops of the form. Build up the center of the bed with 4″-6″ of gravel (for drainage) and cover the gravel with plastic sheeting to make a vapor barrier. Cover the sheeting with 2″ of sand; put a little sand down in the trenches to hold the edges of the vapor barrier down. When you've finished spreading the sand, there should be 4″ between the top of the sand layer and the tops of the forms. Finally, put down steel reinforcing mesh. Hold the mesh 1″-2″ above the sand with rocks or scrap wood. (See Figure 2.)

5 **Prepare the bed.**

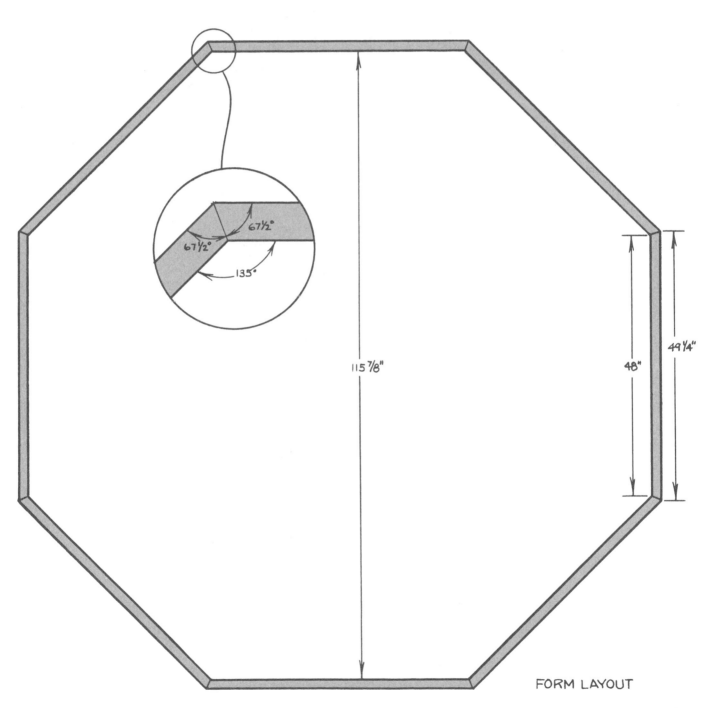

FORM LAYOUT

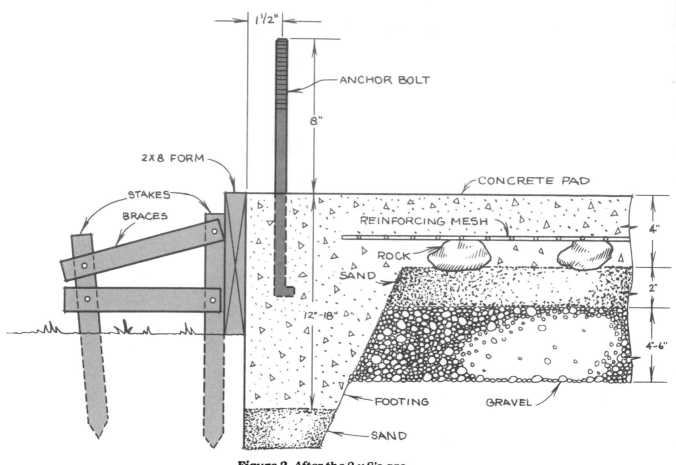

Figure 2. After the 2 x 8's are staked and braced to the ground, dig a footing trench and prepare the bed with gravel (for drainage), a sheet of plastic (for a vapor barrier), sand (to hold the vapor barrier in place), and wire mesh (for reinforcement). Pour the concrete over the bed and set the anchor bolts in place.

6 **Pour the pad.**

Pour the concrete in the form, then "screed" it off level with the top of the form by dragging a 10-foot 2 x 4 across the tops of the form members. With a helper, 'saw' this 2 x 4 screed back and forth across the concrete as if you were operating an old two-man saw. This motion will result in the smoothest possible surface.

TIP If you order the concrete delivered, and the truck will not be able to back up to your site, then arrange for extra wheelbarrows or a pump truck.

Set one anchor bolt at each corner of the pad, as shown in the *Anchor Bolt Placement Detail* drawing. The bolts should protrude about 8″ from the top of the concrete pad. Smooth out the concrete with a "darby" or a trowel, then let the concrete cure for at least 48 hours before you continue with any further construction.

7 **Set the anchor bolts.**

TIP To make the concrete as hard as possible, lightly sprinkle it with water every 4-6 hours while it cures. Continue this for 48 hours.

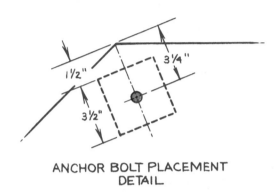

ANCHOR BOLT PLACEMENT
DETAIL

Building the Wall Frames

Cut the posts 90″ long, then drill holes for the anchor bolts in the bottom of each post. Make sure the holes are ¹⁄₁₆″ larger than the diameter of the bolts. This will give the wood some room to swell during wet weather. Place metal post ends on the concrete pad, then slip the posts over the anchor bolts. The anchor bolts will hold the posts upright. (See Figure 3.)

8 **Set the posts.**

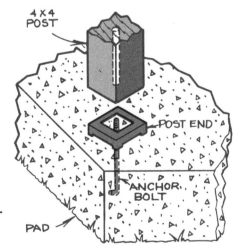

Figure 3. Drill holes in the bottom of the posts ¹⁄₁₆″ larger than the anchor bolts. Set metal post ends over the bolts and slip the posts in place over the anchor bolts. The posts should rest on the post ends.

TIP To bore deep holes in the ends of the posts, use an 'aircraft drill', or a 'drill bit extender' and an auger bit.

9 **Attach the top plate.**

Cut top plate members 48″ long and miter the ends 67½° so they will butt together, as shown in the *Top Plate, Top View* drawing. Put the top plate in place and nail it to the posts with 16d nails.

10 **Attach the cap plate.**

Cover the top plate with a cap plate, as shown in the *Cap Plate, Top View* drawing. This cap plate consists of 2 x 6 spacers and a framework that ties the top corners together and keeps them from flexing. This frame is made with two 'long' and four 'short' joists. The 'outside' ends of all the joists are cut square. (See Figure 4.) The 'inside' ends of the short joists are mitered, and the long joists are lapped in the middle. (See Figure 5.) With a friend, raise the joists and put the spacers in place, making sure all the parts fit properly. When you're sure, attach the cap plate parts to the top plate using 16d nails.

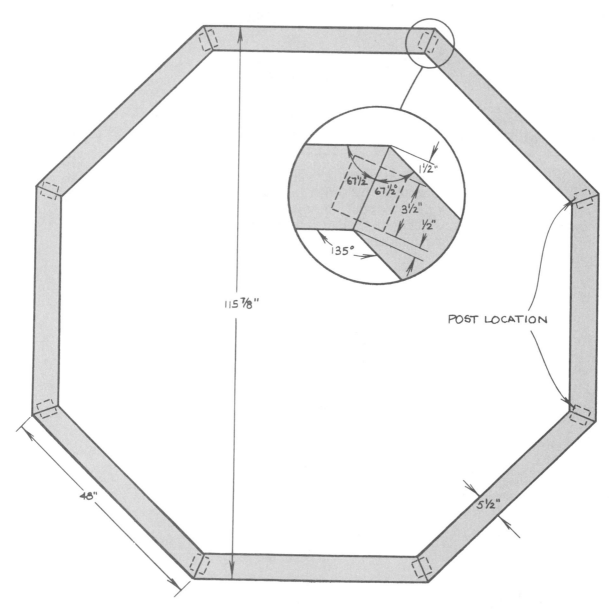

TOP PLATE, TOP VIEW

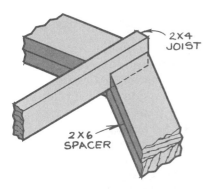

Figure 4. *The cap plate joists sit on the top plate. These joists are kept from shifting by 2 x 6 spacers.*

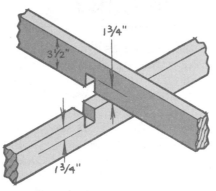

Figure 5. *Cut notches in the long joists where they overlap to create a lap joint.*

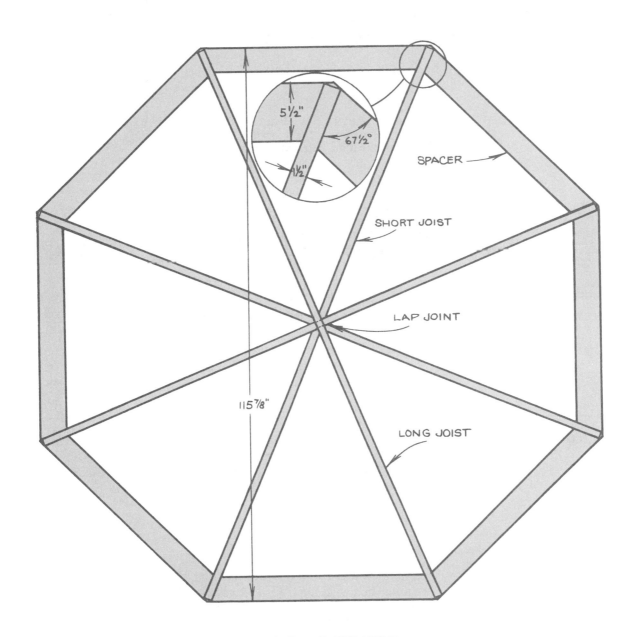

CAP PLATE, TOP VIEW

11 **Cut the rafters and the key.**

Cut eight rafters from 2 x 8 stock, making four of the rafters slightly longer than the other four, as shown in the *Long* and *Short Rafter Layout* drawings. Cut the curves with a bandsaw or saber saw. Miter the ends of the rafters at 45°. Then bevel the 'inside' ends of the short rafters at 90°, as shown in the *Key Detail,* and the outside ends of *all* the rafters at 135°, as shown in the *Rafter End Detail* drawing. Finally, cut a 'bird's mouth' notch in each rafter where it will fit over the cap plate. Also, cut the 2 x 2 key to size. (See Figure 6.)

Note: The rafters should overhang the top plate by 8″. If you have changed the dimensions of the gazebo, adjust the length of your rafters to allow for the overhang.

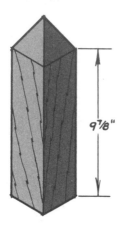

Figure 6. *Cut 2 x 2 stock to make a 'key'. This key is the 'meeting place' for all the rafters at the peak.*

9⅞″

TIP You can also cut the rafters from two 2 x 4's, glued edge to edge. Use waterproof resorcinol glue, then reinforce the glue joint with 16d nails *after* you cut the curve.

---

12 **Erect the rafters.**

With a friend, lift the first two rafters in place and brace them with scrap lumber. Put the key in place between the rafters, as shown in the *Roof Frame, Cutaway View* drawing. Attach the rafters to the key with 16d nails, and to the cap plate with metal gussets and 4d nails. Put the remaining long rafters in place, then the short rafters.

---

13 **Attach the facing strips.**

Cut facing strips from 'one-by' stock to fit across the rafters at the overhang of the roof. Bevel one edge of the facing strips 35°, as shown in the *Facing Detail* drawing, to match the slope of the roof where the curve begins. Miter the ends at 67½°, so they butt together where they are attached to the rafters. Nail the strip to the rafters with 6d nails.

---

14 **Install roof slats.**

Instead of sheathing the roof with large sheets of plywood, this roof is sheathed with horizontal slats. The slats will follow the curve over the rafters. Cut the slats for one section of the roof, put them in place to be sure they fit properly, then use them as a guide for cutting the remaining slats. Nail the slats to the rafters with 6d nails.

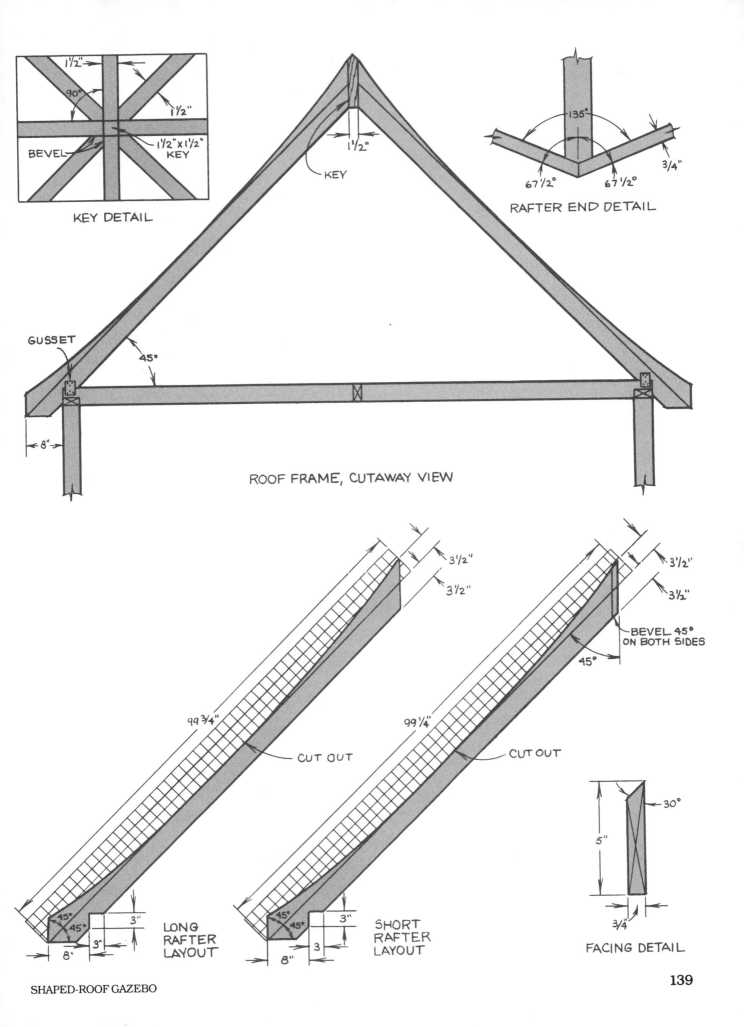

KEY DETAIL

1/2"

90°

BEVEL

1 1/2"

1 1/2" x 1/2" KEY

KEY

1/2"

RAFTER END DETAIL

135°

67 1/2°

67 1/2°

3/4"

GUSSET

45°

8"

ROOF FRAME, CUTAWAY VIEW

3 1/2"

3 1/2"

99 3/4"

CUT OUT

45°

45°

8"

3"

3"

LONG RAFTER LAYOUT

3 1/2"

3 1/2"

BEVEL 45° ON BOTH SIDES

45°

99 1/4"

CUT OUT

45°

45°

8"

3

3"

SHORT RAFTER LAYOUT

FACING DETAIL

30°

5"

3/4"

SHAPED-ROOF GAZEBO

139

15 **Install the drip edge, tarpaper, and shingles.**

Install metal drip edge around the edges of the roof. Once the drip edge is in place, cover the entire roof with a double layer of tarpaper. Then install the shingles. (See Figure 7.)

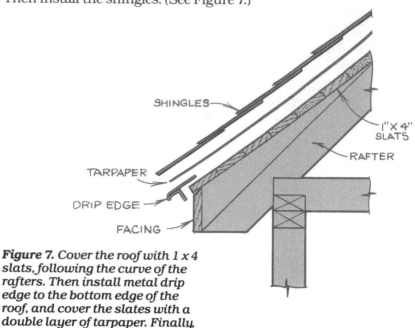

Figure 7. Cover the roof with 1 x 4 slats, following the curve of the rafters. Then install metal drip edge to the bottom edge of the roof, and cover the slates with a double layer of tarpaper. Finally, install the shingles.

---

16 **Install the peak cap and ridge caps.**

Create a ridge cap at each 'hip' by 'double-wrapping' the shingles, just as you would on the ridge of an ordinary roof. (See Figure 8.) Finish the roof by installing an aluminum cap over the peak.

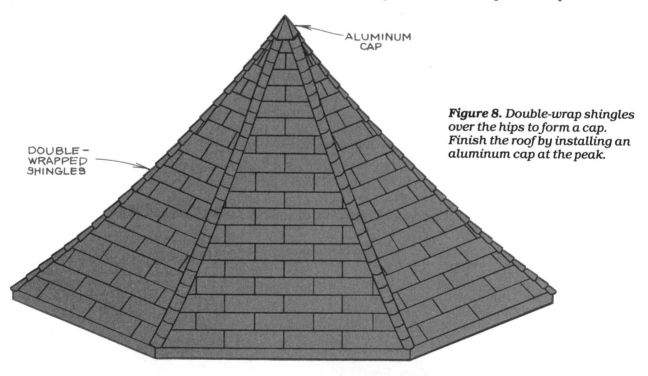

Figure 8. Double-wrap shingles over the hips to form a cap. Finish the roof by installing an aluminum cap at the peak.

TIP If you can't order an eight-sided peak cap from your local building supplier, have a tinsmith create one for you. Most heating and cooling contractors do their own tinsmithing.

Installing the Lattice

Make the lower frames first: Cut the frame pieces to size, as shown in the *Lower Lattice Frame Layout* drawing. Join the pieces at the corners with lap joints. (See Figure 9.)

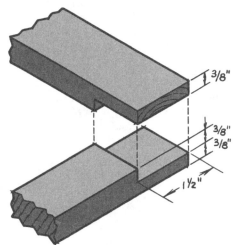

Figure 9. *Join the lattice frame members at the corners with lap joints.*

The upper frames are a bit trickier. The straight parts are made and joined in the same manner as on the lower frames. But the curved part is made in sections, as shown in the *Upper Lattice Frame Layout* drawing.

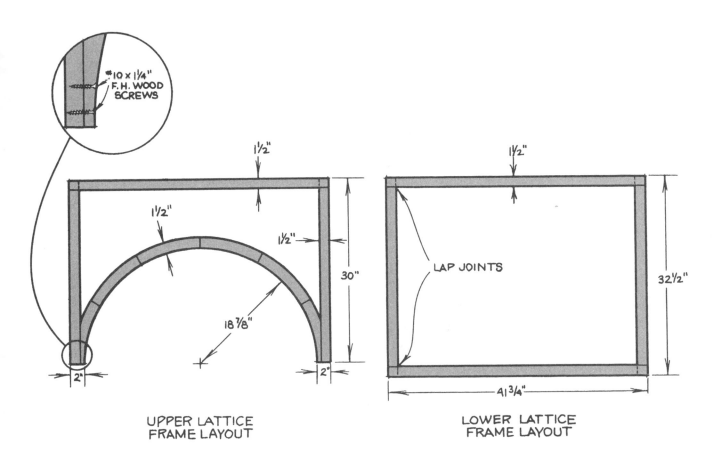

UPPER LATTICE
FRAME LAYOUT

LOWER LATTICE
FRAME LAYOUT

Cut these sections from ¾" x 2½" x 11" boards, as shown in the *Curved Lattice Frame Detail* drawing. First, miter the ends of these boards at 60°, then rout a groove in the ends, as shown in the *Groove Section* drawing. Cut the curves with a bandsaw or saber saw, and join the boards with splines to form an arch with a radius of 18⅞". (See Figure 10 and the *Curved Lattice Frame Assembly* drawing.) Trim the sides of the arch to fit between the vertical members of the upper frame, and attach the arch to the frame with screws.

Note: Once you get to this point, it's wise to take your own measurements rather than rely on ours. Construction-grade lumber isn't all that accurate (even if you are), and the spacing between the posts may be off slightly. To avoid unsightly gaps or buckled lattice sections, custom-fit each frame to its opening.

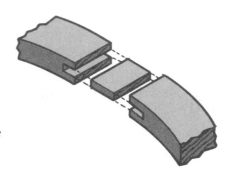

Figure 10. *Join the curved members of the upper lattice frames with splines to form an arch.*

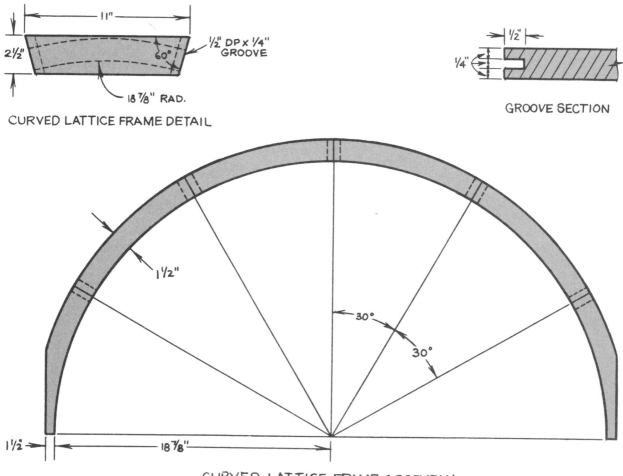

CURVED LATTICE FRAME DETAIL

GROOVE SECTION

CURVED LATTICE FRAME ASSEMBLY

Cover the frames with lattice, as shown in the *Finished Upper Lattice Section* and *Finished Lower Lattice Section* drawings. When the lattice is in place, attach 1 x 2 horizontal trim to the top of the upper section, and to the top and bottom of the lower section. Miter the ends of this trim at 67½°, as shown in the *Trim Detail* drawing, and nail the trim to the lattice sections with the miter angled in toward the lattice. Use 4d nails to attach the trim. Trim the arch with ¼" thick, 1" wide lath. This thin stock will bend easily to fit the arch. Tack the lath to the curved frame with #6 x 1" flathead wood screws.

18 Cover and trim the frames.

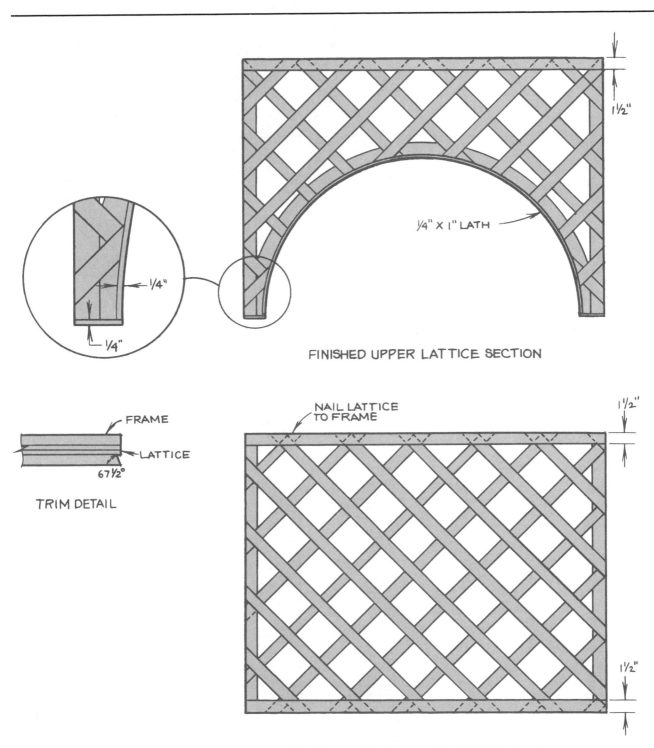

¼" X 1" LATH

FINISHED UPPER LATTICE SECTION

TRIM DETAIL

FRAME

LATTICE

67½°

NAIL LATTICE TO FRAME

FINISHED LOWER LATTICE SECTION

19 Install the lattice sections.

Nail the 'inside' set of ¾″ x ¾″ cleats to the sides of the posts, toward the inside edge, as shown in the *Wall Joinery Detail* drawing. Tack the upper and lower lattice sections in place against these cleats, where shown in the *Wall Layout, Side View* drawing. Toenail the sections to the posts with 6d nails, keeping the trim to the outside. Then nail the 'outside' set of cleats in place. The cleats will hold the lattice sections to the posts. Leave one wall section open, to create an entrance, as shown in the *Wall Layout, Top View* drawing.

20 Install the handrails.

Carefully measure the space for the handrails and cut them to length from 2 x 4 stock. Miter and cut ¾″ long, 1¼″ wide tabs in the ends of the handrails, where they will fit between the cleats, as shown in the *Handrail Detail* drawing. Put the handrails in place over the lower lattice sections. Nail them to the posts and the lattice sections with 8d nails.

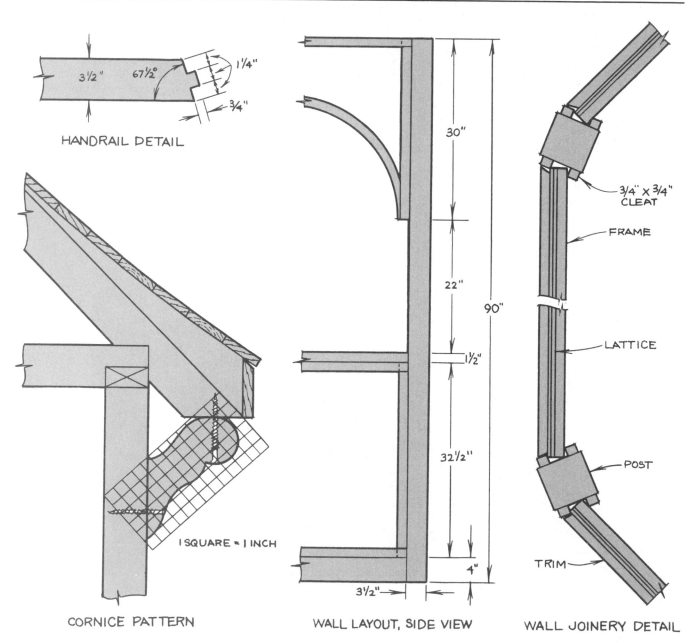

HANDRAIL DETAIL

CORNICE PATTERN

I SQUARE = I INCH

WALL LAYOUT, SIDE VIEW

WALL JOINERY DETAIL

Finishing Up

Cut the cornices as shown in the *Cornice Pattern* drawing. Fit the cornices in place and attach them to the rafters with long lag screws. Countersink the head of the lag screws so they aren't so obvious. If you wish, hide the heads with wood plugs.

21 Install the cornices.

Use exterior paint or waterproofing stain to protect the wood from water and insect damage. Paint or stain all exposed wood surfaces.

22 Paint or stain all exposed wood.

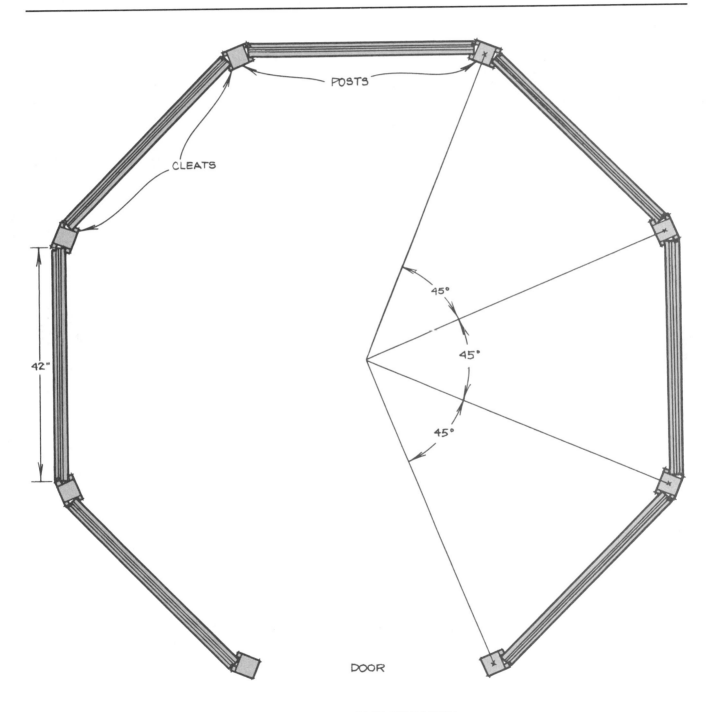

POSTS

CLEATS

42"

45°

45°

45°

DOOR

WALL LAYOUT, TOP VIEW

Summer House

There's no more "civilized" way to enjoy the great outdoors than from the inside of a screen house. Forget assaults from insects, sudden summer showers, and an overdose of sunshine. A screen house is a perfect structure from which to enjoy an unobstructed view of nature every day of the warm-weather season—come rain or shine.

This 12' x 16' summer house is set on a concrete pad. The concrete is impervious to water, and helps keep the structure cool during hot weather. The walls are framed, but covered with screen to give you a panoramic view in all directions. The gabled roof overhangs the walls by a foot in all directions to help keep the structure dry, even during a summer storm. Cover the roof with traditional roofing materials, as shown here, or use corrugated fiberglass panels.

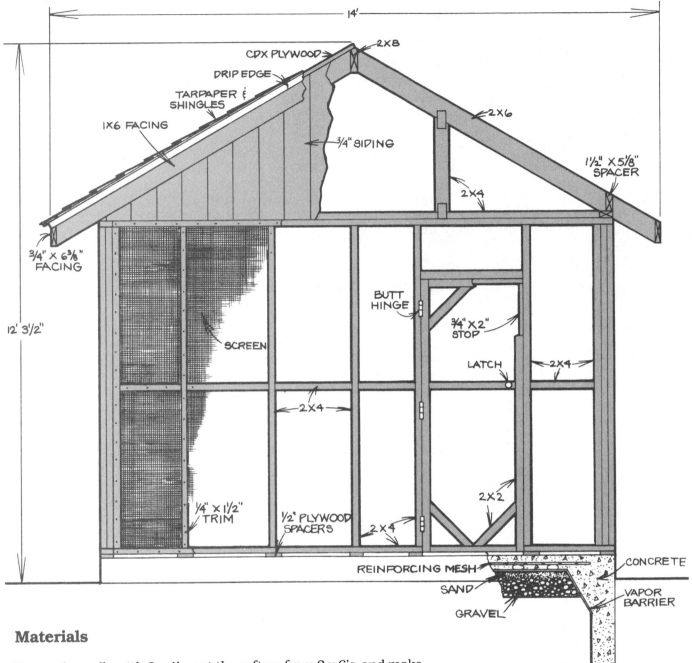

14'

CDX PLYWOOD
2X8
DRIP EDGE
TARPAPER &
SHINGLES
1X6 FACING
3/4" SIDING
2X6
1 1/2" X 5 1/8"
SPACER
3/4" X 6 3/8"
FACING
12' 3 1/2"
2X4
BUTT
HINGE
3/4" X 2"
STOP
LATCH
2X4
SCREEN
2X4
2X2
1/4" X 1 1/2"
TRIM
1/2" PLYWOOD
SPACERS
2X4
REINFORCING MESH
CONCRETE
SAND
VAPOR
BARRIER
GRAVEL

FRONT ELEVATION

Materials

Frame the walls with 2 x 4's, cut the rafters from 2 x 6's, and make the ridgeboard from a 2 x 8. You'll also need some 'one-by' (3/4" thick) stock to make the facing strips, and wood siding to cover the gable ends.

To cover the roof, you'll need 1/2" CDX (exterior) plywood for sheathing, as well as tarpaper, drip edge, and roofing nails. If you'd rather cover the roof with fiberglass panels, then you'll need redwood 'closure' and 'caps', aluminum nails with neoprene washers, silicone sealant, and corrugated aluminum ridgecap.

To pour the concrete pad, you'll need concrete (of course), 2 x 8's and some scrap lumber to make the form, sand, gravel, reinforcing wire mesh, and anchor bolts. You'll also need a supply of galvanized nails, staples, hinges, a door latch, and screen.

Before You Begin

1 **Adjust the size of the summer house.**

As shown in the working drawings, this summer house is 12' wide, approximately 10' high (at the peak), and stretches 16' long. However, this may be smaller or larger than you need. If you want a smaller house, reduce one or more of the dimensions. The design can be enlarged in either direction simply by adding more studs to the wall frames and stretching the rafters. You can raise the roof by using longer wall studs.

2 **Check the building codes.**

Because this summer house is attached to a permanent foundation, it may be affected by building codes. Check your local codes for any restrictions that apply to outbuildings. You may find that, unless you apply for a variance, you'll have to locate this summer house several yards back from your property line. If needed, obtain a building permit before you begin work.

Pouring the Concrete Pad

3 **Stake out the foundation.**

Use stakes and a string to lay out the location of your pad, as shown in the *Concrete Pad Layout* drawing. Place the stakes outside of the foundation lines and fasten the string between them so that the points where the horizontal and vertical strings cross mark the exact location of the corners of your summer house. (See Figure 1.) Measure carefully—if your pad isn't square, your building won't be either.

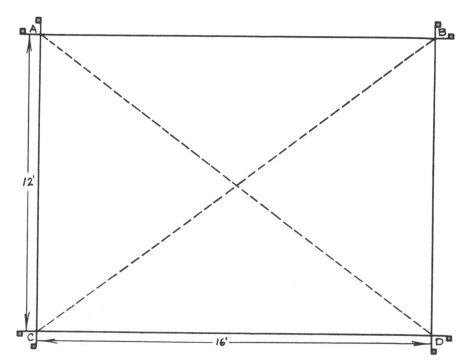

Figure 1. Use stakes and strings to mark the foundation lines for your shed. Measure diagonally from corner to corner to be sure your layout for the foundation is square. AD should equal BC.

Peel up the sod and remove any large rocks, debris, or plants from the area where you will pour the pad. Stake 2 x 8's around the outside edge of the pad as marked by your string. Be sure the forms are level and that all four corners are square. (Check by measuring the forms diagonally, from corner to corner.) Use stakes and scrap lumber to brace the forms. Since the edges of the pad will serve as the footings for the walls, dig a trench inside the forms that is 12″ wide and 24″ deep. Level the bed inside the trench with a flat shovel; then tamp the bed down. *It's important that the top of the bed be 6″-8″ below the tops of the forms.* Build up the center of the bed with 4″-6″ of gravel (for drainage) and cover with plastic sheeting to make a vapor barrier. Next, cover the sheeting with 2″ of sand; put a little sand down in the trenches to hold the edges of the vapor barrier down. When you've finished spreading the sand, there should be 4″ between the top of the sand layer and the tops of the forms. Finally, put down steel reinforcing mesh. Hold the mesh 1″-2″ above the sand with rocks or scrap wood. (See Figure 2.)

4 **Set the forms and prepare the bed.**

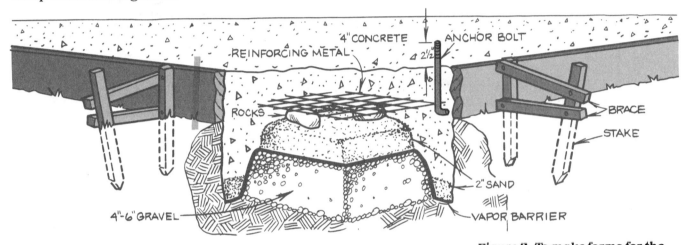

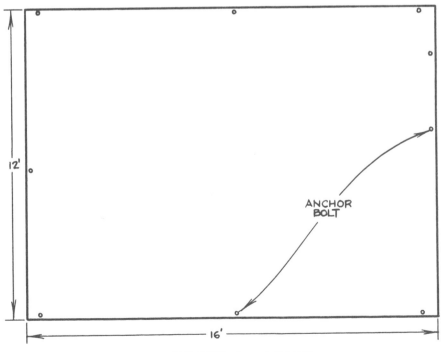

CONCRETE PAD LAYOUT

Figure 2. To make forms for the slab, stake 2 x 8's to the ground. Dig a footing trench and prepare the bed with 4″-6″ of gravel (for drainage), a sheet of plastic (for a vapor barrier), 2″ of sand, and wire mesh held 1″-2″ off the sand (for reinforcement). Pour the concrete over the bed and set the anchor bolts in place.

5 Pour the concrete.

For this foundation, you'll probably want to order a ready-mix delivery. Arrange for extra wheelbarrows or a pump truck if the concrete truck cannot back up to the site. Once the concrete is poured, have a friend help you level it with the tops of the forms by dragging a 'screed' —long 2 x 4—back and forth with a sawing motion across the tops of the form boards.

6 Set the anchor bolts.

Set the anchor bolts where shown in the drawings. The bolts should protrude 2½″ from the top of the pad. Smooth out the concrete with a 'darby' or a trowel, then let the concrete cure for at least 24 hours before you continue with any further construction.

Building the Wall Frame

7 Cut and drill the sole plates.

Cut the sole plates from pressure-treated or rot-resistant lumber, because they will draw moisture from the concrete. Line up the sole plate along the edge of the foundation and mark the position of the anchor bolts. Then drill the holes about ¼″ larger than the diameter of the bolts so the wood will fit easily over the bolt but still be snug enough to prevent shifting. For example, drill a ¾″ hole to fit over a ½″ bolt. Lay the sole plate in place to make sure the edges are straight and the holes fit over the bolts.

8 Build the wall frames.

Cut the frame parts to length. Put the top plate next to the sole plate, bottom side up. Using a framing square, measure where each stud will be placed and draw a pencil line across both plates. Use the *Side Wall Frame, Back Wall Frame*, and *Front Wall Frame* drawings to determine the location of each stud. Once you have marked the location of each stud, lay out the parts for each wall in turn on the concrete pad, and nail them together with 16d nails. From ½″ exterior plywood, cut 3½″ x 3½″ blocks. Nail these blocks to the bottoms of the sole plate, directly beneath each stud. The blocks will keep the wooden wall frame slightly raised off the concrete so that little water can collect under the frame. This will help to prevent rot.

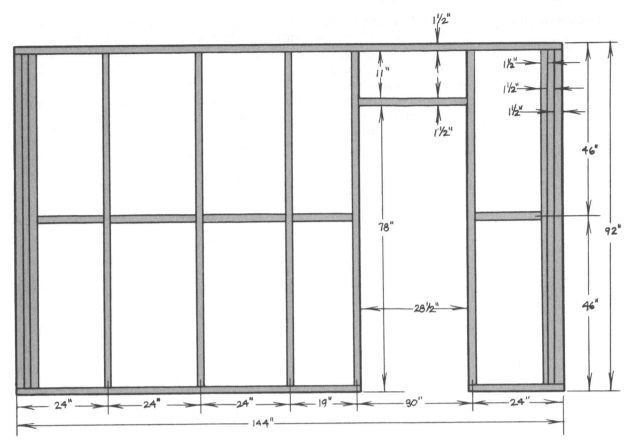

FRONT WALL FRAME

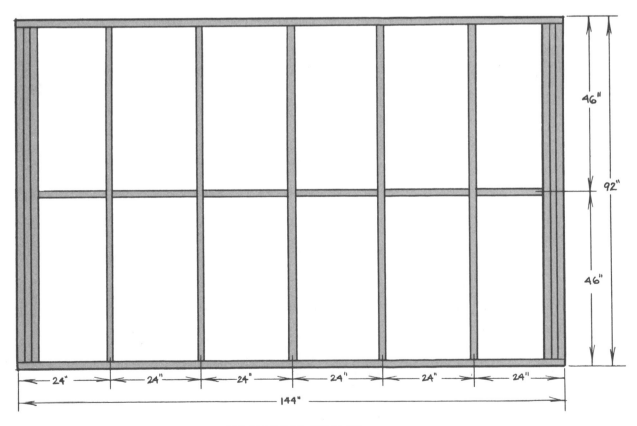

BACK WALL FRAME

9 **Raise the walls.**

With the help of a friend positioned at one end of a wall frame and you at the other, gently lift each wall into place. Temporarily brace them upright as you go. Refer to the *Stud Layout* drawing to see how the walls go together. Fasten the sole plates to the pad with the anchor bolts. Connect the walls to each other at the corners using 16d nails spaced every 24". Use a carpenter's level to make sure all the studs are plumb vertically and the top plates are level horizontally.

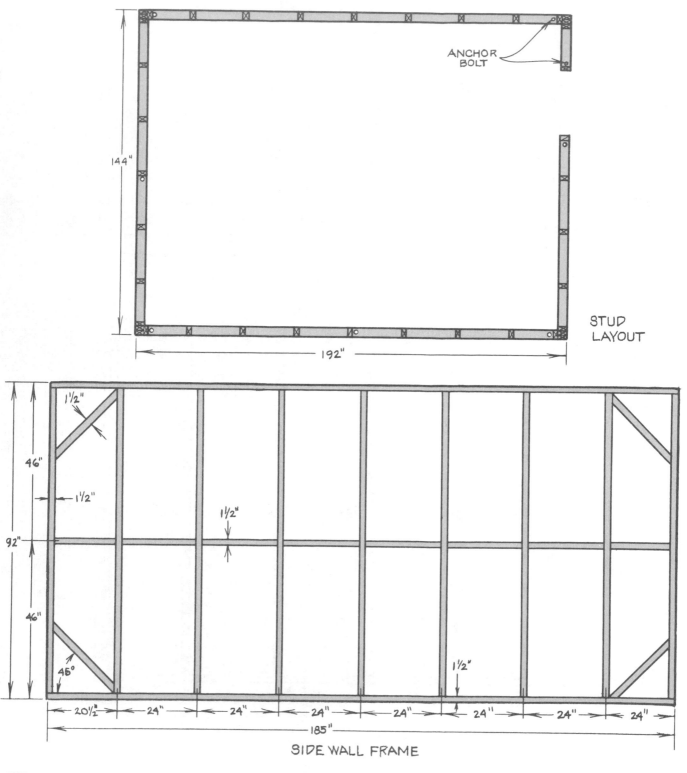

ANCHOR BOLT

144"

192"

STUD LAYOUT

1½"

46"

1½"

1½"

92"

46"

45°

1½"

20½" 24" 24" 24" 24" 24" 24" 24"

185"

SIDE WALL FRAME

To tie the entire assembly together, nail a second top (or 'cap') plate down onto the first top plate with 16d nails. Be sure the ends of the cap plate lap the joint between the top plate members. (See Figure 3.)

10   **Attach the cap plate.**

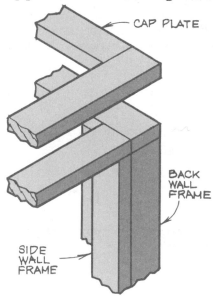

Figure 3. The cap plate is nailed to the top of the wall frames and ties the wall assembly together. The ends of the cap plate must overlap the joints between the wall frames.

Building the Roof

Use 2 x 6 stock for the rafters and cut them to the desired length. (This summer house has a 12″ overhang on each side to allow the rain to drain off away from the building, so remember the overhang if you have adjusted the length of the rafters.) After you have cut the rafters, notch them to fit over the top plate and miter the ends as shown in the *Rafter Layout* drawing.

11   **Cut the rafters.**

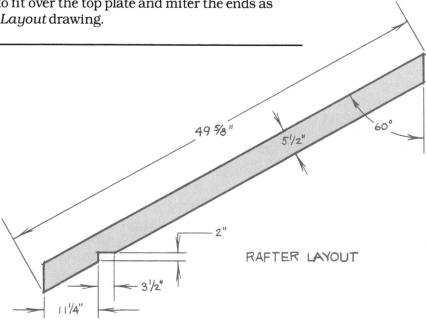

RAFTER LAYOUT

12 Set the ridgeboard in place.

Cut the 2 x 8 ridgeboard to size, allowing for the 11¼″ overhang at the gable ends. Bevel the top edge of the ridgeboard 120° as shown in the *Ridgeboard Detail* drawing. Temporarily, support the ridgeboard above the wall frame. Make temporary supports by nailing 2 x 4's to the side wall frames so that they stick up about 5′ above the frames. Clamp (don't nail) the ridgeboard to these supports. (See Figure 4.) The clamps will allow you to adjust the height of the board, if needed.

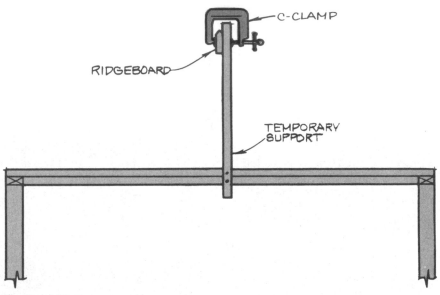

Figure 4. *Nail 2 x 4's to the cap plate as temporary supports for the ridgeboard. Clamp (don't nail) the ridgeboard to these supports.*

13 Nail the rafters in place.

Nail the rafters to the ridgeboard and the top plate, every 2′ on center, as shown in the *Roof Frame, Side View* and *Roof Frame, End View* drawings. Use 16d nails to secure the rafters in place.

TIP Save the time and trouble of notching rafters by using metal rafter ties to attach the rafters to the ridgeboard and top plate. The one drawback to using these ties is that they don't look as good as plain wood.

14 Nail the gable braces to the end rafters.

Cut two 2 x 4's for the braces for each gable end and miter the top edges to fit flush against the rafters, as shown in the working drawings. Attach the braces at each gable end, between the rafters and the cap plate, with metal truss plates or gussets. Put these gussets on the *outside* so that they won't be seen when the structure is complete—they'll be covered by siding. Remove the temporary ridgeboard braces.

Cut 2 x 4 spacer blocks and bevel the top edge 30°, as shown in the *Spacer Block Detail* drawing. Attach the spacer blocks flush with the top of the rafters so bugs can't get in. Use 16d nails.

15 **Install spacer blocks between the rafters.**

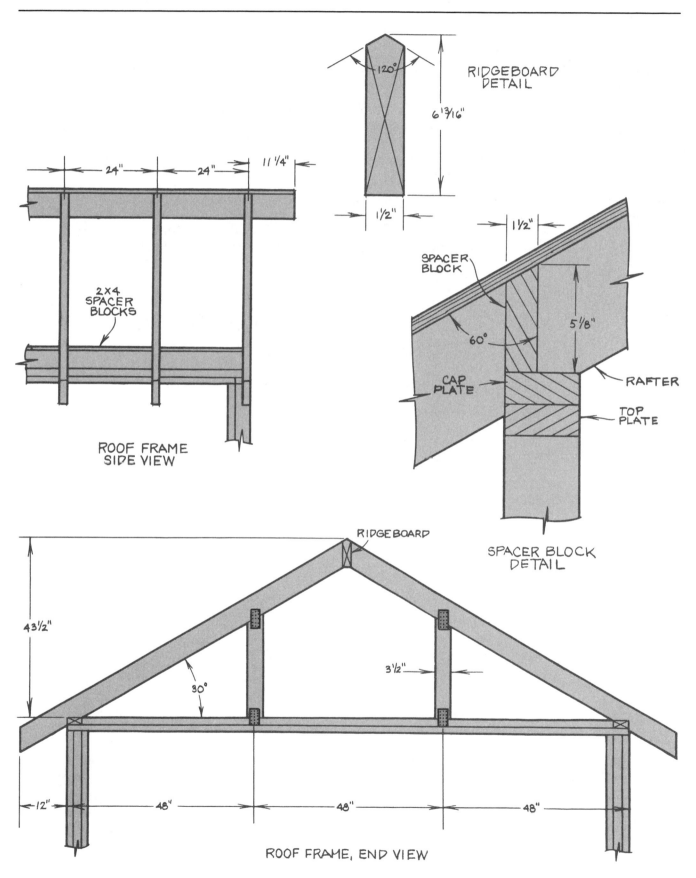

RIDGEBOARD DETAIL

120°

6 13/16"

1 1/2"

24"   24"   11 1/4"

2 X 4 SPACER BLOCKS

ROOF FRAME SIDE VIEW

1 1/2"

SPACER BLOCK

60°

5 1/8"

CAP PLATE

RAFTER

TOP PLATE

SPACER BLOCK DETAIL

RIDGEBOARD

43 1/2"

30°

3 1/2"

12"   48"   48"   48"

ROOF FRAME, END VIEW

16 **Attach facing strips to the sides and gable ends.**

Cut the side facing strips from ¾" stock, and bevel the top edge to match the slope of the rafters. Cut the gable end facings to size, then miter each strip so they fit together at the peak. Attach the side facing strips first, nailing them to the rafter ends with 8d nails. Then attach the gable end facing strips, nailing them to the ridgeboard and the side facing strips, as shown in the *Facing Strip Detail* drawing.

17 **Install the roof sheathing.**

Cover the roof frame with ½" CDX (exterior) plywood roof sheathing. Since you won't be installing a ceiling, be sure to place the good side of the plywood down. Nail the sheathing to the rafters using 6d nails spaced every 6" along the edges and 12" elsewhere.

18 **Install the roofing materials.**

Run metal drip edge around all the sides of the roof. This drip edge will keep the rain from collecting under the shingles at the edge of the roof and possibly rotting out the sheathing. Then cover the entire roof with tarpaper and shingles. (See Figure 5.) Put a double layer of tarpaper across the sheathing and 'double-wrap' the shingles at the peak to make a ridgecap.

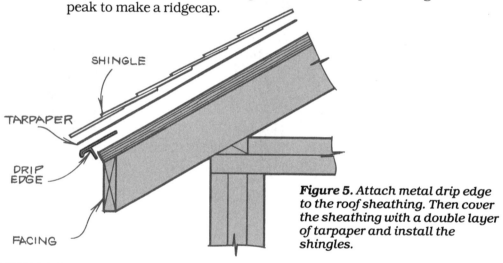

Figure 5. Attach metal drip edge to the roof sheathing. Then cover the sheathing with a double layer of tarpaper and install the shingles.

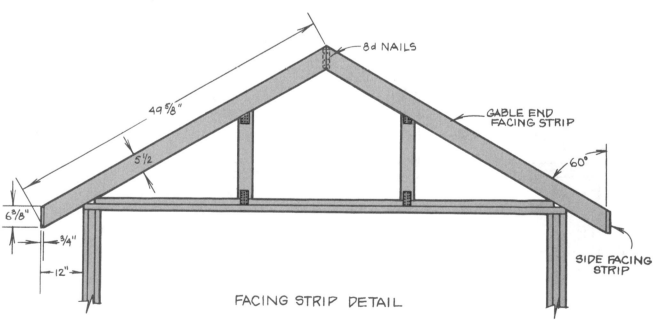

FACING STRIP DETAIL

Cut siding to fit the gable ends, and nail it to the rafters, gable braces, and cap plate. The siding should fit flush against the sheathing at the top, and lap the cap plate at the bottom but *not* the top plate. Leave the top plate uncovered—you'll need it as a nailing surface to attach the screen and the trim. Since there won't be any siding on the sides, trim the gable ends with short lengths of 1 x 2. (See Figure 6.)

19 **Install siding on the gable ends.**

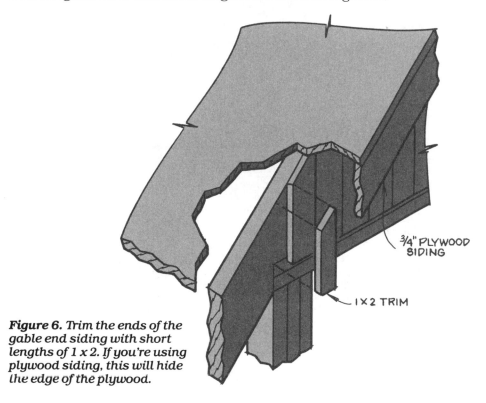

¾" PLYWOOD SIDING

1 X 2 TRIM

Figure 6. Trim the ends of the gable end siding with short lengths of 1 x 2. If you're using plywood siding, this will hide the edge of the plywood.

Cover all the exposed wood surfaces with paint or stain. You'll find it much easier to do this now, before you install the screen.

20 **Paint or stain all exposed wood surfaces.**

Cover the entire wall frame with screen, from the top plate to the sole plate. Leave the door uncovered, of course. Staple the screen in place, then cover the edges of the screen with ¼" x 1½" trim. (See Figure 7.) Rip this trim from pressure-treated 'two-by' stock, and nail it in place with 4d finishing nails.

21 **Install screen and trim.**

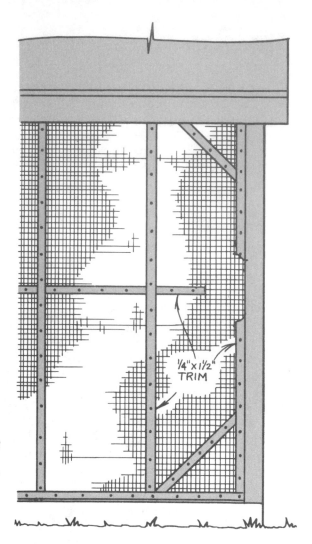

Figure 7. *Attach trim along the screen openings to hold the screen in place. Nail the trim to the framing members, or use ¾″ flathead wood screws. The screws will allow you to easily remove the trim and replace the screens.*

¼″ x 1½″ TRIM

TIP If you have small children or pets, and you think that you might have to repair or replace the screens from time to time, use #8 x ¾″ flathead wood screws instead of nails to attach the trim.

22 **Build the door.**

Cut 2 x 2's to frame the door, as shown in the *Door Frame* drawing. To strengthen the door, install braces in the corners. Assemble the door frame using #10 x 3″ flathead wood screws. Paint or stain the door frame to match the summer house, then staple screening to the frame. Install trim in the same manner that you trimmed the wall frames. To help strengthen the door, lap the ends of the trim over the joints between the members of the door frame.

Cut 2″ x ¾″ strips for the door stop. Place the strips flat against the framing studs so the strips are flush with the back of the opening, leaving a 1½″ space for the door, as shown in the *Door Stop Detail*. Attach the stops using 8d nails. Install hinges and a latch on the door. Put the door in place and attach the hinges to the frame. Chisel out the frame member where the latch will strike and install a striker plate.

23 **Hang the door.**

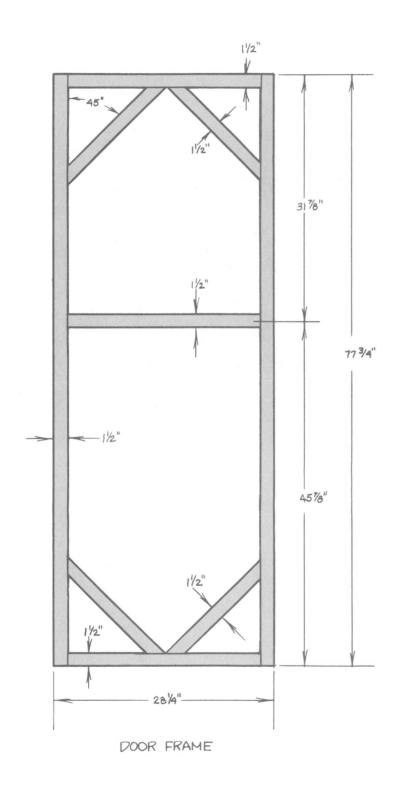

DOOR FRAME

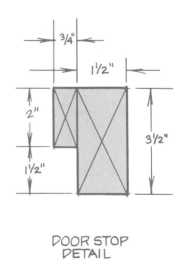

DOOR STOP
DETAIL

Sunspace

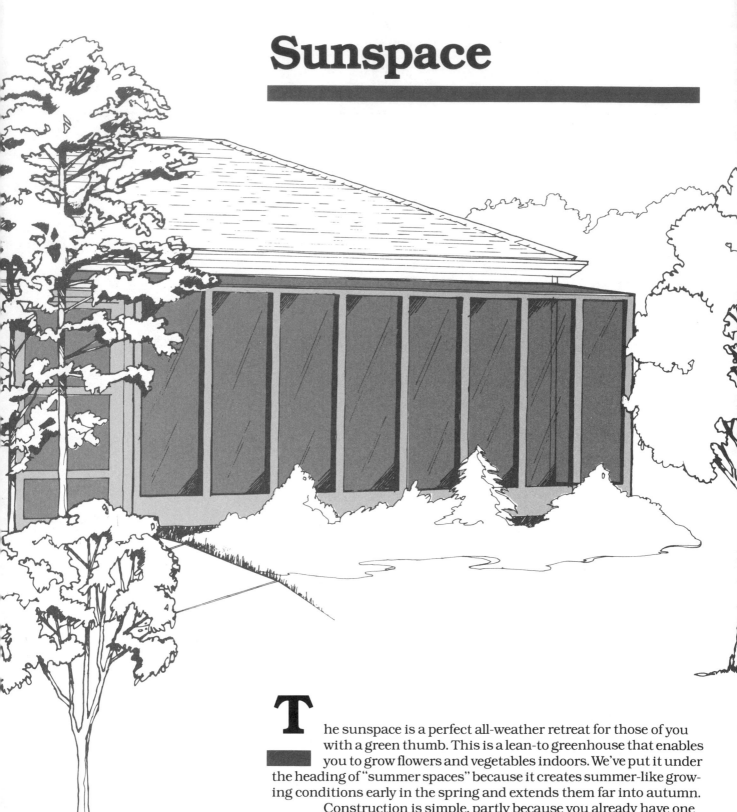

The sunspace is a perfect all-weather retreat for those of you with a green thumb. This is a lean-to greenhouse that enables you to grow flowers and vegetables indoors. We've put it under the heading of "summer spaces" because it creates summer-like growing conditions early in the spring and extends them far into autumn.

Construction is simple, partly because you already have one wall built. The structure leans against the side of your home. The sunspace is the size of a large room, and sits on a concrete footer-and-wall foundation. The structure is framed with 2 x 4's, like many of the projects in this book, but the walls and roof are covered with transparent glazing to let the sunshine in.

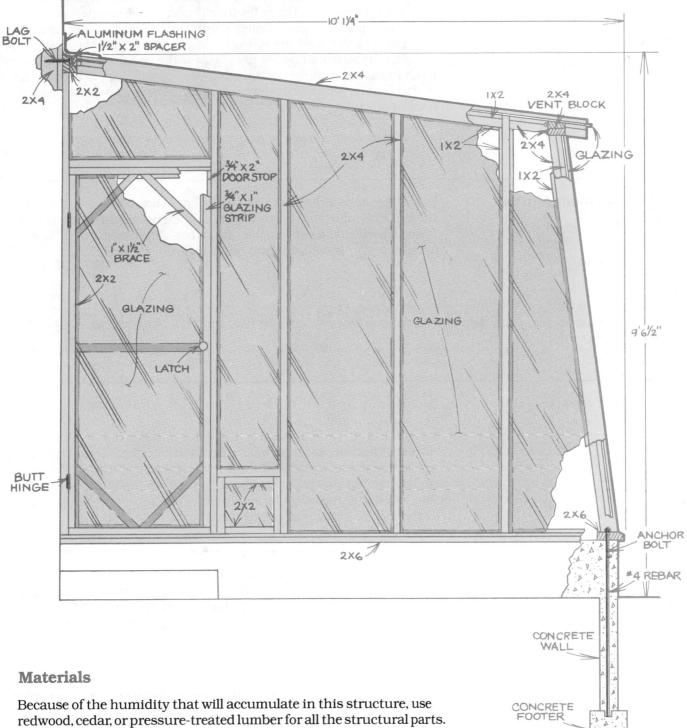

LAG BOLT

ALUMINUM FLASHING

1½" x 2" SPACER

2X4

2X2

10' 1¼"

2X4

1X2

2X4 VENT BLOCK

2X4

GLAZING

1X2

¾" x 2" DOOR STOP

¾" x 1" GLAZING STRIP

2X4

1X2

1" x 1½" BRACE

2X2

GLAZING

GLAZING

LATCH

9' 6½"

BUTT HINGE

2X6

ANCHOR BOLT

2X2

#4 REBAR

2X6

CONCRETE WALL

CONCRETE FOOTER

SIDE ELEVATION

Materials

Because of the humidity that will accumulate in this structure, use redwood, cedar, or pressure-treated lumber for all the structural parts. Moisture can rot other woods. If you build from redwood, select kiln-dried heartwood. The sapwood (light colored part) of redwood is not rot-resistant.

Use 2 x 4's for the entire frame, with the exception of the corner posts and sole plates. The posts are cut from 4 x 4 stock. Purchase 2 x 6 stock for the sole plates. The 2 x 2's for the ledger strip and door frame can be ripped from larger stock, or purchased already cut. You'll also need some 'one-by' (¾″ thick) stock to make glazing strips and the facing strip.

To pour the foundation, you'll need concrete, #4 reinforcing bar (rebar), 12 stakes, 1 x 2 spreaders, wire ties, and scrap lumber to build the form. You'll also have to purchase 1″ rigid insulation for the concrete walls and anchor bolts. In addition to these materials, purchase *flat*, clear fiberglass panels, galvanized nails, aluminum nails with rubber seals, flashing, silicone sealant, screws, and lag bolts.

Note: If you would rather use glass for the sunspace 'glazing', you won't need the aluminum nails. However, you will need to beef up the roof so that it can support the extra weight. Use 2 x 6's for the rafters instead of 2 x 4's.

Before You Begin

1 Take advantage of the sun.

Your sunspace's ability to catch the sun depends on its orientation to the sun's rays. For that reason, build your structure against a south-facing wall. If that's not possible, build it against an east wall. West should be your last choice, and north is no choice at all.

TIP Facing south, a sunspace will act as a passive solar collector, and help cut your heating bills in the winter.

2 Adjust the size of the sunspace.

Because this structure is a lean-to, you'll have to adjust the size to fit against an existing wall. As shown in the working drawings, this sunspace is 10′ wide, approximately 8′ high, and stretches 16′ long. However, this may be larger or smaller than you need. If you want a smaller sunspace, simply reduce the width or length. The design can be elongated or shortened by adding or subtracting framing studs. You can change the width by lengthening or shortening the rafters. Remember, keep the rafters and studs spaced 24″ on center.

3 Check the building codes.

Because this greenhouse is attached to a permanent foundation, it may be affected by building codes. Check your local codes—they may also have restrictions on how you may attach it to your home. If needed, secure a building permit before you start work.

Pouring the Foundation

4 Lay out the foundation.

Use stakes and string to lay out the foundation. The front line must be precisely parallel with the side of the house, so measure carefully. Drive in stakes outside the foundation lines—the intersection of the strings will mark the corners. (See Figure 1.) Measure diagonally from corner to corner to be sure the layout is square. Next, mark the borders of the trench footing, as shown in the *Foundation Layout* drawing.

162

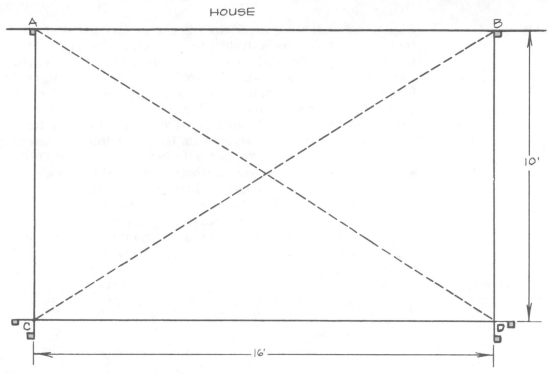

Figure 1. *Use stakes and string to lay out the foundation lines. Measure diagonally from corner to corner to be sure your layout is square. AD must equal BC.*

TIP Use a trickle of sand to show where to dig the trenches.

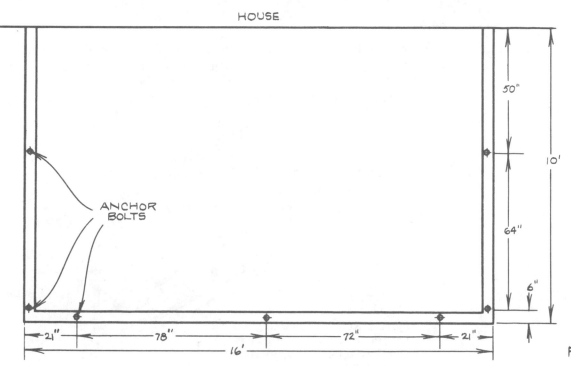

5 Dig the trench.

Dig a trench that is 24" to 36" deep, below the frost line. Make the trench wide enough on both sides to allow you to stand in it. When the trench is completed, dig another trench that is 6" deep and 8" wide for the footer, as shown in Figure 2. At 3' intervals along the footer trench, drive 12" stakes a few inches into the ground along each side. Select a stake that looks like it is in a high spot in the trench and use string and a string level to mark all the stakes at the same level. Next to each stake, drive an 18" long reinforcing bar into the ground until its top is even with the mark on the stake. The bars will serve as the grade pegs for the footer. Remove the stakes.

Note: The trench footings used here are 30" to 42" below the ground, so you may want to hire a professional with a backhoe to do the digging. If you dig them yourself, rent a gasoline-powered trencher. As you dig, keep in mind that the footings must rest on undisturbed earth. Don't dig too far down—you can't fill the soil in again.

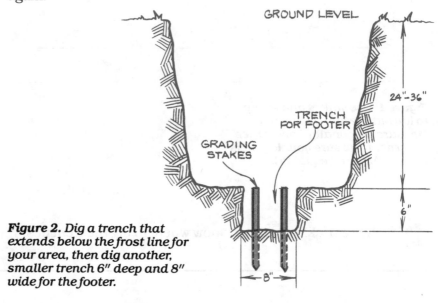

Figure 2. Dig a trench that extends below the frost line for your area, then dig another, smaller trench 6" deep and 8" wide for the footer.

6 Lay rebar in the footer trench.

Set #4 reinforcing bars along the inside of each row of grade pegs and support the bars on bricks or stones so they are about 3" above the ground, as shown in Figure 3. Overlap the bars 12"-15" where they meet and tie them together with wire ties. Also tie the rebar to the grade pegs.

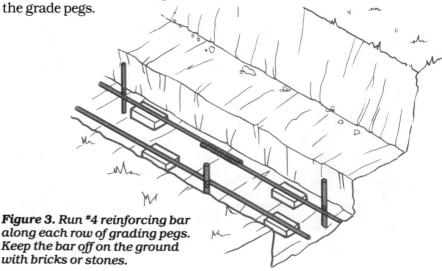

Figure 3. Run #4 reinforcing bar along each row of grading pegs. Keep the bar off on the ground with bricks or stones.

With the help of a friend, pour the concrete into the trench and spread it with square-tipped shovels. 'Slice' into the concrete with the shovels or with a yardstick to break up air pockets. As you pour, level the concrete with 'floats'. (These floats are nothing more than scraps of 2 x 4.) Using the floats, pat down the concrete, then zigzag horizontally across the concrete until the surface is fairly smooth. Sweep the surface with the trailing edge of the float, dragging it diagonally toward you. When you are finished, the tops of the grade pegs should be barely visible above the concrete. Press 1 x 1's on edge in the center of the concrete to make a notch or key. Allow the concrete to cure for 24 hours and remove the 1 x 1's.

7 **Pour the concrete for the footer.**

Use 2 x 4 studs and ¾″ plywood sheathing to build the forms for the walls, as shown in Figure 4. This form must extend at least 12″ above the ground. Cross brace the form to tie the sides together. In addition to the braces, use wire ties and spreaders to keep the sides of the forms from bowing while you're pouring the concrete. To insert the wire ties, drill holes on each side of the studs and pass the wire through the holes. Use a 1 x 2 spreader near each tie so you won't pull the form sides out of alignment when you tighten the wire. Then twist the wire with a screwdriver or block of wood until it is tight.

8 **Build the forms for the foundation walls.**

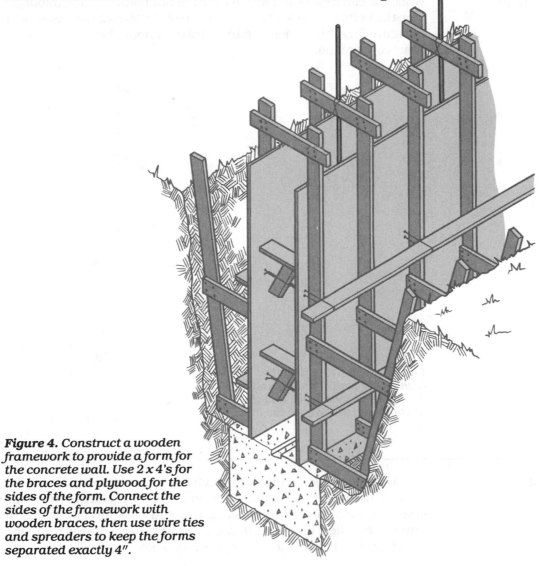

Figure 4. Construct a wooden framework to provide a form for the concrete wall. Use 2 x 4's for the braces and plywood for the sides of the form. Connect the sides of the framework with wooden braces, then use wire ties and spreaders to keep the forms separated exactly 4″.

9   **Set the rebar vertically in the forms.**

Place lengths of No. 4 reinforcing bar vertically inside the forms and hold them in place with wire ties. Space the rebar about 24" apart throughout the foundation wall. Make sure the bars extend about 1' above the forms. The bars can be cut flush with the top of the wall when the concrete has cured.

10   **Pour the concrete wall.**

With the help of a friend, pour the concrete into the forms. As the concrete is being poured, remove the spreaders. Leave the wire ties in place. (Cut them after the concrete has cured.) After the concrete has been poured, use a concrete vibrator to settle the concrete and remove any air pockets. You can rent a concrete vibrator at a building supply store.

TIP   If the concrete truck can't back up to your building site, arrange for extra wheelbarrows or a pump truck.

11   **Set the anchor bolts.**

While the concrete is still wet, set the anchor bolts in place, making sure the bolts extend 2½" above the top of the foundation, as shown in Figure 5. Let the concrete cure at least 24 hours before you continue construction.

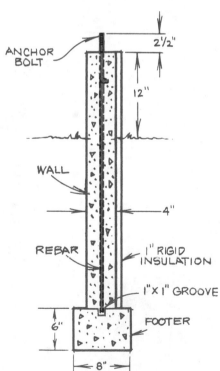

Figure 5. When finished, the foundation wall should stand 12" above ground level, and the anchor bolts 2½" above the top of the wall.

12   **Insulate the wall.**

Remove the wooden forms, cut the wire ties and the rebar flush with the wall, then place 1" rigid insulation along the insides of the wall to keep ground moisture from soaking into and weakening the concrete. This will also help keep heat in the sunspace. Throw gravel into the trench and fill it with earth, tamping as you go.

                                                           GAZEBOS AND SUMMER SPACES

Building the Frame

Cut the 2 x 6 sole plate to length, and bevel one edge 30°, as shown in the *Sole Plate Detail* drawing. Line up the sole plate along the edge of the foundation and mark the position of the anchor bolts. Drill the holes ¼″ larger than the diameter of the bolts, then lay the sole plate in place to make sure the edges are straight and the holes fit over the bolts. Cut 2 x 4's and 4 x 4's for the studs, then miter each end at 82½°, as shown in the *Front Wall Stud Detail* drawing.

Cut a 2 x 4 top plate and place it next to the sole plate, bottom side up. Using a framing square, measure where each stud will be placed. Draw a pencil line across both plates to mark the stud positions. Assemble the wall, positioning the studs ¾″ from the back of the sole plate, as shown in the working drawings. Space the studs 24″ on center, as shown in the *Front Wall, Front View* drawing.

13 **Build the front wall.**

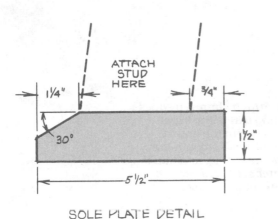

SOLE PLATE DETAIL

FRONT WALL STUD DETAIL

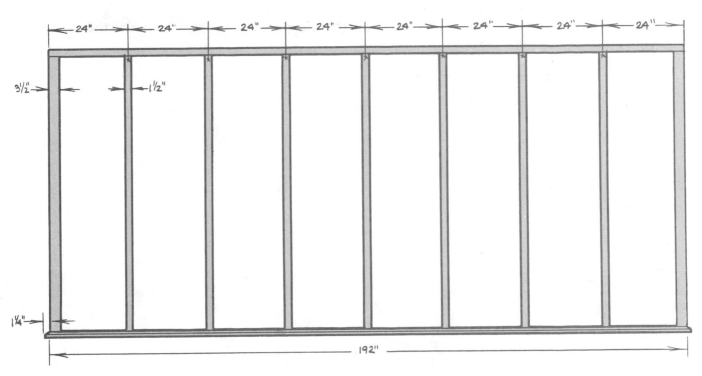

FRONT WALL, FRONT VIEW

14 Cut the rafters.

Cut the rafters from 2 x 4 stock, as shown in the *Rafter Layout* drawing. Miter each end as shown, and notch the rafters so they fit over the ledger strip and the top plate.

15 Prepare the side of the house where you will attach the sunspace.

If you're attaching the sunspace to a brick, block, or flat wall, you can skip this step. You don't need to specially prepare the wall where the frame will butt up against it. However, if the wall is covered in lapped siding, you must somehow make the surface flat. The easiest way to do this is by cutting shaped wedges, 3½″ wide, to 'shim' the places where the siding laps. Carefully measure and mark where the frame will meet the house. Then nail the wedges in place with 4d nails. (See Figure 6.)

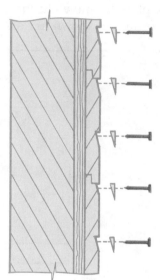

Figure 6. *Attach shaped wedges to the siding where the frame members will join the house. These wedges will smooth the surface of the siding, and you'll get a tighter seal between the sunspace and the house.*

TIP For an airtight seal between the house and the sunspace frame, spread caulk on the wall before you nail the wedges in place.

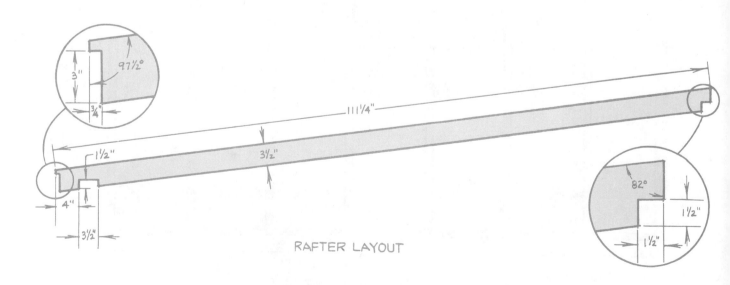

RAFTER LAYOUT

Cut a header and a ledger strip exactly as long as the top plate. Nail the two pieces together with 12d nails, passing the nails through from both the front and back of the assembly. This will strengthen the joint. The ledger strip should be flush with the bottom of the header, as shown in the *Header Detail* drawing. Attach the completed assembly to the house with lag screws. These screws must bite into the frame studs so that the ledger strip can support the weight of the roof. (See Figure 7.)

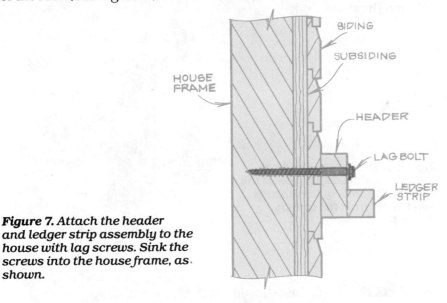

Figure 7. Attach the header and ledger strip assembly to the house with lag screws. Sink the screws into the house frame, as shown.

16 Attach header and ledger strip to the house.

TIP   If you are attaching your sunspace to a masonry wall, use ½″ lead expansion shield and bolts. Space shield and bolts up to 36″ on center.

Cover the top of the foundation wall with a sheet of plastic, to keep moisture from the concrete from rotting the wood frame. With the help of a friend, gently lift the front wall into place. Temporarily brace it upright at the proper angle, then fasten the sole plate to the anchor bolts. Check to be sure the wall is square. If necessary, drive a wooden wedge under the sole plate until the wall is properly positioned.

17 Erect the wall and brace in place.

Carefully measure along the ledger strip and the top plate, marking where the rafters will go. They should be spaced every 24″ on center. Fit the rafter 'bird's mouths' over the top plate at one end and the ledger at the other, as shown in the *Wall Frame, Side View* drawing. Nail the rafters in place with 16d nails. Cut a facing strip to fit in the front end of the rafter, as shown in the working drawing. Attach the facing strip to the rafters with 8d nails.

18 Attach the rafters and facing strip.

19 **Frame the side walls.**

Cut 2 x 6 sole plates for the sides. Mark the location of the anchor bolts and drill holes in the sole plates to fit over the anchor bolts. Put the sole plates in place and fasten them down. Cut 2 x 4 studs for the side walls and notch the upper ends, as shown in the *Wall Stud Joinery Detail* drawing. Attach the studs to the frame using 16d nails, where shown in the *Side Wall Frame* and *Side Wall Frame with Door* drawings. The notched end is attached to the rafters. (See Figure 8.) Frame in the door and vent openings by nailing 2 x 4's between the studs.

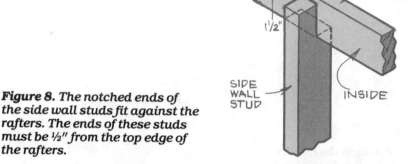

Figure 8. The notched ends of the side wall studs fit against the rafters. The ends of these studs must be ½" from the top edge of the rafters.

TIP Before you attach the frame members that fit against the side of the house, apply a heavy bead of caulk to the house to seal the joint between the sunspace and the house.

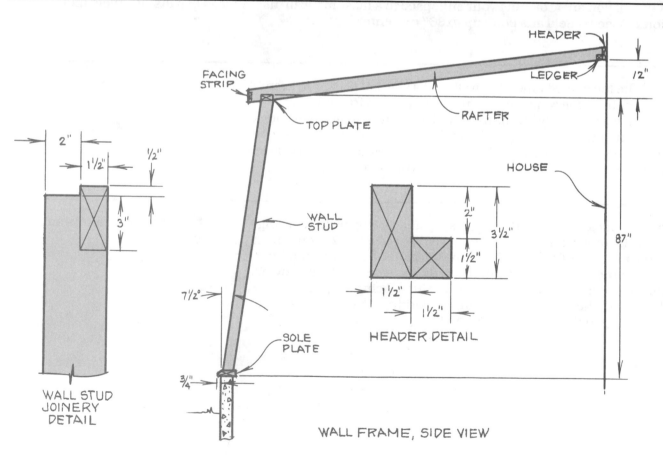

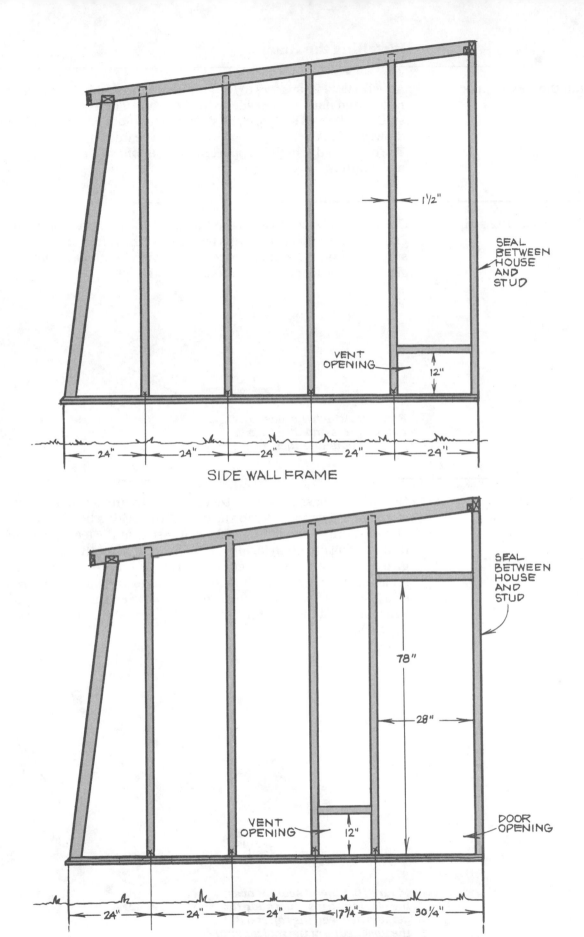

1½"

SEAL
BETWEEN
HOUSE
AND
STUD

VENT
OPENING

12"

24" 24" 24" 24" 24"

SIDE WALL FRAME

SEAL
BETWEEN
HOUSE
AND
STUD

78"

28"

VENT
OPENING

12"

DOOR
OPENING

24" 24" 24" 17¾" 30¼"

SIDE WALL FRAME WITH DOOR

SUNSPACE

171

20 **Install the roof supports.**   Attach 1½" x 2" spacers to the header. Fit the spacers between the rafters and nail to the header with 12d nails. Then cut 1½" x 3" stock for the roof supports, as shown in the *Roof Frame, Top View* drawing. Nail the roof supports between the rafters with 16d nails. The bottom edge of the support must be flush with the bottom edge of the rafters.

21 **Install the glazing strips.**   Cut 1½" x ¾" stock to form the glazing strips for attaching your glazing panels. Place the strips flat against the rafters and the wall studs, ½" from the outside edge, as shown in Figure 9. Attach the glazing strips to rafters and studs with 6d nails.

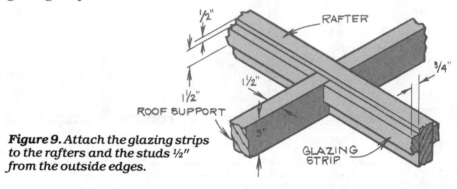

Figure 9. Attach the glazing strips to the rafters and the studs ½" from the outside edges.

22 **Install the glazing.**   Cut the fiberglass panels to size with a contractor's saw and a plywood blade. Use some scrap sheathing to provide a backup to the blade. Before you cut, flip the blade over so that the teeth are pointing in the wrong direction (against the direction of rotation)—this will give you a much smoother cut in the thin fiberglass material. Spread sealant along the tops of the glazing strips to provide a waterproof seal. Slide the fiberglass panels in place between the rafters or studs. Near the top of the roof, butt the glazing against the spacers. (See Figure 10.)

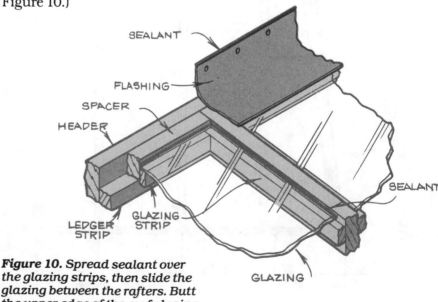

Figure 10. Spread sealant over the glazing strips, then slide the glazing between the rafters. Butt the upper edge of the roof glazing against the spacers.

172

The bottom edge of the roof panels overhang the facing strip slightly, and do not need to be sealed on this one edge. (See Figure 11.)

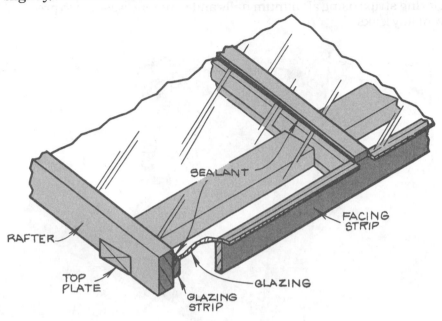

Figure 11. *The bottom edge of the roof glazing must overhang the facing strip slightly.*

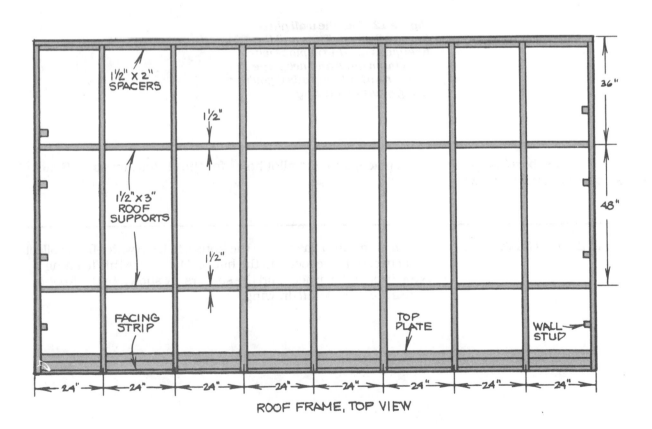

ROOF FRAME, TOP VIEW

The wall panels must be sealed on all four sides; they don't overhang any part of the frame. (See Figure 12.) Attach the panels to the glazing strips using aluminum nails and neoprene washers to prevent any leaks.

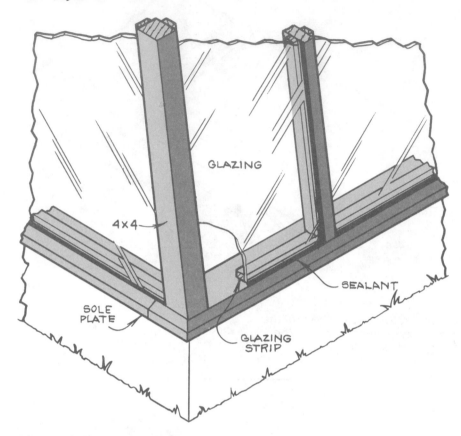

GLAZING

4×4

SEALANT

SOLE PLATE

GLAZING STRIP

Figure 12. Seal the wall glazing on all edges. Then nail the fiberglass glazing in place with aluminum nails and neoprene washers. (Omit the nails if you're using glass for glazing.)

TIP When the fiberglass panels are in place, drill 5/32″ pilot holes for the aluminum nails. Space the nail holes about every 12″.

23 **Install the roof flashing.** Attach aluminum flashing to the house, to keep water from collecting between the header and the house. Make sure this flashing extends past the header and the spacers, as shown in the *Roof-to-House Joinery Detail* drawing.

Finishing Up

Cut 2 x 2's to frame the door and vents, as shown in the *Door Frame* and *Vent Frame* drawings. To strengthen the door, cut 1½″ x 1″ stock to form braces and miter the ends to fit flush against the frame. Screw the frame members together using #8 flathead wood screws. Cut 1″ x ¾″ glazing strips and attach to the door and vent door frames, as shown in the *Section A* detail. Attach the strips flush with the back of the frames so there is a ½″ space at the front. (This is where you will attach the glazing.)

24 Build the door and vent frames.

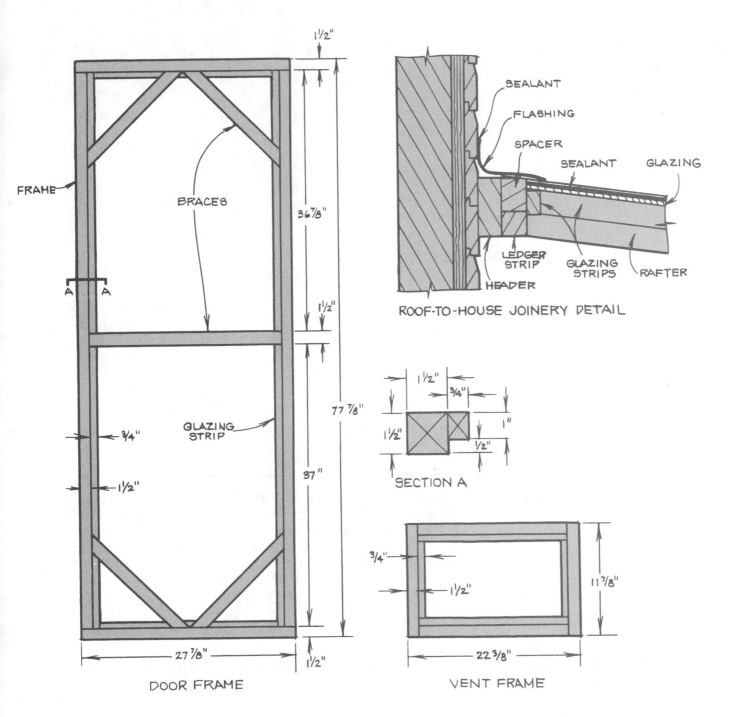

ROOF-TO-HOUSE JOINERY DETAIL

SECTION A

DOOR FRAME

VENT FRAME

25 **Install the glazing in the door and vents.**

Cut glazing to size, then spread sealant over the glazing strips, as you did when you glazed the sunspace frame. Put the glazing panels in place, as shown in Figure 13. Attach the panels to the glazing strips using aluminum nails and neoprene washers to prevent any leaks. Drill pilot holes for the nails, to prevent the wood from splitting.

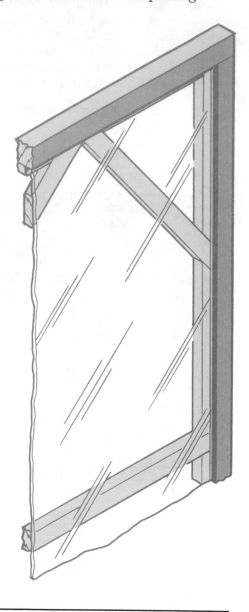

Figure 13. The glazing in the door fits over the bracework, as shown.

26 **Hang the door and vents.**

Cut 2″ x ¾″ strips to make door and vent stops. Place the door and vent stops flat against the framing studs, as shown in the *Door/Vent Stop Detail* drawing. The stops should be flush with the back of the opening, leaving a 1½″ space at the front. Attach the stops using 6d nails. Mortise the door, vents, and sunspace frame for hinges, then hang the door and vents in the frame. Install a latch in the door and turnbuttons on the vents to keep them closed.

TIP Put the hinges on the bottom of the vent frame, and use a small chain to hold the vent in the open position without letting it flop on the ground.

The vents near the bottom of the side walls let air *into* the sunspace. This air escapes through the spaces between the roof glazing and the top plate. There will be times when you want to block off the air flow. To do this, make vent blocks, as shown in the *Vent Block Detail* drawing. Cut the wooden blocks slightly smaller than needed, then glue felt around the edges. This will give you an airtight friction fit when the blocks are in place.

27 **Make vent blocks for the roof.**

Use waterproofing paint or stain to seal the exposed wood surfaces and to keep the humidity in the sunspace from rotting the wood. Be sure to paint or stain wood surfaces on the inside of the sunspace as well as the outside.

28 **Paint and stain exposed wood surfaces.**

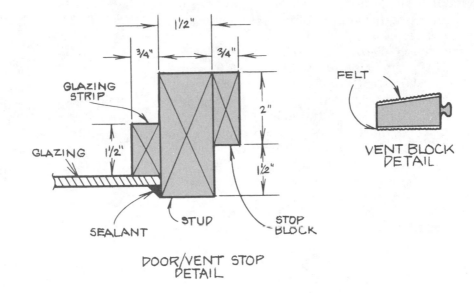

Arbors and Trellises

Adding an arbor or trellis to your yard will enhance your landscaping and increase the aesthetic value of your property, if not its real value. Today's arbors and trellises are descended from the ancient grape arbors of the Old World, though they're seldom used just for grapevines anymore. Instead, a modern arbor usually serves as a decorative entrance to a flower garden or a patio. Add benches and it will make a simple, but cozy, garden shelter. A trellis, either freestanding or attached, can be used to line a walk, define a property line, or decorate a wall.

The biggest attraction to arbors and trellises is that they support vining plants. This particular outdoor structure is meant to be 'lived upon', so to speak. With some careful planting and pruning, they can become showcases for your gardening skills. A trellis is lighter in construction than an arbor, and won't support as much weight. But it still provides adequate support for a wide variety of plants and flowers. Ever-blooming roses such as Dream Girl, American Pillar, or Summer Snow grow well on trellises, as well as perennial vines like Akebia, Silver Lace vine (sometimes called the Chinese Fleece vine), and one of the varieties of Bittersweet vines. An arbor will support all these, plus the heavier vines, such as grapes. Check with your local nursery to see which vines grow well in your area.

Before You Begin

Plan an arbor or trellis that suits your tastes and needs. We show you how to build four common types of arbors and trellises. You can mix and match among the designs and features you see here to create an arbor or trellis that is uniquely suited to your outdoor setting. All of the projects except the attached trellis use simple pole foundations, so it's easy to adjust the size of the arbor or trellis —all you do is alter the number and placement of the posts. In the case of the attached trellis, just adjust the size and the number of panels.

Here are some other considerations:

Materials. Carefully choose your materials. Be careful if you work with pressure-treated lumber. Some of the chemicals used to treat wood can be harmful to plants. For instance, one popular preservative, pentachlorophenol (or just 'penta'), is similar to Agent Orange, the defoliant used in the Vietnam War. Creosote is also harmful to plants. The telephone companies use this preservative not only to keep their poles from rotting, but to control the weeds that might climb up them and become entangled in the wires. Your best bet is to use wood that's been treated with chromated copper arsenate (CCA). This chemical stains the wood a light green. Of course, if you want to be completely safe, use naturally rot-resistant woods such as cedar or redwood.

178

Orientation to the sun. When you build your arbor or trellis, select a site that will provide at least four hours of sunlight a day—otherwise you won't be growing much on it. Try to place the structure so that it gets a southern or eastern exposure.

Grid patterns. You can choose from many different grid patterns for your arbor or trellis. The type of vine you plan to grow will partially determine the type of grid you want. For example, heavy vines require a sturdy grid, so use wide slats or interlocking slats. The more interlocking boards you have laced in your grid, the stronger it will be. Likewise, twining vines or vines with tendrils need a trellis with lots of interlocking supports so they can climb properly. Fragile vines such as morning glories ask for only a minimum of support from a grid.

There may be factors besides the plants that will determine your grid pattern. If you want to use a trellis as a privacy screen, use a closely-woven grid pattern so when the green leaves fall away your yard will still be shielded from view. Mix and match the grid patterns we've illustrated to suit your site, your plants, and any other needs.

Paints and stains. The types of plants you grow may also determine whether or not you can paint or stain your arbor or trellis. If you want to paint it, don't grow roses, grapes, or other perennials on the structure. You'll have to chop those vines down in a few years to repaint it. Instead, use redwood or cedar that doesn't require painting—or select annual vines.

Local restrictions. Finally, check with your local building inspections office before you begin work to be sure your trellis or arbor meets code requirements. You may have to alter the site and the size to meet local codes.

ARBORS AND TRELLISES

Ladder Arbor

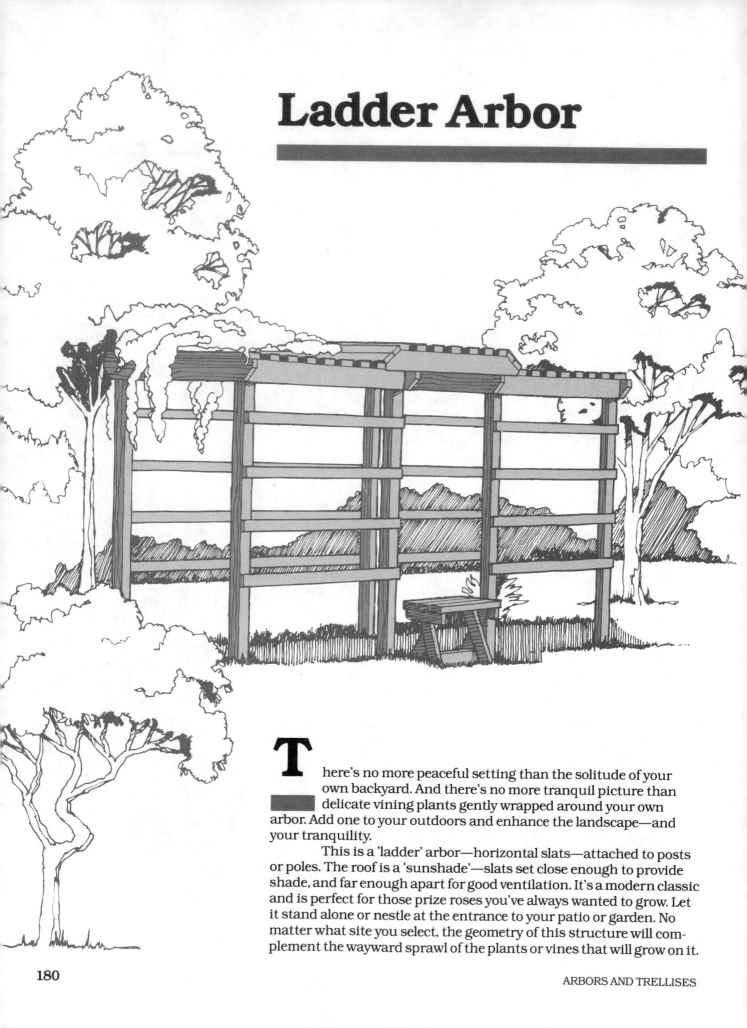

There's no more peaceful setting than the solitude of your own backyard. And there's no more tranquil picture than delicate vining plants gently wrapped around your own arbor. Add one to your outdoors and enhance the landscape—and your tranquility.

This is a 'ladder' arbor—horizontal slats—attached to posts or poles. The roof is a 'sunshade'—slats set close enough to provide shade, and far enough apart for good ventilation. It's a modern classic and is perfect for those prize roses you've always wanted to grow. Let it stand alone or nestle at the entrance to your patio or garden. No matter what site you select, the geometry of this structure will complement the wayward sprawl of the plants or vines that will grow on it.

ARBORS AND TRELLISES

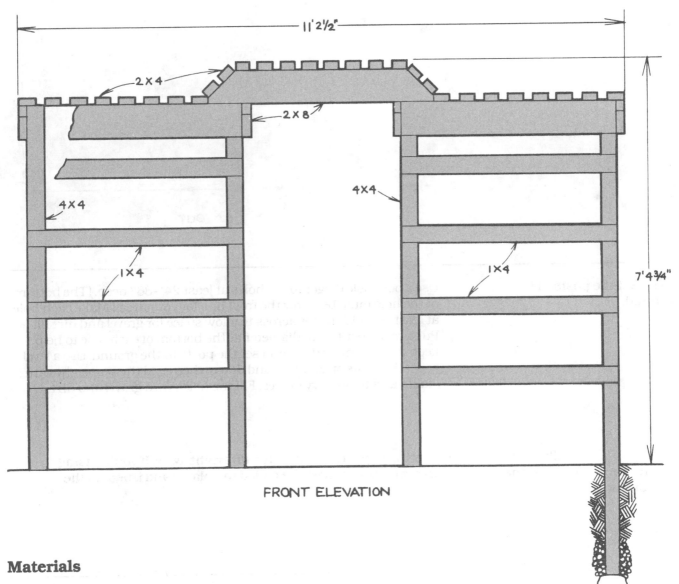

FRONT ELEVATION

Materials

Use redwood, cedar, or lumber that has been chemically treated with chromated copper arsenate (CCA) for this project. You'll be able to spot this wood in the lumberyard right off because CCA turns the wood a dull green. Other chemicals used in pressure-treatment may be harmful to plant life and could stunt the growth of sensitive flowers or vines growing on the arbor.

Use 4 x 4's for the posts, 2 x 4's and 2 x 8's for the sunshade roof, and 1 x 4's for the 'ladder' slats. You'll also need some galvanized nails.

LADDER ARBOR   **181**

1 **Lay out the posts.**

Use stakes and strings to mark the location of the posts. The easiest way to get your layout square is to mark the location of the four outside posts first, then measure diagonally, from corner to corner. Both diagonal measurements should be the same. When you're satisfied that the corner posts are properly located, measure along the strings to locate the inside posts, as shown in the *Post Layout* drawing.

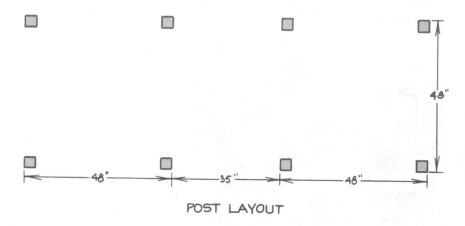

POST LAYOUT

2 **Set the posts in the ground.**

Use a post hole digger to dig holes at least 24"-36" deep. (The bottom of the hole must be below the frost line for your area.) Make each hole at least 10 to 12 inches across to allow space for gravel and dirt fill. Put a rock about 8" in diameter at the bottom of each hole to help keep them from settling, and set the posts in the ground. Use a level to set the posts straight up and down, then hold them upright with stakes and temporary braces. Fill the holes with gravel and dirt.

TIP If you're filling the holes with earth, do *not* tamp down the dirt right away. If you're using concrete, do not pour the concrete just yet. Wait until you've added top plates and trued up the arbor frame.

Building the Frame

3 **Mark the tops of the posts.**

Pick a corner post and measure 80" up from the ground. Using a string level, find the tops of the other posts, using this one corner post as a reference. (See Figure 1.)

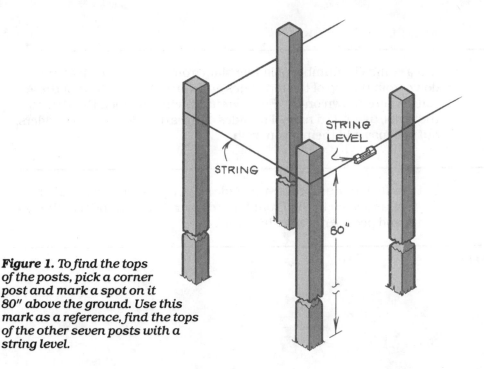

Figure 1. *To find the tops of the posts, pick a corner post and mark a spot on it 80" above the ground. Use this mark as a reference, find the tops of the other seven posts with a string level.*

Cut the top plates and the pediments to the dimensions shown in the *Front View* drawings. Miter the ends of the side top plates and the pediments as shown in the *Side View* drawings and Figure 2. Nail the top plates in place on the posts with 16d nails, lining the top of the top plates up with the top marks. Using a handsaw, cut off the part of the posts that sticks up above the top plates. Then nail the pediments in place on the top plates, using 16d nails.

4   **Cut and attach the top plates to the posts.**

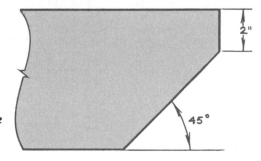

Figure 2. *Miter the ends of the side top plates and the pediments as shown here.*

Using a level, check that the arbor frame is square. If it isn't, push firmly against one or more posts until all the posts are straight up and down. Tamp down the earth at the foot of the posts (or fill the holes with concrete) to set the posts permanently in the ground. Remove the temporary braces.

5   **Square the arbor frame.**

TIP   To pack the earth as tightly as possible, use the end of a 4 x 4 or a large, capped iron pipe as a tamp.

Finishing the Arbor

6 **Attach the 'ladder'.**

Cut a sufficient number of ladder slats from 1 x 4 stock. Measure down from the top of the top plates and mark the positions of these slats, where shown on the *Front View* drawings. Attach the slats to the posts, using 12d nails. The sides of the arbor do not need ladders, but you may add them if you wish.

7 **Attach the sunshade.**

Cut a sufficient number of sunshade slats, and space them out on the roof, as shown in the *Front View* drawing. Attach them to the top plates and pediments with 16d nails.

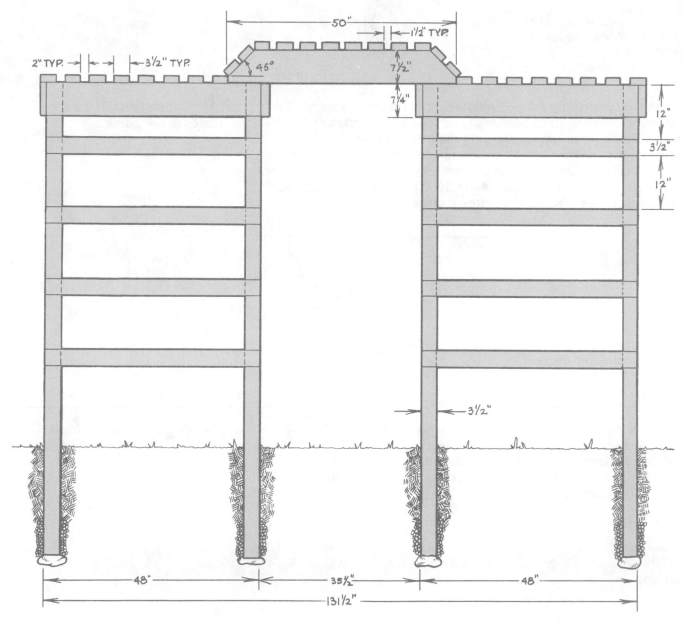

FRONT VIEW

A Place to Rest

This arbor just wouldn't be complete without a place to sit down and admire it, so we've added the plans for a small bench. (See the *Bench Front View*, *Bench Side View*, and *Bench Top View*.) This can be made from 2 x 8 and 2 x 4 scraps from the arbor. Nail the top and the crossed legs together with 16d nails, then bolt the legs to the top with ⅜" x 3½" and ⅜" x 5" carriage bolts.

8 Make a bench.

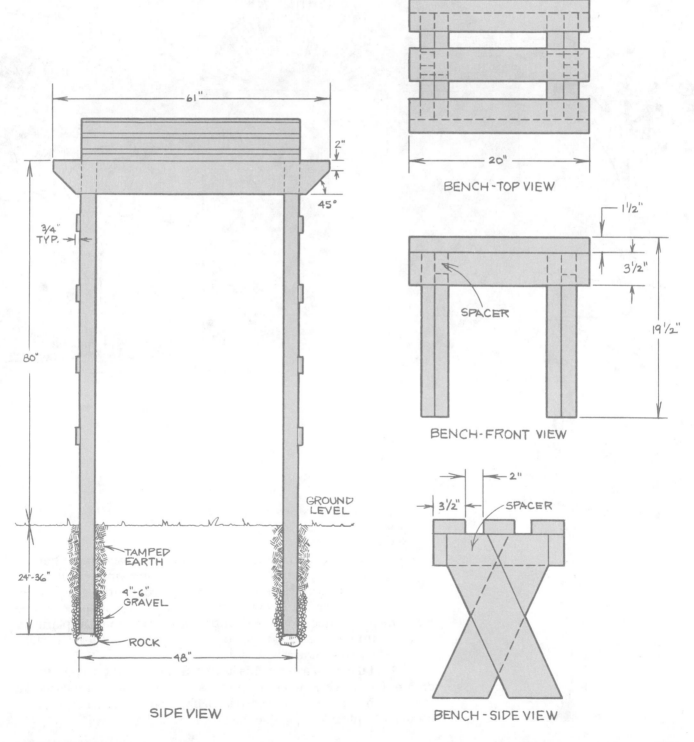

SIDE VIEW

BENCH-TOP VIEW

BENCH-FRONT VIEW

BENCH-SIDE VIEW

Lattice Arbor

While ladder arbors are good for grapes, roses, and other climbing plants with thick stalks, some vines require an arbor with a closer weave. Lattice arbors are a good place to grow delicate flowering vines and ivy. Not only is the space between slats smaller; lattice provides a *continuous* support for the plants to twine around. There are no gaps between rungs, and the lattice fills in quickly with greenery.

Like most arbors, this lattice arbor is built on a pole foundation. The space between the poles is filled with diamond-weave lattice, and the roof is covered with lath. If you so wish, there's even room for built-in seats inside the arbor.

ARBORS AND TRELLISES

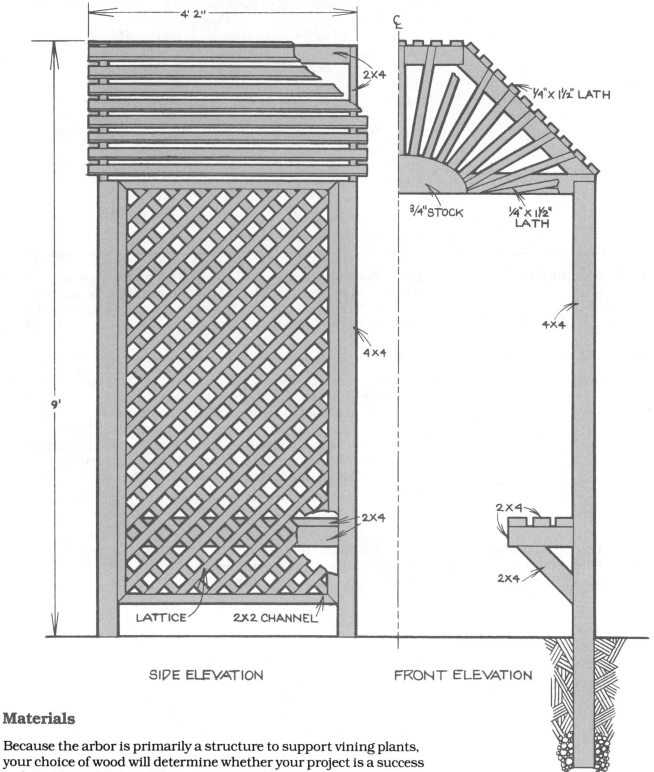

4' 2"

2X4

¼" x 1½" LATH

¾" STOCK

¼" x 1½" LATH

4X4

9'

4X4

2X4

2X4

2X4

LATTICE   2X2 CHANNEL

SIDE ELEVATION   FRONT ELEVATION

Materials

Because the arbor is primarily a structure to support vining plants, your choice of wood will determine whether your project is a success or failure. Some lumber is treated with chemicals that make it poisonous to plants. For example, lumber that has been treated with black creosote will kill any foliage it comes in contact with.

Look for Southern yellow pine, Western hemlock, Douglas fir, and Ponderosa pine lumber that is kiln-dried and pressure treated with chromated copper arsenate (CCA) for this project. You'll be able to spot this wood in the lumberyard right off because CCA turns the wood a dull green. CCA is relatively harmless to plants. Redwood and cedar are also good choices. The tendrils of ivy and many other vines penetrate the wood and cause it to rot, but redwood and cedar have natural oils which make them rot-resistant.

Use 4 x 4's as the supporting posts for the arbor, and 2 x 4's for top frame and optional seats. For the grid, you can purchase ready-made 4' x 8' sheets of 'diamond weave' lattice at most lumberyards and home centers. Or you can make your own grid from ¼" x 1½" lath strips. Frame the lattice in 2 x 2 'channel' stock. Like the lattice, this channel is available commercially. Or you can make your own by cutting a ½" wide, ¾" deep groove down one side of 2 x 2 stock. Cover the roof with ¼" thick, 1½" wide lath.

In addition to this lumber, you'll also need some galvanized nails and lag bolts.

Setting the Posts

1 **Lay out the posts.**

Use stakes and strings to mark the location of the posts, as shown in the *Post Layout* drawing. Mark the location of the four corner posts, then check that the layout is square by measuring diagonally, from corner to corner. Both diagonal measurements should be the same.

2 **Set the posts in the ground.**

Use a post hole digger to dig holes at least 24"-36" deep. (The bottom of the hole must be below the frost line for your area.) Make each hole at least 10 to 12 inches across to allow space for gravel and dirt fill. Put a rock about 8" in diameter at the bottom of each hole to help keep them from settling, and set the posts in the ground. Use a level to set the posts straight up and down, then hold them upright with stakes and temporary braces. Fill the holes with gravel and dirt.

TIP Do *not* tamp down the dirt right away. Wait until you've added top plates and trued up the frame. When you do tamp the dirt, be sure that the distance between the inside edges of the posts at the sides is precisely 41". Otherwise, the lattice won't fit correctly.

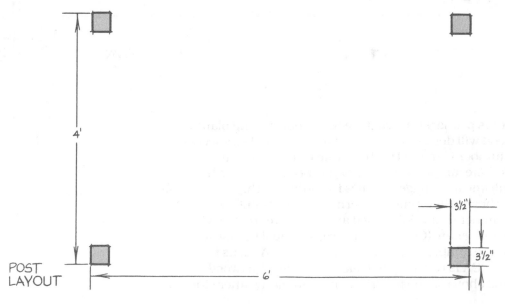

POST LAYOUT

4'

6'

3½"

3½"

Pick a corner post and measure 84″ up from the ground. With a string and string level, find the tops of the other posts, using the first corner post as a reference. (See Figure 1.) Cut the corner posts off at the proper height with a handsaw.

3 **Cut the corner posts to the proper height.**

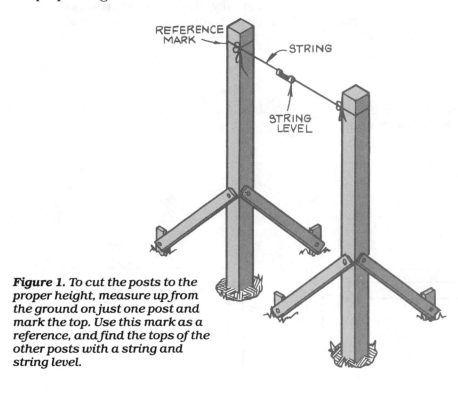

Figure 1. To cut the posts to the proper height, measure up from the ground on just one post and mark the top. Use this mark as a reference, and find the tops of the other posts with a string and string level.

Building the Top Frame

Cut the top plates to the dimensions shown in the *Arbor Frame, Front View* and *Side View* drawings. Cut notches in the ends of the front and back top plate pieces. (See Figure 2.) Check the fit of all the parts, and nail the top plate to the posts with 16d nails.

4 **Cut and attach the top plates.**

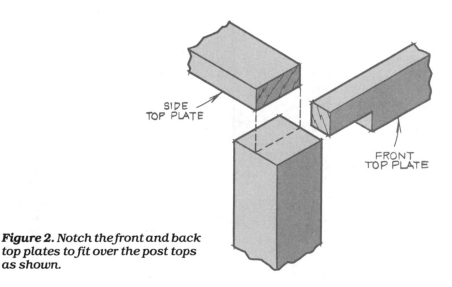

Figure 2. Notch the front and back top plates to fit over the post tops as shown.

5 **Build the roof frame.**

Cut the parts for the roof frame, as shown in the *Roof Frame, Top View* and *Side View* drawings. Miter the bottom ends of the diagonal boards and notch the other ends as shown. Nail the diagonal and horizontal boards together first, then attach them to the top plate, right above each corner post. Finally, attach the ridge boards between the two halves of the roof frame. Use 16d nails for all this framework.

TIP The roof frame will be a bit wobbly until you install the lath. You may want to install a few temporary braces until it's time to put the roof on.

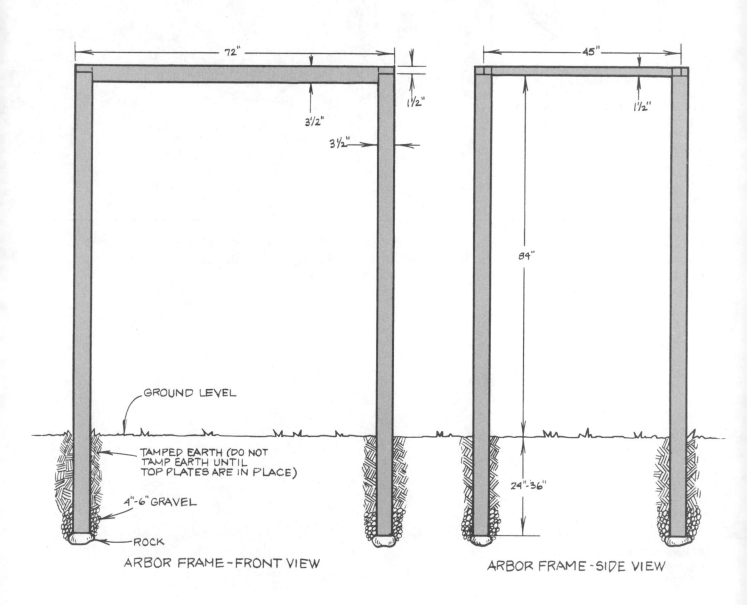

ARBOR FRAME – FRONT VIEW

ARBOR FRAME – SIDE VIEW

Filling in the Walls

If you're making your own lattice, put together two large sheets, at least 4' x 7'. Make the lattice from ¼" thick, 1½" wide lath, stapled together at each junction. You can use the design that we show here or invent your own. You'll also need some 'channel' stock, to frame the lattice. Make the channel from 2 x 2 stock, and cut a groove down the center of one side. This groove should be ¾" deep and as wide as the lattice you've made is thick.

6 Make the lattice and 'channel', if necessary.

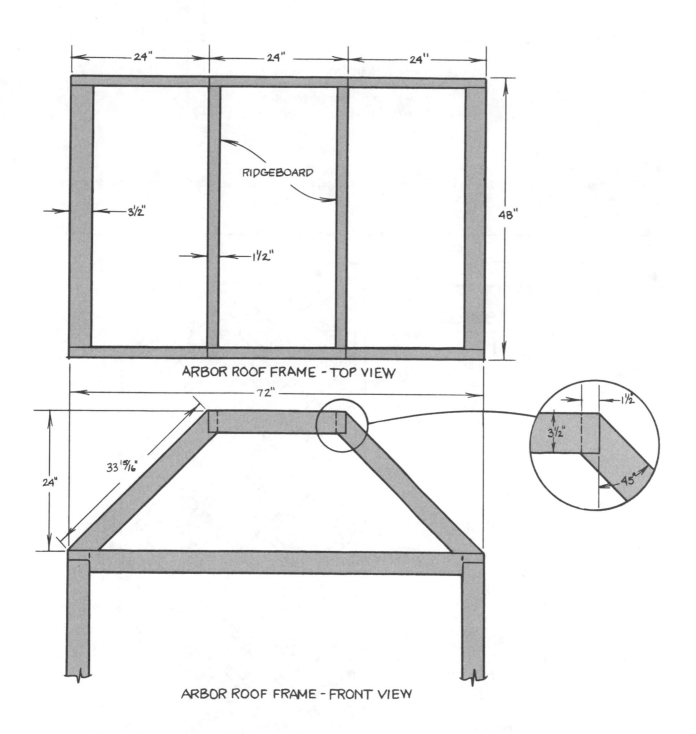

ARBOR ROOF FRAME - TOP VIEW

ARBOR ROOF FRAME - FRONT VIEW

7 **Frame the lattice.**

Cut two sheets of lattice, 39½″ x 76½″. Cut the frame parts from 'channel' stock, as shown in the *Side Frame Layout* drawings. Miter the ends of the channel at 45°. Assemble the frame so that the lattice rests in the groove in the channel, as shown in the *Section A* drawing. Nail the frame together with two 6d finishing nails at each corner. Drive the nails in at right angles to one another, to strengthen the miter joint. (See Figure 3.)

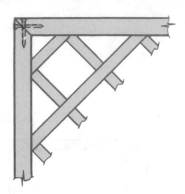

Figure 3. Reinforce the lattice frame miter joints with nails. Drive the nails at right angles to each other.

TIP Before driving nails into the corner joints, drill pilot holes. This will prevent the wood from splitting.

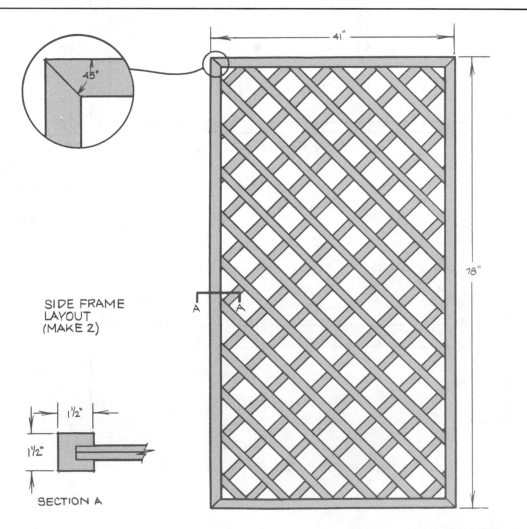

SIDE FRAME
LAYOUT
(MAKE 2)

SECTION A

Put the lattice frames in place in the side walls, as shown in the *Finished Arbor, Side View* drawing. The frames should be centered on the posts and butted up against the top plates. When they are properly positioned, nail them in place with 6d finishing nails. Put the nails in the channel groove, in between the lattice. (See Figure 4.) Pound them in as far as you can with a hammer, then drive them home with a nail set.

8 **Attach the lattice to the posts.**

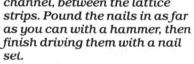

Figure 4. *To attach the lattice frames to the posts, drive finishing nails in the grooves of the channel, between the lattice strips. Pound the nails in as far as you can with a hammer, then finish driving them with a nail set.*

Installing the Roof

Cut a sufficient number of lath strips to cover the roof. If you can't buy these ready-cut from your local lumberyard, you can rip ¼" thick, 1½" strips from 2 x 4 stock. Space the strips 1" apart, and nail them to the roof frame with 4d nails.

9 **Attach lath to the top of the roof frame.**

TIP To help you space the lath quickly and evenly, rip a board 1" wide and use this as a spacing gauge.

10 Attach lath to the sides of the roof frame.

Also, attach lath to the front and back side of the roof frame. As shown in the *Finished Arbor, Front View* drawing, this lath fans out from the center of the front and back top plates to make a design. It takes some extra time to lay out and cut the lath strips to make this design, but the end result is worth the bother. Work out from the center, angling each strip 10° more to the left or right than the proceeding strip. After you've made the rays, attach a 'sun' on top of the rays. This sun is actually just half of an oval, cut from wood, as shown in the *Sun Detail* drawing.

Option: If this sun-with-rays pattern doesn't suit you, you can also install the lath vertically on the front and back of the roof frame. Space the lath just as you did on top of the roof.

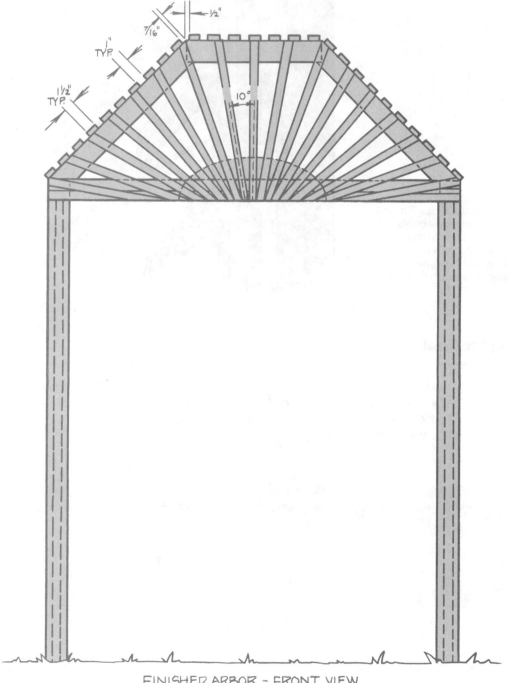

FINISHED ARBOR - FRONT VIEW

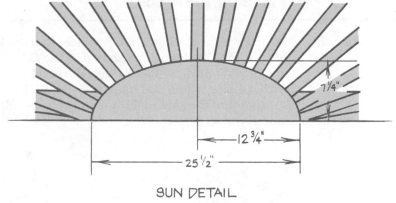

SUN DETAIL

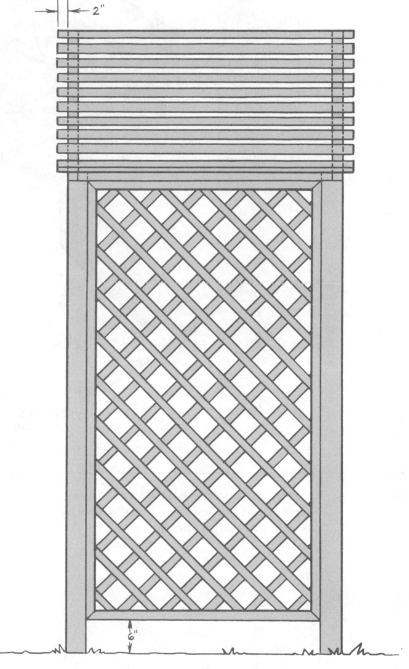

FINISHED ARBOR - SIDE VIEW

Finishing the Arbor

11 Install seat in the arbor, if you wish.

No arbor is complete without a place to sit. Make two seats and four braces from 2 x 4 stock, as shown in the *Optional Seat, Bottom View* and *Side View* drawings. Attach the seats and braces to the posts with lag screws and 16d nails. (See Figure 5.) The seats should line the inside of the side walls, but you can also place them on the outside, if you so desire.

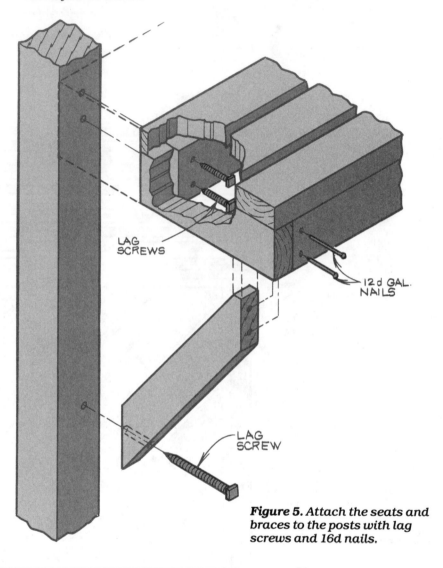

Figure 5. *Attach the seats and braces to the posts with lag screws and 16d nails.*

12 Paint or stain the arbor.

Paint or stain the exposed wood surfaces of the arbor, if this suits you. With all the cracks and crevices in this structure, you'll save time by using a paint sprayer.

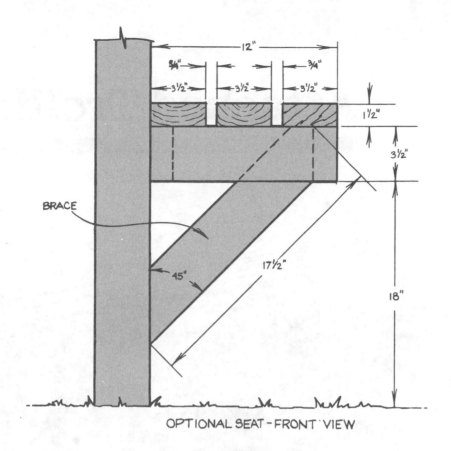

BRACE

12"

3/4"   3/4"

3½"   3½"   3½"

1½"

3½"

45°   17½"

18"

OPTIONAL SEAT-FRONT VIEW

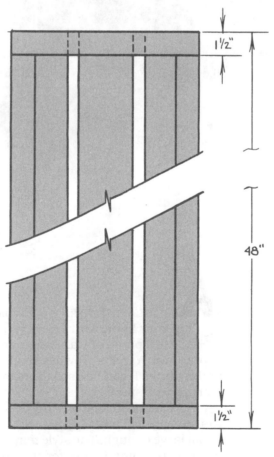

1½"

48"

1½"

OPTIONAL SEAT- BOTTOM VIEW

Attached Trellis

An attached trellis is a simple, economical way to spruce up your home with greenery and flowers. But attaching a trellis to a porch or wall does more than enhance your home's appearance. A trellis provides shade in the summer and a wind break in the winter. It can be used to enclose an open porch and create a more private space.

There are many, many designs for trellises, and the design you see here will probably not work for you. It may be too long, too tall, or not large enough. The style may not work well with the architecture of your home. No matter, it's a project that can be easily adapted to your needs. Change the dimensions and the style to suit yourself, and use our instructions as a guide.

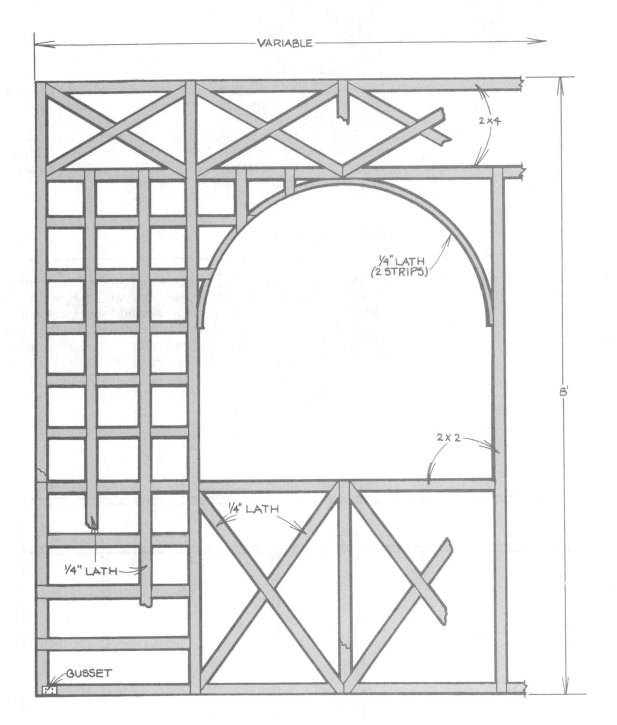

VARIABLE

2×4

¼" LATH
(2 STRIPS)

8'

2×2

¼" LATH

¼" LATH

GUSSET

FRONT ELEVATION

Materials

Because your trellis will be used to grow plants, your choice of wood is critical. Some lumber is treated with chemicals that are poisonous to plants. For example, lumber that has been treated with black creosote will kill any foliage it comes in contact with.

Look for Southern yellow pine, Western hemlock, Douglas fir, and Ponderosa pine lumber that is kiln-dried and pressure-treated with chromated copper arsenate (CCA) for this project. You'll be able to spot this wood in the lumberyard right off because CCA turns the wood a dull green. CCA is relatively harmless to plants. Redwood and cedar are also good choices. The tendrils of ivy and many other clinging vines penetrate the wood and cause it to rot, but redwood and cedar have natural oils which make them rot-resistant.

Use 2 x 2's to frame the trellis; you can rip these from 2 x 4 stock. Cover the frame with ¼" thick, 1½" wide lath. If pressure-treated or rot-resistant lath isn't available at your lumberyard, this can also be ripped from 2 x 4's. If you wish, you can also use 'diamond weave' lattice to cover the trellis. This is available at most lumberyards and home centers in 2' x 8' and 4' x 8' sheets.

In addition to this lumber, you'll also need some galvanized nails to assemble the trellis. If you're building a large trellis, buy some hanger bolts to attach it to your house. If you're building a small trellis, you can hang it on L-hooks.

Before You Begin

1 Adjust the design to fit your home.

As we mentioned at the beginning of this chapter, the trellis you see in the working drawings will probably not work for your home without modification. Depending on the size and style you want, you may design something that looks markedly different. But the building techniques will remain essentially the same. Our trellis includes all the major variations of a trellis—side panels, top and bottom panels, archways, openings, horizontal, vertical, and diagonal members. Study what we've done here, then borrow the features you need for your own trellis.

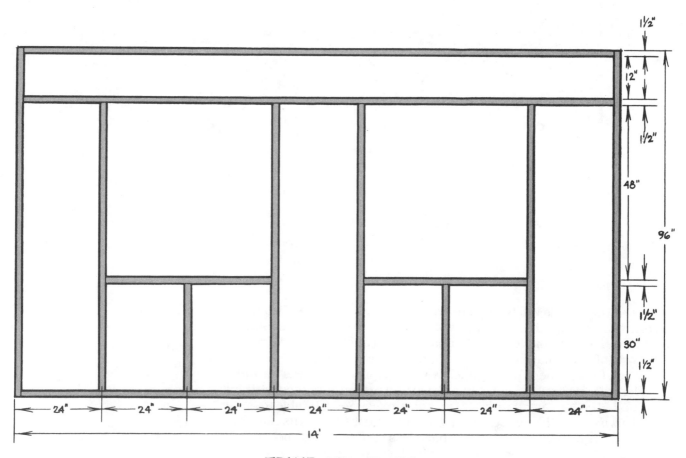

FRAME-FRONT VIEW

Building the Trellis

Cut the 2 x 2's you need to make the trellis frame. Nail them together with 12d nails, as shown in the *Frame, Front View* drawing. 12d nails won't split the wood, but neither will they keep the frame joints from pulling apart after a few seasons in the weather. To strengthen the joints, cut metal gussets from galvanized sheet metal. Nail these gussets over each joint with roofing nails. (See Figure 1.) These gussets will barely show when the trellis is completed.

2 Lay out and build the trellis frame.

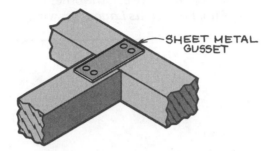

Figure 1. Reinforce the joints between the frame members with sheet metal gussets and roofing nails.

TIP If your trellis is very large, like the one shown here, nail gussets on *both* sides of the frame.

To make an arch, you must bend some wood. For a large arch (3' or more in diameter), you can use ¼" thick lath to make the arch. Lath, made from softwoods, is usually limber enough to be bent without steaming or other special processing. Bend two pieces, face to face (as shown in Figure 2 and the *Arch Layout* drawing), to make a single arch, ½" thick. Glue the two pieces together with waterproof resorcinol glue, and clamp the arch in the frame while the glue dries. After the glue has set up, the arch will retain its shape with no need of clamping. Remove the arch from the frame and chamfer the ends, as shown in Figure 2. Then put it back in place and attach the arch to the frame with 4d nails.

3 Make the arches (if your design includes arches).

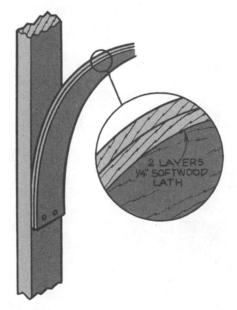

Figure 2. To make an arch, bend two long pieces of lath to the proper radius and glue them together. When the glue sets up, attach the arch to the frame with 4d nails.

4 **Attach the horizontal and diagonal lath.**

Attach the horizontal and diagonal lath to the frame with 2d nails. Miter the lath as shown in the *Horizontal and Diagonal Lath Layout* drawing, so that all the ends butt together. When attaching lath to the arches, drill pilot hole for the nails. This will keep the wood from splitting.

5 **Attach the vertical lath.**

Attach the vertical lath to the trellis assembly, atop the horizontal and diagonal lath. The vertical lath should cover all or most of the butt joints between the other pieces of lath, as shown in the *Finished Trellis Layout* drawing.

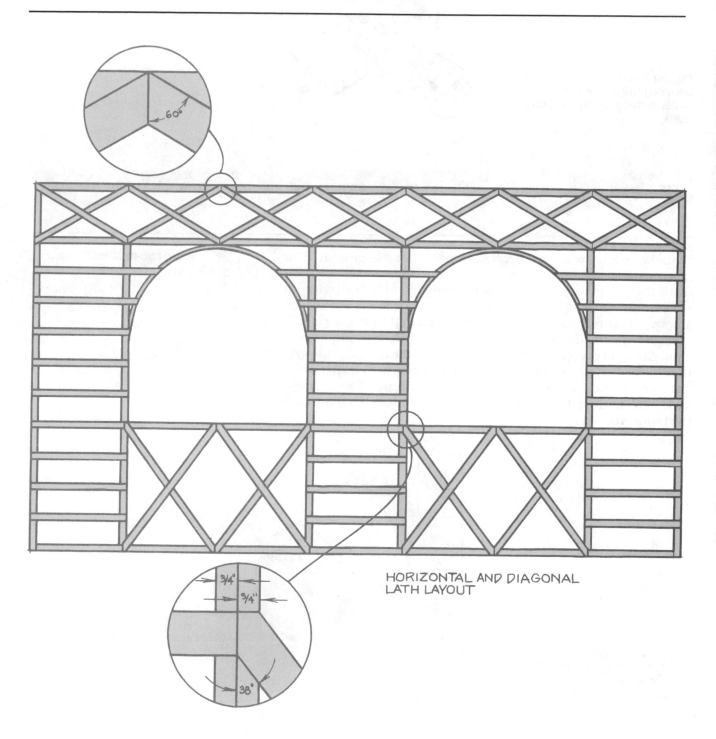

HORIZONTAL AND DIAGONAL LATH LAYOUT

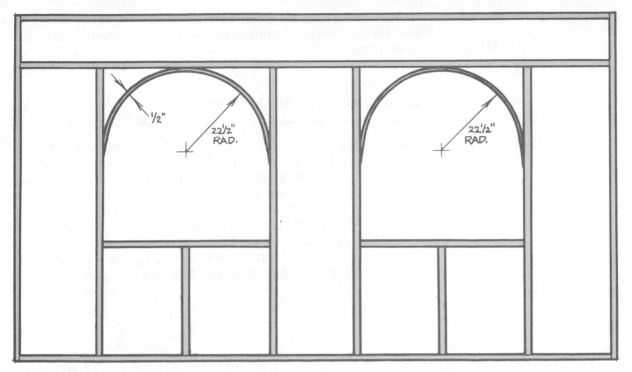

ARCH LAYOUT

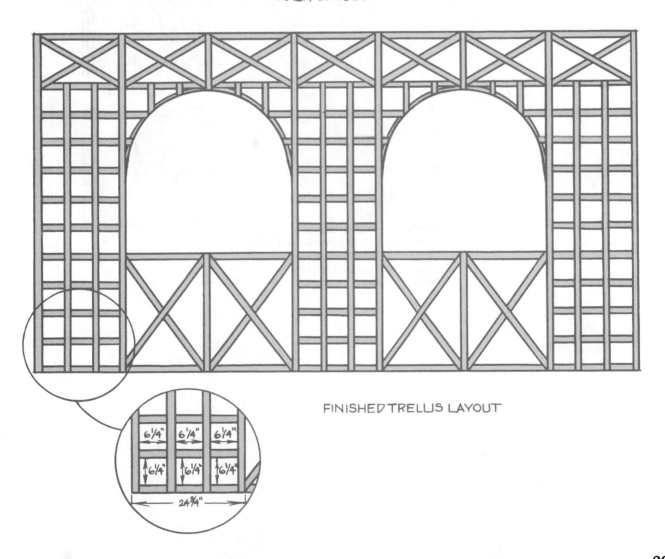

FINISHED TRELLIS LAYOUT

6 **Paint or stain the trellis.**

If you're going to paint or stain this trellis, do it *before* you hang it on your home. Use a paint sprayer to get into all the cracks and crevices in this project.

Attaching the Trellis to Your Home

7 **Install the hangers on your house.**

The most important thing to remember about hanging a trellis on your home is that you want to be able to take it down fairly easy. There will be times when you'll need to remove it to paint the house or repair the wall. So don't nail the trellis in place permanently. Instead, hang it on hardware that will hold it securely in place, but release when you need to take it down. If you're hanging a small trellis, use the same system many folks use for hanging shutters—screw L-hooks into the wall. Then drill corresponding holes in the *horizontal* members of the trellis frame. (See Figure 3.)

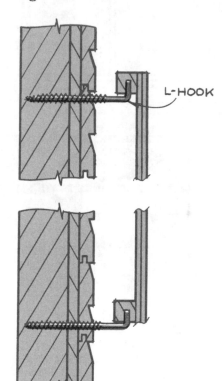

Figure 3. If you're hanging a small trellis, use L-hooks to keep it in place. Drill corresponding holes in the horizontal trellis frame members.

If your trellis is very large (larger than 32 square feet), use hanger bolts. These bolts come in larger sizes than L-hooks, and will support more weight. Screw the hanger bolts into the house and drill corresponding holes in the *vertical* frame members. (See Figure 4.)

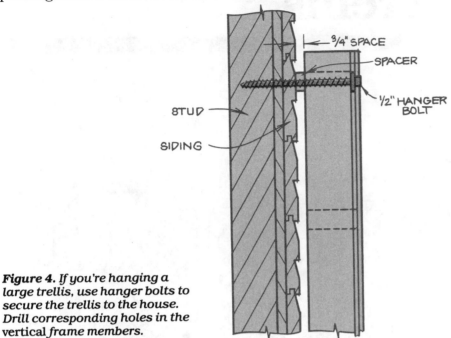

3/4" SPACE

SPACER

1/2" HANGER BOLT

STUD

SIDING

Figure 4. If you're hanging a large trellis, use hanger bolts to secure the trellis to the house. Drill corresponding holes in the vertical frame members.

8 **Lift the trellis into place.**

Lift the trellis into place and rest it on the hangers. If the trellis is very large, you may need one or more helpers to put it up. As you raise the trellis, be careful that you don't twist the frame. If you're using hanger bolts to attach the trellis to the house, secure it with washers and nuts.

Stand-Alone Trellis

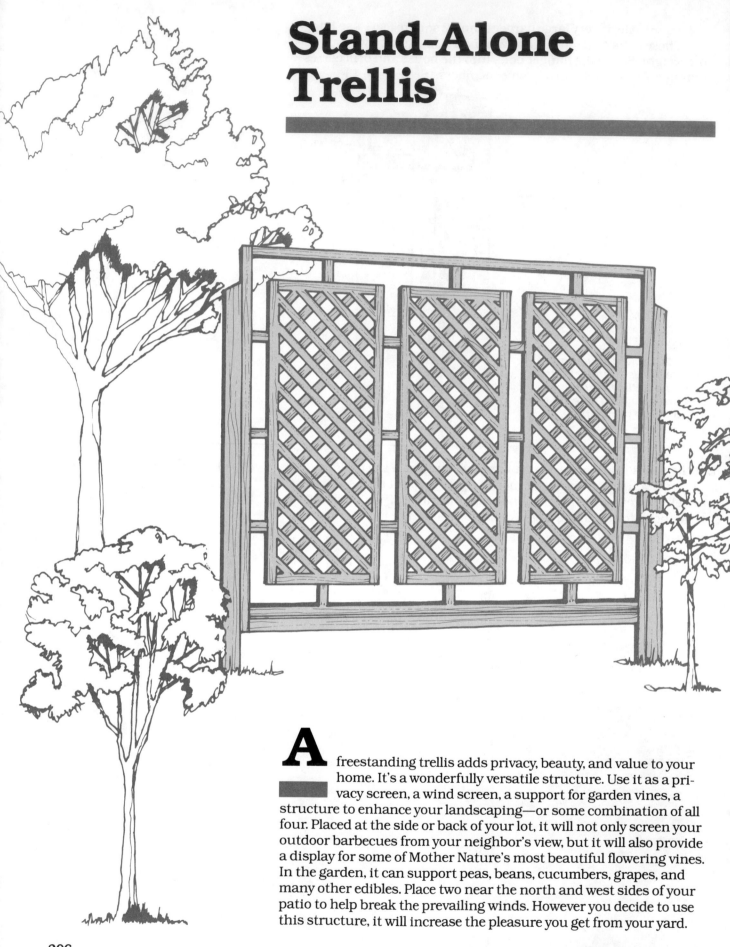

A freestanding trellis adds privacy, beauty, and value to your home. It's a wonderfully versatile structure. Use it as a privacy screen, a wind screen, a support for garden vines, a structure to enhance your landscaping—or some combination of all four. Placed at the side or back of your lot, it will not only screen your outdoor barbecues from your neighbor's view, but it will also provide a display for some of Mother Nature's most beautiful flowering vines. In the garden, it can support peas, beans, cucumbers, grapes, and many other edibles. Place two near the north and west sides of your patio to help break the prevailing winds. However you decide to use this structure, it will increase the pleasure you get from your yard.

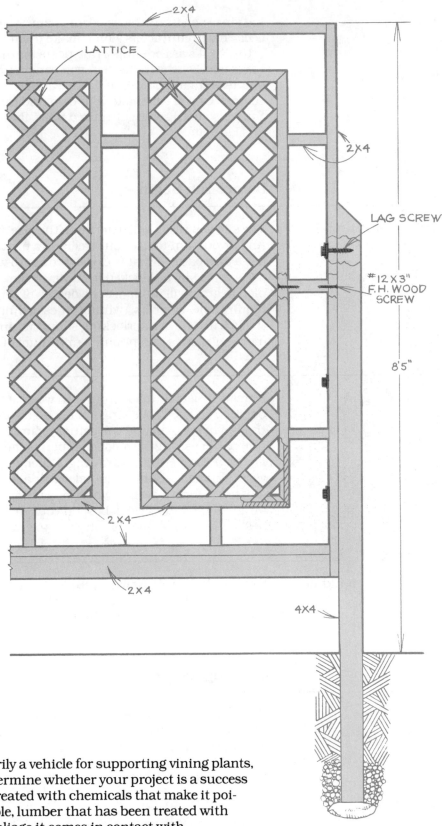

2X4

LATTICE

2X4

LAG SCREW

#12 X 3"
F.H. WOOD
SCREW

8'5"

2X4

2X4

4X4

FRONT ELEVATION

Materials

Because the trellis is primarily a vehicle for supporting vining plants, your choice of wood will determine whether your project is a success or failure. Some lumber is treated with chemicals that make it poisonous to plants. For example, lumber that has been treated with black creosote will kill any foliage it comes in contact with.

Look for Southern yellow pine, Western hemlock, Douglas fir, and Ponderosa pine lumber that is kiln-dried and pressure treated with chromated copper arsenate (CCA) for this project. You'll be able to spot this wood in the lumberyard right off because CCA turns the wood a dull green. CCA is relatively harmless to plants. Redwood and cedar are also good choices. The tendrils of ivy and many other cling-

STAND-ALONE TRELLIS

ing vines penetrate the wood and cause it to rot, but redwood and cedar have natural oils which make them rot-resistant.

Use 4 x 4's as the supporting posts for the trellis, an 2 x 4's for the frame. For the trellis grid, you can purchase ready-made 2' x 8' sheets of "diamond weave" lattice at most lumberyards and home centers. Or you can make your own grid from ¼" x 1½" lath strips. You'll also need some galvanized nails, flathead wood screws, staples, and lag bolts.

Before You Begin

1 Determine the size and location of your trellis.

The site you select will determine the size and design of your trellis. And if you want to use your trellis as a privacy screen, you'll want to use a grid pattern that is closely woven so when the green leaves fall away you'll still be blocked from view. Likewise, the type of vines you plan to grow will influence the design of your trellis. Heavy vines require a heavy trellis, twining vines or vines with tendrils need a trellis with lots of interlocking supports, and fragile vines such as morning glories ask for only a minimum of support from a trellis.

2 Determine the grid pattern.

A trellis is usually made from thin wood strips, with open spaces between strips. Those strips make up the grid, or the part of the trellis where the plants will actually climb. There are many different grid patterns, but the most common is the diamond, as shown in the *Different Grid Patterns* drawing. Other grid patterns include simple vertical, diagonal, or horizontal slats nailed in a frame, or criss-crossed to form a 'wagon wheel' or a checkerboard pattern. Determine which grid will look best on your trellis—or which will work best for the type of plants you want to grow.

3 Check your local building codes.

You should also check with your local building inspector to make sure your trellis complies with any building code restrictions for your area. You may also need to obtain a building permit before you begin.

Building the Grid

4 Rip the lath strips.

If you haven't purchased ready-made lattice or lath, rip the ¼" thick lath strips you'll need from 2 x 4 stock. Try to select 2 x 4's with as few knots as possible—knots will weaken the lath strips. If all the 2 x 4 stock at the lumberyard is pretty knotty, you may have to pay a premium for 'clear' 2 x 4's.

Cut the lath a little oversize and arrange it in a grid pattern. Tack the grid together by stapling at every intersection with ⅝″ galvanized staples. Use an anvil or a thick sheet of iron to 'peen' the tips of the staples—after you've stapled the entire grid together, go back and put the anvil under each joint. Give the staples one or two solid blows with a hammer. The points of the staple will bend over and the staples will be locked in the wood. If you're making a grid in which the strips don't overlap (such as the vertical, diagonal, or horizontal patterns shown in the *Different Grid Patterns* drawing), simply cut the strips and lay them out.

5 Cut the strips and assemble the grid.

TIP To help hold the strips in place while you assemble the grid, temporarily tack the bottom layer to a sheet of plywood. Arrange the top layer, staple both layers together, and remove the completed grid from the plywood. If you're making a grid that doesn't overlap, tack the strips to the plywood and leave them there to mark them for cutting.

Carefully measure the size of the grid you need and mark the strips for cutting. With a circular saw or a saber saw, carefully cut the grid. Hold the grid down firmly on a cutting surface to keep the lifting action of the saw from pulling apart the stapled joints. If you're making a grid that doesn't overlap, you can save time by cutting the strips on a bandsaw.

6 Trim the grid to size.

TIP Use a chalk 'snap line' to quickly mark the grid strips for cutting.

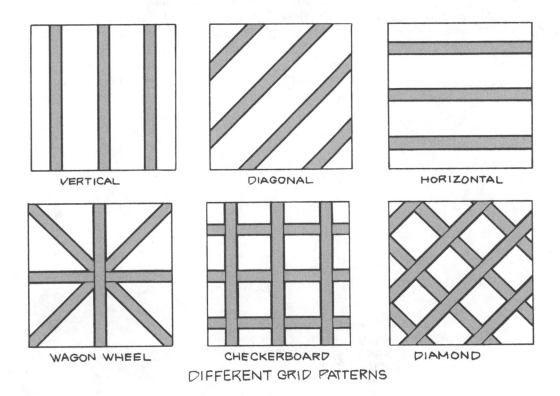

VERTICAL   DIAGONAL   HORIZONTAL

WAGON WHEEL   CHECKERBOARD   DIAMOND

DIFFERENT GRID PATTERNS

7 Build the grid (inside) frames.

Cut 2 x 4 stock to length, mitering the corners as shown in the *Grid Frame Layout* drawing. With a dado blade or a router, cut a ½" wide x ¾" deep groove down the center of the inside face of these frame members, as shown in the *Section A* drawing. (Make the groove ¼" wide if the strips on your grid don't overlap.) With a helper, assemble the grid frames, slipping the grids into the grooves. Hold each corner together with 2-3 #12 x 3" flathead wood screws. To make the corners as strong as possible, drive these screws at right angles to each other, as shown in Figure 1.

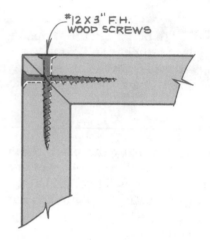

Figure 1. Join the corners of the grid frames with wood screws. Drive the screws at right angles to each other to strengthen the miter joint.

Optional Method: Build the grid frame up in layers, with butt-joint corners. Cut stock 1½" wide by 1½" thick. (Cut stock 1½" wide by 1⅝" thick if the strips of your grid don't overlap.) Lay out the 'bottom' frame, screwing the corners together. To this frame, tack two layers of ¼" thick by ¾" wide lath all around the perimeter, flush to the outside edges. (Use just one layer if the strips of your grid don't overlap.) Tack the grid in place, then lay out the 'top' frame and screw the corners together. Make sure that the top frame parts overlap the butt joints on the bottom frame—this will add strength to the overall assembly. (See Figure 2.) Attach the top and bottom frames together with 12d nails.

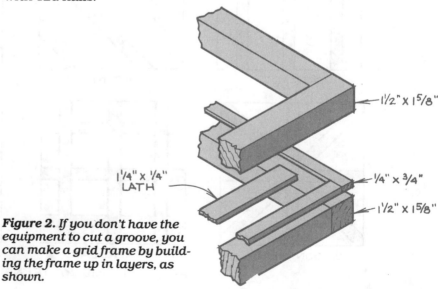

Figure 2. If you don't have the equipment to cut a groove, you can make a grid frame by building the frame up in layers, as shown.

Cut 2 x 4 stock to make the trellis frame, as shown in the *Trellis Frame Layout* drawing. Assemble this frame with 16d nails. The bottom of the frame gets *two* frame members, arranged in a 'T' shape, as shown in the *Section B* drawing. This helps to support the weight of the grids and grid frames.

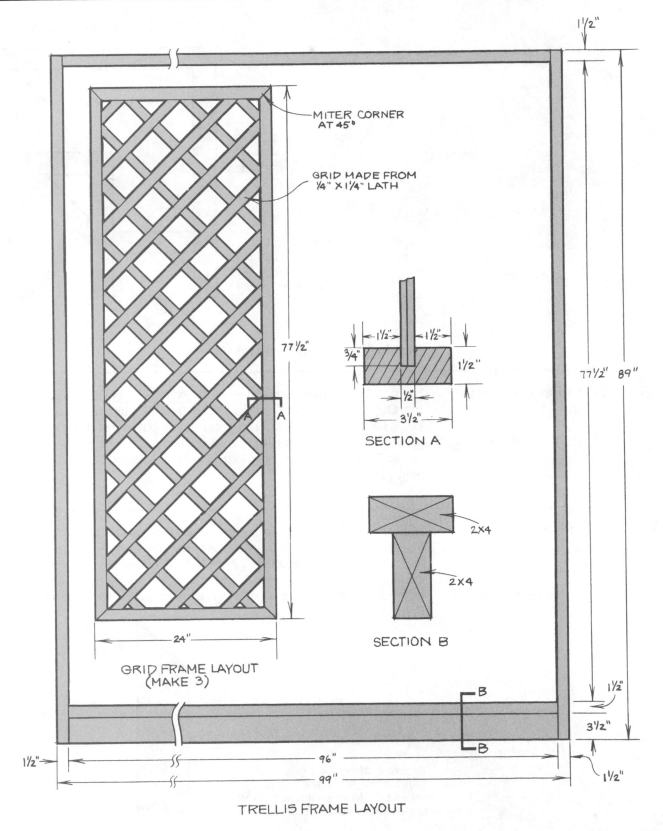

1½"

MITER CORNER
AT 45°

GRID MADE FROM
¼" X 1¼" LATH

77½"

A A

1½" 1½"

¾" 1½"

½"

3½"

SECTION A

2X4

2X4

SECTION B

77½" 89"

24"

GRID FRAME LAYOUT
(MAKE 3)

B

B

1½"

3½"

1½"

1½"

96"

99"

1½"

TRELLIS FRAME LAYOUT

9 Assemble the grid frames to the trellis frame.

Cut 6″ lengths of 2 x 4 from the scrap. Lay out the grid frames inside the trellis frame, using these short lengths as spacers, as shown in the *Finished Trellis* drawing. Assemble the trellis frame, grid frames, and spacers with #12 x 3″ wood screws, as shown in Figure 3. Each of the butt joints between the spacers and the frames should be reinforced with at least *two* screws—one at the front of the trellis, and one at the back.

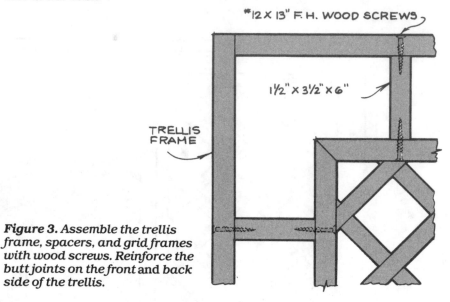

Figure 3. Assemble the trellis frame, spacers, and grid frames with wood screws. Reinforce the butt joints on the front and back side of the trellis.

Setting Trellis in Ground

10 Set the supporting posts in the ground.

Lay out the location of the posts as shown in the *Post Layout* drawing. Dig holes 24″-36″ deep, below the frost line in your area. Put a large, flat rock, about 8″ in diameter, at the bottom of these holes to provide a solid base for the posts. Set the posts in the holes, but do *not* fill in the holes just yet. You may need to adjust the location of the posts 1″-2″.

11 Cut the top of the posts at the same height.

Hold the posts upright with temporary braces and stakes. (Don't attach the braces to the *inside* face of the posts—they will interfere with the next step.) Measure 72″ to the top of one post. With a string and string level, find the top of the other post, using the mark as a reference. Cut the posts off with a handsaw, mitering the tops. (See Figure 4.)

Figure 4. Miter the tops of the supporting posts at 45°. The miter should slope down from the inside face to the outside face.

With a helper, place the trellis in between the posts. Hold it 12″ off the ground with blocks and scrap wood. Check to see if the trellis frame is level. If not, adjust one side or the other by adding or removing scrap. When the frame is level, move the support posts in so they butt up against the side frame members. Attach the trellis frame to the posts with ⅜″ x 3½″ lag bolts. Check that the trellis is straight up and down, then fill in the post holes. Tamp down the dirt and remove the temporary braces, stakes, blocks, and scrap.

12 **Attach the trellis to the supporting posts.**

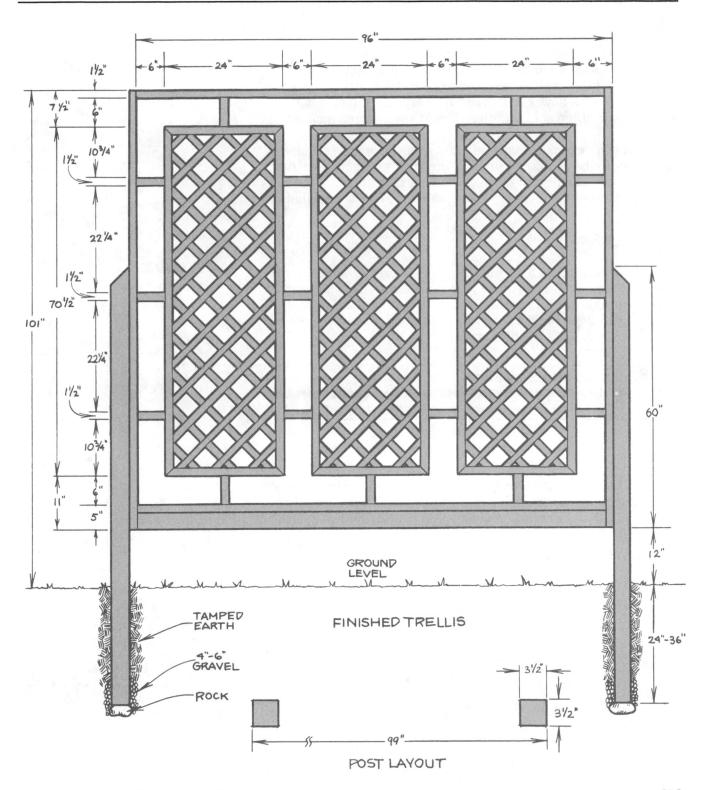

FINISHED TRELLIS

POST LAYOUT

Fences and Privacy Screens

A fence not only embellishes the beauty of your home, but serves you in many practical ways as well. A fence can define your property line, keep the dog out of your garden, provide a background for a flower bed. In the drab days of winter a fence provides visual interest to brighten those dismal, barren months of the year.

A fence can take many different forms. From the simple picket fence to the contoured privacy screen, this outdoor structure can be modified to suit your needs and your architectural preferences. All it takes is some basic woodworking skills and a little imagination.

Choosing a Design

Because the techniques for building a fence or privacy screen in the ground are similar, you can mix and match designs freely. Perhaps the most difficult part of building your fence will be deciding which of the many designs you'll want to use. The chart below may help you determine which fence is right for you.

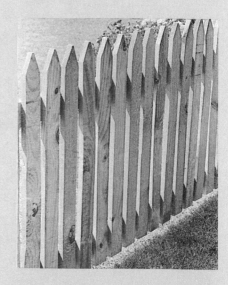

Uses of Fences and Screens

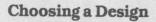

Design	Primary Uses	Skill Needed
Picket Fence	Establish property line, keep kids in, animals out	Simple to build, uses basic skills, requires little time
Solid Privacy Screen	Frame property, privacy, set off garden or lawn, keep pets in	Simple to build, uses basic skills, requires little time
Ventilated Privacy Screen	Accent property, allows air to flow across yard, but still provides privacy	Moderate skills, requires special techniques
Lattice Fence	Masks patio, lawn, or garden, provides support for climbing plants, allows for ventilation	Moderate skills, requires special techniques
Contoured Screen or Fence	Used to add visual interest to flat landscapes, or to follow the contours of rolling landscapes	Moderate skills, requires special measuring and carpentry skills

The styles listed here aren't the only types of fences you could build, of course. But they cover the most common varieties. Use the chart to help decide which type of fence comes closest to meeting your particular needs, then modify it as you need to. Use the procedures in the chapters in this section to help plan and build your fence project, no matter what the final design.

Common Fence Building Techniques

As mentioned earlier, many of the techniques for setting and building a fence are common from fence to fence. Laying out the posts, setting the posts, and cutting the posts to the proper height are precisely the same for every fence and privacy screen project in this section. Rather than repeat ourselves five times, we'll explain those common techniques just once, in this introductory chapter.

Before You Begin

1 **Check your local building codes.**

Before you dig the post holes, saw the lumber, or even settle on the design for your fence, you should check with your local building inspections office to see if there are any ordinances or regulations concerning the height, location, or materials of fences. You may also need to obtain a building permit.

2 **Be certain of your property boundaries.**

If you will be framing your yard with your fence, it's wise to have your property line checked by a surveyor. If you build your fence so that it encroaches on your neighbor's side of the line, it becomes his property.

Plotting Your Fence or Screen

3 **Gather the proper tools to survey the fence line.**

Unless your lot is as smooth as a baseball diamond, you'll have to pay special attention to laying out your fence. The first step is to locate the exact course your fence or screen will run and mark the line with stakes and string. To do this you'll need the following tools:
- 50-foot or 100-foot steel tape measure
- Ball of mason's twine or any nonstretchable, nylon string
- Stakes
- Large (8d-16d) box nails
- Some scraps of paper
- Piece of colored chalk
- Plumb line

Plotting a straight fence. Mark end or corner post locations with stakes, driven solidly into the ground. Run mason's twine or string between stakes, draw the string tight and tie it firmly to the stakes. (See Figure 1.) If bushes or other low obstructions are in the way, use tall stakes to clear them. Locate the remaining posts by measuring along twine and marking post centers on twine with chalk. Most fence posts are set 4'-6' apart. Transfer post marks from twine to ground by sticking nails through scraps of paper, then driving them in the ground directly below the string marks.

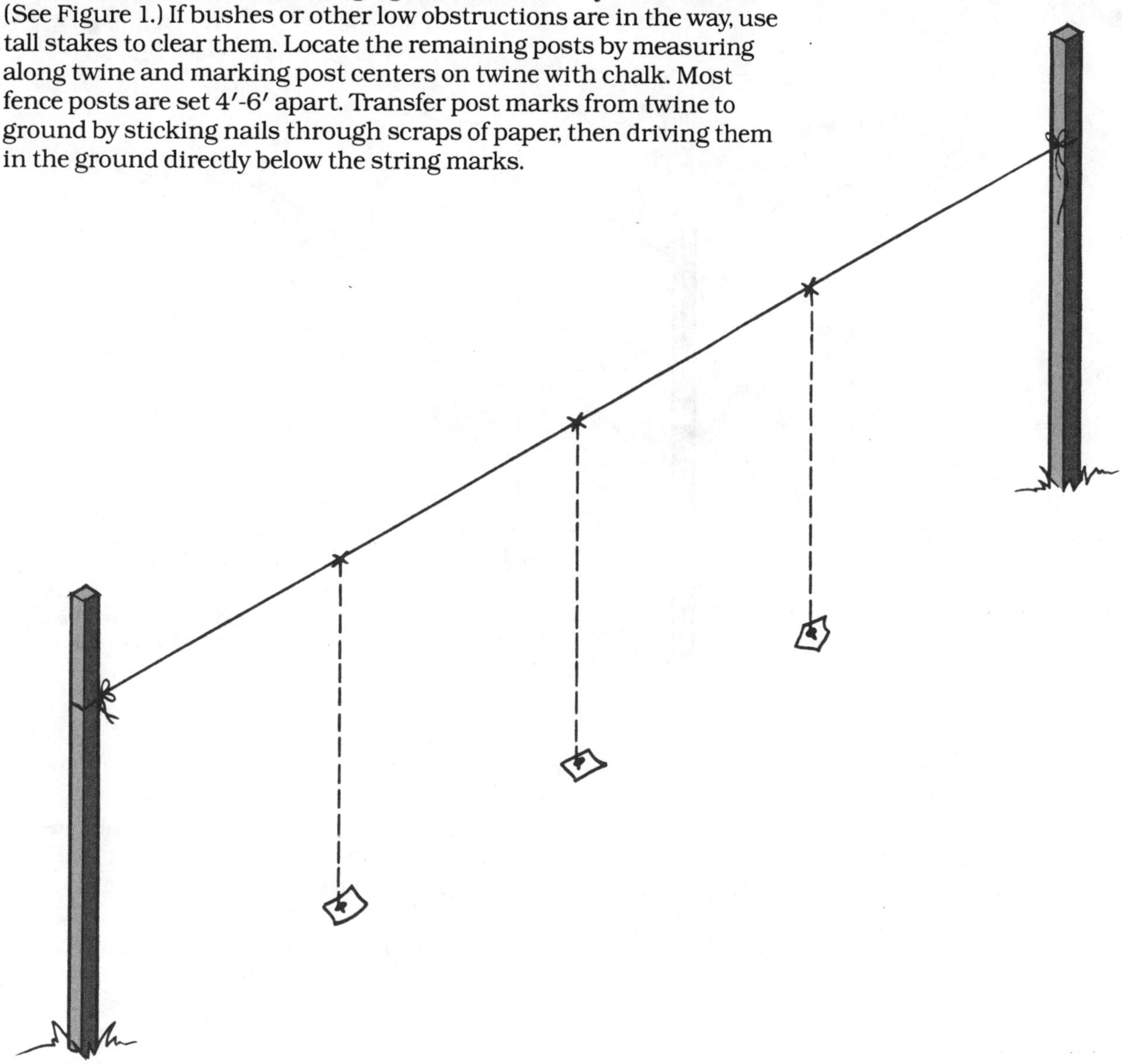

4 **Lay out the fence line.**

Figure 1. *To lay out the fence posts, stretch a string between two stakes. Measure along this string to locate each post, and mark the position of the post in the ground with a nail and a scrap of paper.*

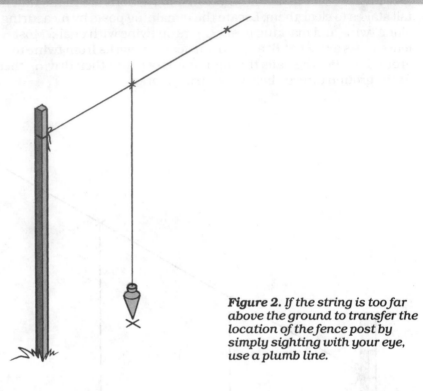

Figure 2. *If the string is too far above the ground to transfer the location of the fence post by simply sighting with your eye, use a plumb line.*

Plotting a right angle. If your fence layout calls for perfectly square corners, first set the stakes as accurately as you can and tie string between them. Measure out exactly 3′ from the corner stake along one string, and exactly 4′ along the other. Mark the strings with chalk, then measure the diagonal between the chalk lines. (See Figure 3.) The diagonal should be exactly 5′. (Remember the Pythagorean theorem from high school?) If the measurement is *less* than 5′, move the stakes so the corner angle is *larger*. If the measurement is *more* than 5′, move the stakes so that the corner angle is *smaller*.

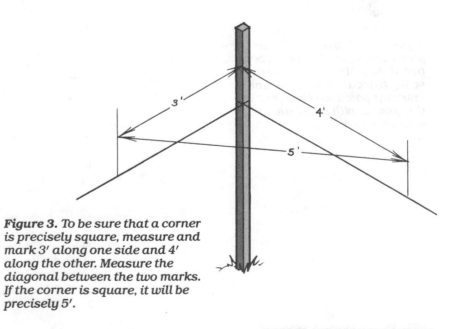

Figure 3. *To be sure that a corner is precisely square, measure and mark 3′ along one side and 4′ along the other. Measure the diagonal between the two marks. If the corner is square, it will be precisely 5′.*

Plotting a curve. For this, you'll need a few extra materials—a stake or pipe about 3′ long for a pivot point, a length of rope, and a straight, pointed stick. Drive stakes where the straight sections of fence will end and you want the curve to begin. (Call these 'end' stakes.) Stretch twine between these stakes, and measure the half-way point along the line. Using the technique for plotting a right angle, run a second string perpendicular to the first, back into the yard from the halfway point. Drive the pivot pipe at a chosen point along the second line. (The farther away from the second line that you drive the pivot, the shallower the arc will be—so keep checking until you have the desired arc.) Measure a cord to reach from the pivot to one end stake. Attach one end of the cord to the pivot and the other to the pointed stick. Then scribe an arc in the ground from one end post to the other, keeping the line taut. (See Figure 4.) Locate the intermediate posts in the arc by bending a flexible steel tape measure around the curve and marking the post locations.

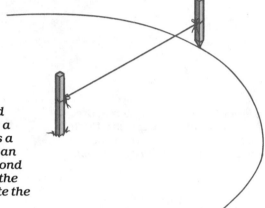

Figure 4. *To lay out a curved fence, use a stake or pipe as a pivot and a length of rope as a compass. Carefully scratch an arc in the ground with a second stake. Then measure along the arc with a steel tape to locate the fence posts.*

Setting the Posts

Be sure that the lumber for the posts has been chemically treated to prevent rotting. You can also paint or dip the posts in wood preservative for protection against insects and decay. If you dip your own, let the posts soak for at least 24 hours. If you want to grow plants near the fences, don't use creosote, pentachloraphenol, or other chemicals that may retard plant life. Your safest choice is chromated copper arsenate (CCA).

5 Select the proper lumber for the fence posts.

6 **Set the posts in the ground.**

Dig holes where you've marked the locations of the posts. Use a post-hole digger to make holes, trying not to disturb the surrounding earth. At least one third of the post should be buried in the ground (one half, if the posts are shorter than 48″). Dig the holes slightly deeper than you think you need to allow for the rock and gravel base. Place a rock in the bottom of the hole to keep the post from settling. (The rock should be twice the diameter of the post.) Put two shovelfuls of gravel in the bottom of the hole for additional drainage, then set the posts perfectly vertical with a level. Temporarily brace the posts to hold them upright, again checking to be sure they are straight and level. (See Figure 5.) Fill the hole with dirt and tamp it down firmly. (See Figure 6.)

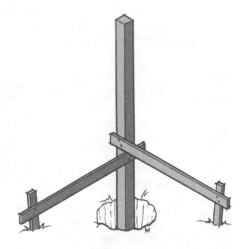

Figure 5. Temporarily brace the posts to hold them upright. Each post needs at least two braces, set at right angles to each other.

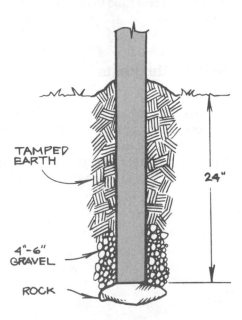

Figure 6. Place the ends of the posts on rocks to keep them from settling. Throw in some gravel to provide drainage, then fill the holes with dirt and tamp down.

Option: In loose soil, it is best to mix concrete and pour it into the hole around the post instead of using dirt. Pour the concrete directly over the gravel base. (*Never* allow the bottom of the post to be set in concrete. This will hasten the decay of the wood.) With a trowel, slope the concrete away from the post to help drain the water away from the wood. (See Figure 7.)

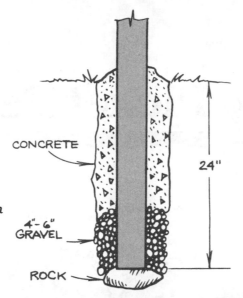

Figure 7. *If you set the posts in concrete instead of earth, slope the top surface of the concrete away from the posts to help drain the water away from the wood. You should still use rocks and gravel for a base. Never set the end of a wood post in concrete; this will cause the wood to rot faster.*

CONCRETE

24"

4"- 6" GRAVEL

ROCK

TIP If you're working with a helper, you can eliminate the hassle of bracing the post upright while you throw in the dirt. Just have your helper serve as a human brace. However, if you're setting the posts in concrete, you'll need to put up the bracework. Even the best of helpers won't stand still for 24 hours while the concrete sets.

The proper height for the posts will vary not only with the design of the fence but whether or not you wish to follow the contour of the land. In most cases, you want to follow the contour. Simply measure up from the ground for each post and cut them all off at the same distance above ground level. On those rare occasions when you want the top of the fence to be perfectly level or follow some other line that doesn't match the contour of the land, stretch a string along the post and use this string to find the tops. (See Figure 8.)

7 Cut the posts to the proper height.

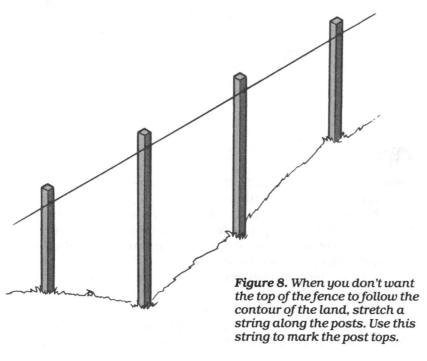

Figure 8. *When you don't want the top of the fence to follow the contour of the land, stretch a string along the posts. Use this string to mark the post tops.*

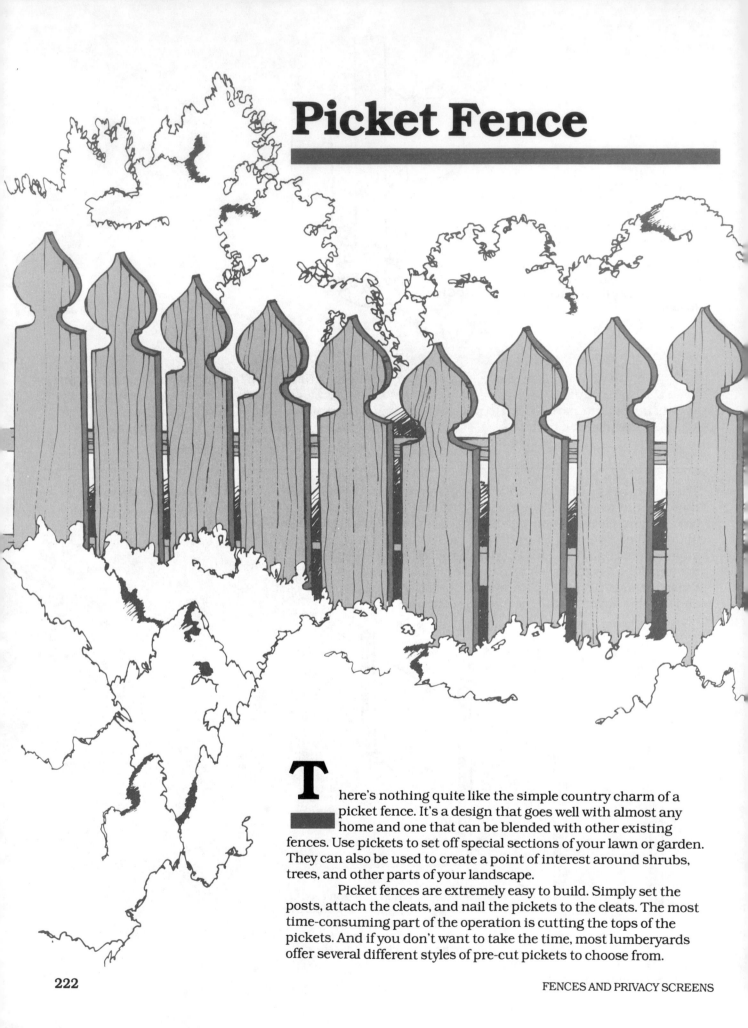

Picket Fence

There's nothing quite like the simple country charm of a picket fence. It's a design that goes well with almost any home and one that can be blended with other existing fences. Use pickets to set off special sections of your lawn or garden. They can also be used to create a point of interest around shrubs, trees, and other parts of your landscape.

Picket fences are extremely easy to build. Simply set the posts, attach the cleats, and nail the pickets to the cleats. The most time-consuming part of the operation is cutting the tops of the pickets. And if you don't want to take the time, most lumberyards offer several different styles of pre-cut pickets to choose from.

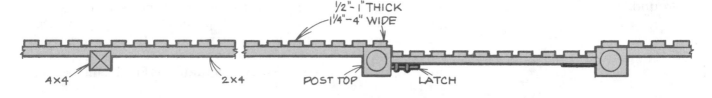

TOP ELEVATION

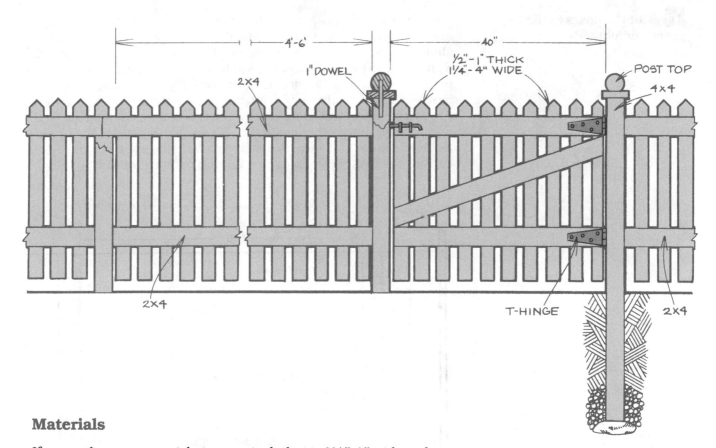

BACK ELEVATION

Materials

If you make your own pickets, use stock that is 1¼"-4" wide and ½"-1" thick. The length of the picket will depend on the height of the fence. As shown in the working drawings, our picket fence is just under 36" high. To make the fence as shown, you need pickets 2"-3" wide and 33½" long. Make the top and bottom rails from 2 x 4's to give the fence the strength it needs to withstand weather, wind, and children. The intermediate, corner, and gate posts should be made of 4 x 4's. All the wooden parts of the fence should be made from redwood, cedar, or pressure-treated lumber, so that your fence will withstand moisture and insect damage. In addition to the wood parts, you'll also need galvanized nails, 'T' hinges, and gate latches.

1 **Lay out the posts.**

Lay out the posts as described in the introductory chapter of the *Fences and Privacy Screens* section. All the posts should be 4'-6' apart, except for the gate posts. These should be precisely 40" apart.

2 **Set the posts in the ground.**

Set the posts as described in the introductory chapter. Dig the post holes deep enough so that at least ⅓ of the posts are below the ground. Set the corner posts and gate posts 6"-12" deeper than the intermediate posts, to provide additional support. Place a large rock in each hole, rest the base of the posts on the rock, brace the post upright, then throw in some gravel for drainage. Finally, fill the hole with dirt and tamp it down firmly.

3 **Cut the posts to the proper height.**

Before you cut the posts off to the desired height, decide whether or not you want the post tops to show. If you don't want the post tops to show, cut the posts off approximately 32" above the ground. If you do want the tops to show, and you plan to add decorative post tops, cut the posts off at 36". Also decide if you want the tops of the pickets to be level with each other, or follow the contour of the land. If you want them to be even, mark one post as a reference, then use a string to find the tops of the other posts. (See Figure 1.) If you want the pickets to follow the contour, measure each post individually. Once you've marked the tops of the posts, cut them off with a handsaw.

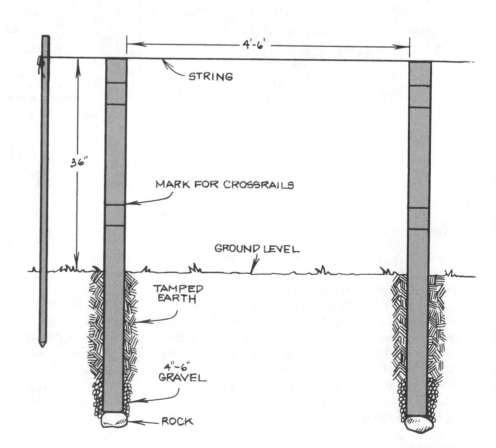

Figure 1. If you want the pickets to be even, use a string to find and mark the post tops. Cut the posts to the proper height, then mark each post where you will attach the rails.

This sort of fence has a 'show' side and a 'neighbor' side. On the show side, all you see is pickets. The neighbors get to see the rails and the other parts of the fence frame. Decide which side of the fence is which before you proceed; the rails must be attached to the show side. When you have decided on the show side, measure down from the tops of the posts and mark where you will nail the rails. The measurements are shown in the *Post Layout* drawings. With a handsaw or circular saw, cut kerfs 1″ deep along the marks. Then chisel out the waste between the kerfs to make a notch or 'dado' for the rails. (See Figure 2.)

4 Notch the posts for the rails.

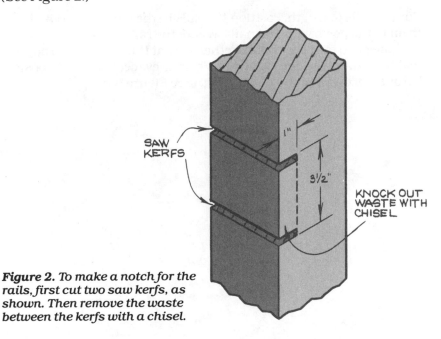

Figure 2. *To make a notch for the rails, first cut two saw kerfs, as shown. Then remove the waste between the kerfs with a chisel.*

Use a wide chisel and have a helper brace the posts while you pound on the chisel with a mallet. The corner posts must be notched on two adjacent sides, as shown in Figure 3.

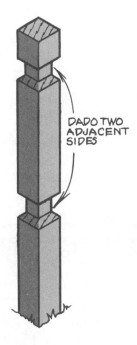

Figure 3. *The corner posts must be notched on two adjacent faces.*

Note: If you don't want the tops of the posts to show, the top notch should start at the very top of the post, as shown in the *Alternate Post Layout* drawing.

Option: Some people may find it easier to notch the posts *before* they set them in the ground. However, if you notch the posts first, you must dig the post holes to precisely the right depth. This can take a great deal of time: Constantly check the depth you dig. If you dig too deep, partially fill the hole with dirt and tamp down.

---

5 **Attach the rails to the posts.**

Cut the rails to length. Position the rails in the notches and nail them to the posts with 16d nails. Where the rails meet on an intermediate post, butt the ends together so that they meet near the middle of the post. (See Figure 4.) Where they meet on corner posts, lap one rail over the end of the other. (See Figure 5.)

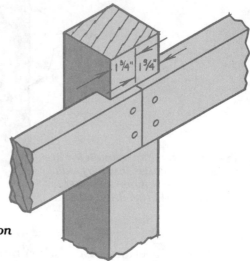

Figure 4. *Where the rails meet on an intermediate post, butt the ends together.*

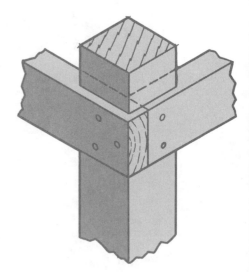

Figure 5. *Where the rails meet on a corner post, lap one rail over the end of the other.*

Attaching the Pickets and Post Tops

Cut the pickets to length. If you've purchased pre-shaped pickets, cut the length off the bottom end. If you're shaping your own pickets, cut the boards all 1″-2″ longer than needed. Cut the shapes in the tops, using a bandsaw, saber saw, or drill. Several possible shapes are shown in the *Picket Designs* drawing. Once the pickets are shaped, set up a 'stop-block' jig on your table saw or radial arm saw to cut all the pickets to precisely the same length.

6 Cut the pickets.

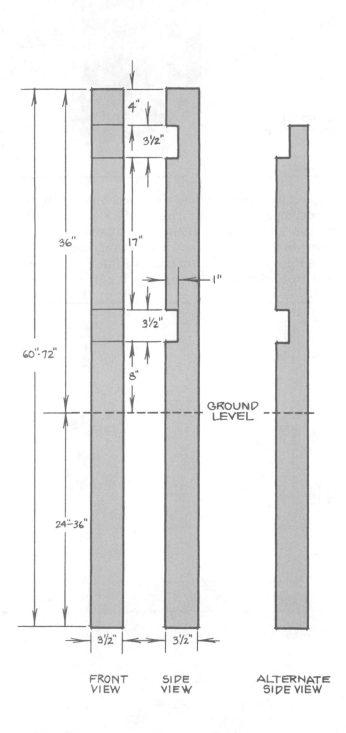

FRONT VIEW   SIDE VIEW   ALTERNATE SIDE VIEW

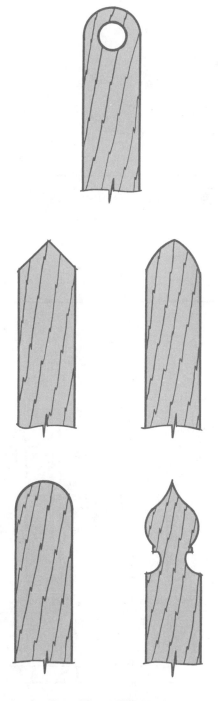

PICKET DESIGNS

7 **Attach the pickets to the rails.**

Nail the pickets to the rails with 6d nails. Position each picket so that it's at least 2″ above the ground, and ½″-1½″ from the adjacent pickets, as shown in the *Fence Assembly* drawing. To evenly space the pickets, make a simple spacing jig from scrap wood, as shown in Figure 6.

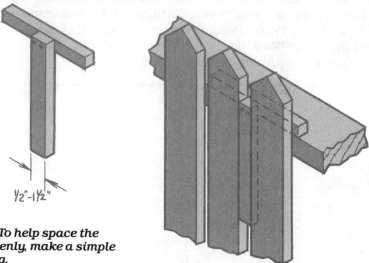

½″-1½″

Figure 6. To help space the pickets evenly, make a simple spacing jig.

8 **Attach the post tops.**

If you want the tops of the posts to show, you'll probably also want some decorative post tops. From short lengths of 4 x 4 stock, saw post tops on your bandsaw, or shape them on your lathe. Several examples are shown in the *Post Top Designs* drawing. You can also purchase post tops from your local lumberyard. Attach these post tops to the posts with 1″ dowels and 6d nails.

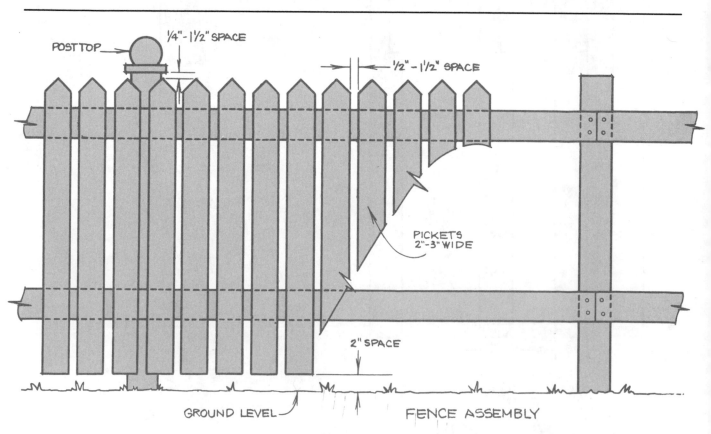

POST TOP

¼″-1½″ SPACE

½″ – 1½″ SPACE

PICKETS 2″-3″ WIDE

2″ SPACE

GROUND LEVEL

FENCE ASSEMBLY

Finishing the Fence

Make as many 40″ wide gates as you need, as shown in the *Picket Gate Layout* drawing. Use 1 x 4's for the cleats and diagonal brace, and assemble the parts with 4d nails. Be sure the parts are square before you nail them together.

9   **Make the gates you need.**

Use two 'T' hinges to hang each gate. Attach the strap of the 'T' to the horizontal cleats with 1¼″ long screws. Then position the gate between the posts so that the gate pickets are even with the fence pickets. The cleats and the diagonal brace should face the 'neighbor' side—the same side as the rails. When the gate is properly positioned, screw the butt side of the 'T' hinges to one post with 3″ long screws. Attach the gate latch to the other post, according to the manufacturer's directions.

10   **Hang the gates.**

If desired, paint or stain all exposed wood surfaces. (You can paint pressure-treated wood—it will hold paint just as well as untreated lumber.) To help paint between the pickets, use a paint sprayer.

11   **Paint or stain the fence.**

TIP   If you're going to paint the fence, it would be easier to paint all the parts *before* you assemble them—even before you set the posts. Then just touch up the paint once the fence is assembled.

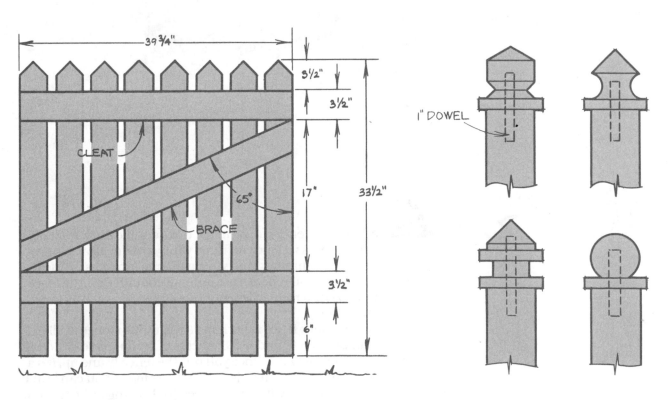

PICKET GATE LAYOUT

POST TOP DESIGNS

Latticework Fence

A latticework fence can be made to do double duty. Not only will it serve as a fence to frame your yard; it will also provide a trellis for climbing and vining plants. It makes an ideal garden fence. The lattice openings are small enough that the fence won't let animals into the garden, but they aren't so small as to restrict ventilation.

A latticework fence is perhaps the simplest fence of all to build. You can buy 2′ x 8′ and 4′ x 8′ sheets of ready-made lattice from your local lumberyard. Once you put the fence frame up, it's a cinch to fill in the sections. It's also a very light fence. You can build each section up to 8′ tall without the need for bracing or heavy framework.

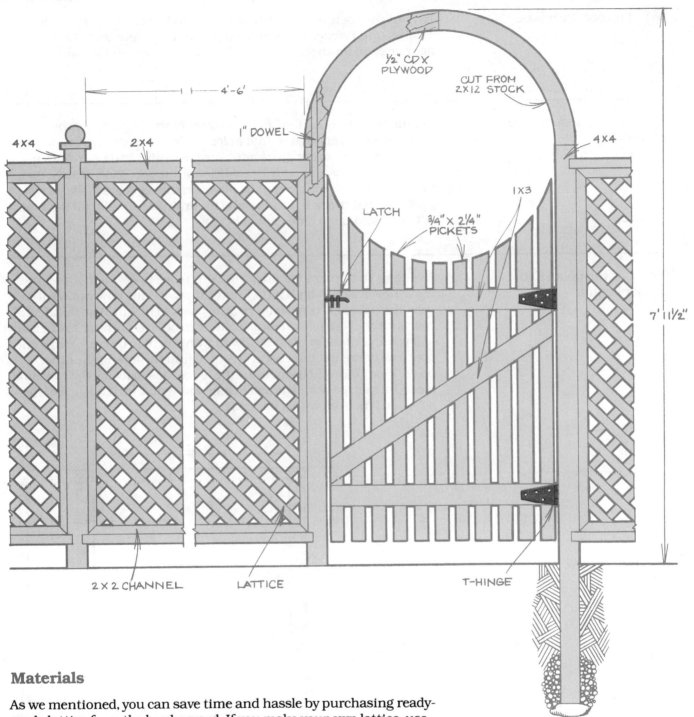

4×4

2×4

½" CDX PLYWOOD

CUT FROM 2×12 STOCK

4'-6'

1" DOWEL

4×4

LATCH

1×3

¾" × 2¼" PICKETS

7' 11½"

2 × 2 CHANNEL

LATTICE

T-HINGE

FRONT ELEVATION

Materials

As we mentioned, you can save time and hassle by purchasing ready-made lattice from the lumberyard. If you make your own lattice, use lath stock that is 1¼"-1½" wide and ¼" thick. The lattice is held in the fence by 1 x 1 cleats, so you'll need some 'one-by' (¾" thick) stock to make these cleats.

Make the top and bottom rails from 2 x 4's to give the fence the strength it needs to withstand weather and wind. The intermediate, corner, and gate posts should be made of 4 x 4's. All the wooden parts of the fence should be made from redwood, cedar, or pressure-treated lumber, so that your fence will withstand moisture and insect damage.

In addition to the wood parts, you'll also need galvanized nails, 'T' hinges, and gate latches.

Building the Fence Frame

1 **Lay out the posts.**

Lay out the posts as described in the introductory chapter of the *Fences and Privacy Screens* section. All the posts should be 4'-6' apart, except for the gate posts. These should be precisely 40½" apart.

2 **Set the posts in the ground.**

Set the posts as described in the introductory chapter. Dig the post holes deep enough so that at least ⅓ of the posts are below the ground. Set the corner posts and gate posts 6"-12" deeper than the intermediate posts, to provide additional support. Place a large rock in each hole, rest the base of the posts on the rock, brace the post upright, then throw in some gravel for drainage. Finally, fill the hole with dirt and tamp it down firmly.

3 **Cut the posts to the proper height.**

As shown in the working drawings, our fence is 72" high if the post tops show, and 68" high if they don't show. Decide how you want to treat the post tops before you cut the posts to height. Also decide if you want the tops of the fence sections to be level with each other, or follow the contour of the land. If you want them to be even, mark one post as a reference, then use a string to find the tops of the other posts. (See Figure 1.) If you want the fence tops to follow the contour, measure each post individually. Once you've marked the tops of the posts, cut them off with a handsaw.

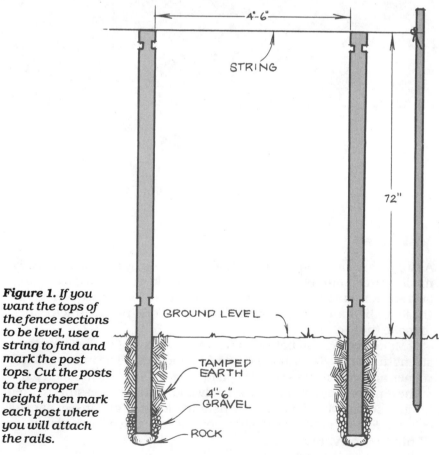

Figure 1. If you want the tops of the fence sections to be level, use a string to find and mark the post tops. Cut the posts to the proper height, then mark each post where you will attach the rails.

The posts must be notched to hold the rails in place. Before you cut these notches, decide whether or not you want the post tops to stick up above the fence. When you have decided this, measure down from the tops of the posts and mark where you will nail the rails. The measurements for a fence post whose top rises above the fence are given in the *Post Layout* drawings. The measurements for fence posts whose post tops don't show are given in the *Alternate Post Layout* drawing. With a handsaw or circular saw, cut kerfs 1″ deep along the marks. Then chisel out the waste between the kerfs to make a notch or 'dado' for the rails. (See Figure 2.) Use a wide chisel and have a helper brace the posts while you pound on the chisel with a mallet. The corner posts must be notched on two adjacent sides, as shown in Figure 3.

Option: Some people may find it easier to notch the posts *before* they set them in the ground. However, if you notch the posts first, you must dig the post holes to precisely the right depth. This can take a great deal of time: Constantly check the depth you dig. If you dig too deep, partially fill the hole with dirt and tamp down.

4 Notch the posts for the rails.

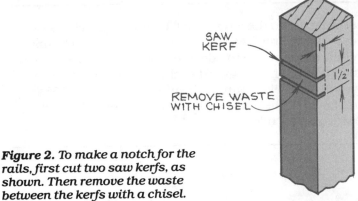

Figure 2. *To make a notch for the rails, first cut two saw kerfs, as shown. Then remove the waste between the kerfs with a chisel.*

Figure 3. *The corner posts must be notched on two adjacent faces.*

5 **Attach the rails to the posts.**

Cut the rails to length. Position the rails in the notches and toenail them to the posts with 16d nails. (See Figure 4.) Where they meet on corner posts, notch the end of one rail to fit around the corner of the other, as shown in the *Fence Frame Assembly* and *Alternate Fence Frame Assembly* drawings.

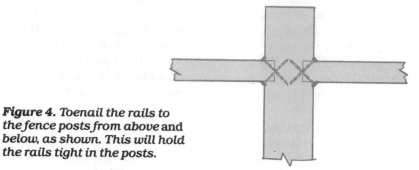

Figure 4. Toenail the rails to the fence posts from above and below, as shown. This will hold the rails tight in the posts.

Attaching the Lattice to the Fence Frame

6 **Attach the first set of cleats to the frames.**

In order to set the lattice in the fence, you'll have to make cleats to keep the lattice in place. From ¾" thick stock, rip a sufficient supply of 1 x 1's. Carefully measure the inside dimensions of each section of the fence frame, and attach a single set of cleats to the insides of the rails and posts with 4d nails. If you want, miter the cleats at 45° at the corners, as shown in the *Fence Assembly* drawing. While this isn't absolutely necessary, it looks classy.

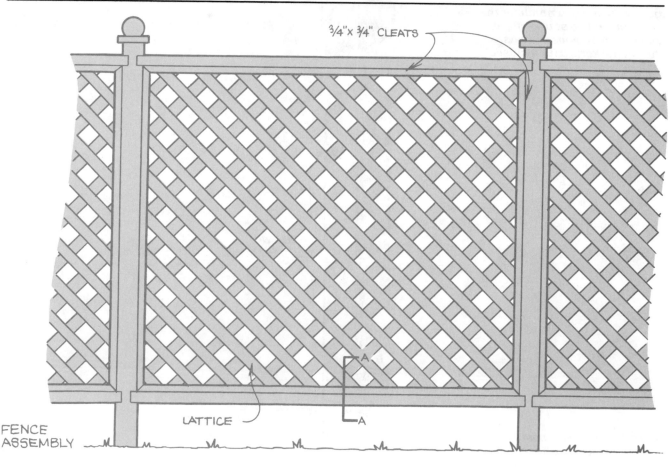

¾" x ¾" CLEATS

LATTICE

A

A

FENCE ASSEMBLY

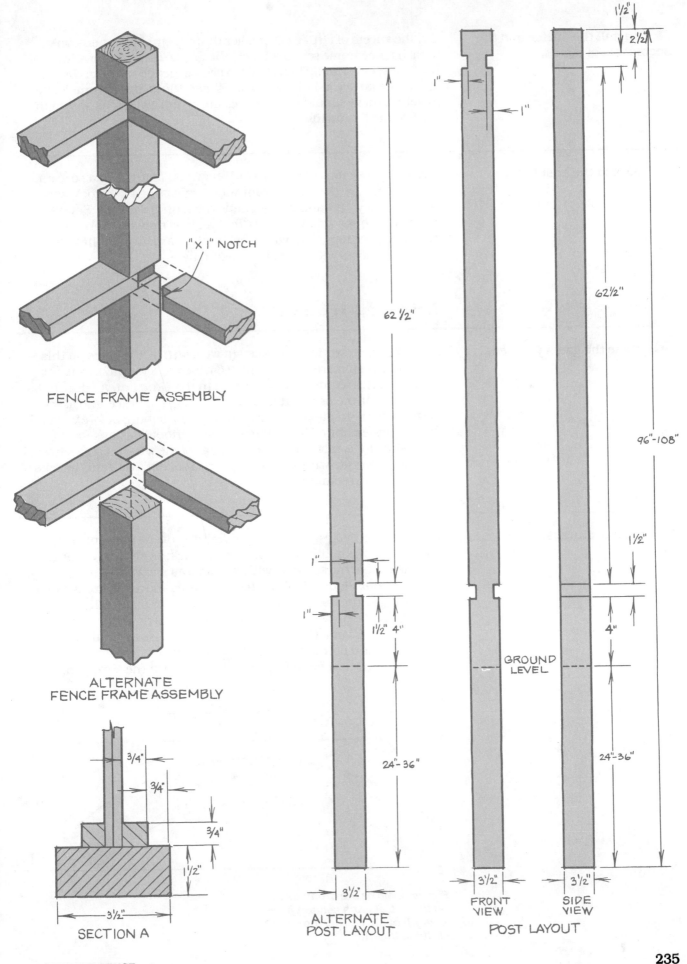

FENCE FRAME ASSEMBLY

1" X 1" NOTCH

ALTERNATE
FENCE FRAME ASSEMBLY

SECTION A

ALTERNATE
POST LAYOUT

GROUND
LEVEL

FRONT
VIEW

SIDE
VIEW

POST LAYOUT

7 **Install the lattice and the second set of cleats.**

Cut the sheets of lattice ⅛" smaller than the inside dimensions of each fence frame section. Insert the lattice in the frames, against the first set of cleats. Then attach a second set of cleats on the other side of the lattice to hold it in place. When the lattice is completely installed, there should be a set of cleats on both sides, as shown in the *Section A* drawing.

8 **Attach the post tops.**

If you want the tops of the posts to show, you'll probably also want some decorative post tops. From short lengths of 4 x 4 stock, saw post tops on your bandsaw, or shape them on your lathe. Several examples are shown in the *Post Top Designs* drawing. You can also purchase post tops from your local lumberyard. Attach these post tops to the posts with 1" dowels and 6d nails.

Finishing the Fence

9 **Make the gates you need.**

While the fence sections are filled in with lattice, the gates on this particular design are made with 1½" wide pickets. Make as many 40" wide gates as you need, as shown in the *Gate Layout* drawing. Use 1 x 3's for the horizontal cleats and diagonal brace, and assemble the parts with 4d nails. Be sure the parts are square before you nail them together. Notice that there are two optional designs for the gates—one for a gate with an archway and one without an archway. If you make the gate with an archway, use a saber saw to cut the tops of the pickets in a semicircle.

10 **Make archways, if you wish.**

If you want to make a semicircular archway over a gate entrance, cut the sections of the arch from 2 x 12 stock, as shown in Figure 5. Then assemble the sections with 8d nails, as shown in the *Archway Layout, Front View* and *Side View* drawings. Place the archway on top of the gate posts and have a helper hold it in place. With a drill and a drill bit extension, drill 1" diameter holes down through the arch, at least 6" into the tops of the gate posts. Insert 1" dowels in these holes and pound them all the way to the bottom with a mallet. Using a coping saw, cut off the dowels flush with the surface of the arch.

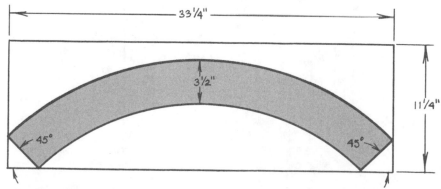

Figure 5. Make the archway in sections. Cut each section from a length of 2 x 12, then nail the sections together.

Use two 'T' hinges to hang each gate. Attach the strap of the 'T' to the horizontal cleats with 1¼" long screws. Then position the gate between the posts so that the gate pickets are at the level you want them. When the gate is properly positioned, screw the butt side of the 'T' hinges to one post with 3" long screws. Attach the gate latch to the other post, according to the manufacturer's directions.

11 Hang the gates.

If desired, paint or stain all exposed wood surfaces. (You can paint pressure-treated wood—it will hold paint just as well as untreated lumber.) To help paint all the cracks and crevices, use a paint sprayer.

12 Paint or stain the fence.

TIP If you're going to paint the fence, it would be easier to paint all the parts before you assemble them—even *before* you set the posts. Then just touch up the paint once the fence is assembled.

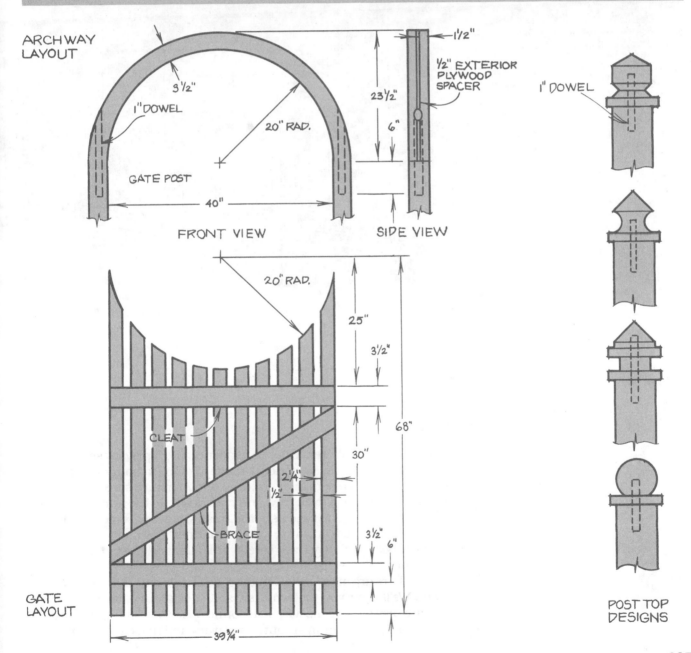

Solid Privacy Screens

The major difference between a fence and a privacy screen is the primary use. A fence is made to define a boundary, frame a yard, keep pets in or children out. A privacy screen shields a portion of your yard from view and cuts down on the noise from the street. They also can be used as 'windbreaks' or wind screens. The two projects are made in a very similar manner. The main difference is that the boards on a privacy screen are taller and closer together than a fence.

There are many different designs and variations of privacy screens. We'll show you the two basic *solid* designs in this chapter. These are privacy screens with little or no space between the boards to give you the maximum amount of privacy and noise reduction.

FENCES AND PRIVACY SCREENS

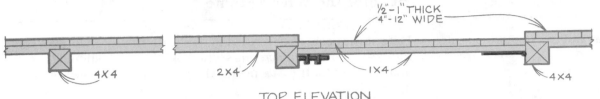

1/2"-1" THICK
4"-12" WIDE

4 X 4

2 X 4

1 X 4

4 X 4

TOP ELEVATION

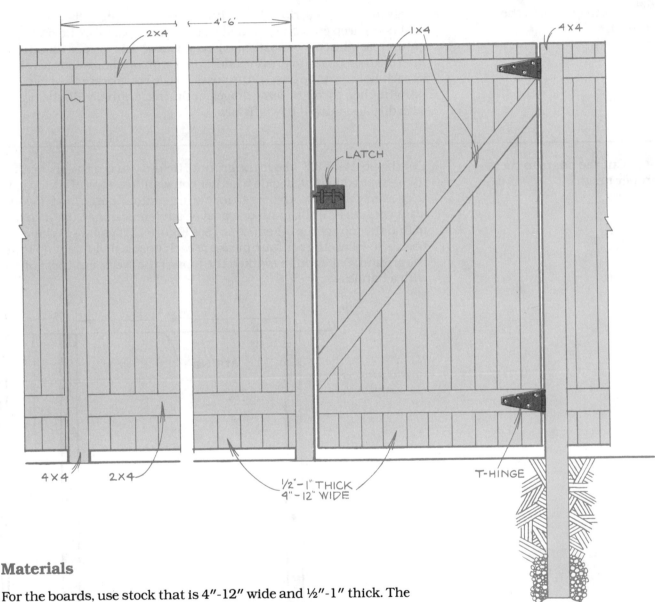

2 X 4

4'-6'

1 X 4

4 X 4

LATCH

T-HINGE

4 X 4   2 X 4

1/2"-1" THICK
4"-12" WIDE

BACK ELEVATION

Materials

For the boards, use stock that is 4"-12" wide and ½"-1" thick. The length of the board will depend on the height of the privacy screen. As shown in the working drawings, our solid screen is 72" high. To make the screen as shown, you need boards 4"-6" wide and 70" long. Make the top and bottom rails from 2 x 4's to give the screen the strength it needs to withstand weather, wind, and children. The intermediate, corner, and gate posts should be made of 4 x 4's. All the wooden parts of the privacy screen should be made from redwood, cedar, or pressure-treated lumber, so that your screen will withstand moisture and insect damage.

In addition to the wood parts, you'll also need galvanized nails, 'T' hinges, and gate latches.

1 **Lay out the posts.**

Lay out the posts as described in the introductory chapter of the *Fences and Privacy Screens* section. All the posts should be 4'-6' apart, except for the gate posts. These should be precisely 40½" apart.

2 **Set the posts in the ground.**

Set the posts as described in the introductory chapter. Dig the post holes deep enough so that at least ⅓ of the posts are below the ground. Set the corner posts and gate posts 6"-12" deeper than the intermediate posts, to provide additional support. Place a large rock in each hole, rest the base of the posts on the rock, brace the post upright, then throw in some gravel for drainage. Finally, fill the hole with dirt and tamp it down firmly.

3 **Cut the posts to the proper height.**

Cut the posts off 72" above the ground. Before you cut the posts off to the desired height, decide whether you want the tops of the boards to be even with each other, or to follow the contour of the land. If you want them to be even, mark one post as a reference, then use a string to find the tops of the other posts. (See Figure 1.) If you want the boards to follow the contour, measure each post individually, up from the ground. Once you've marked the tops of the posts, cut them off with a handsaw.

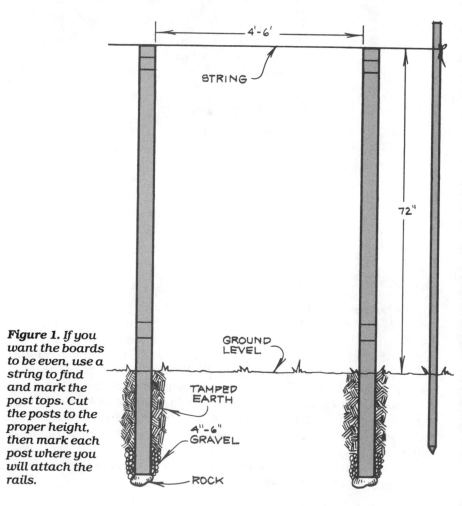

Figure 1. If you want the boards to be even, use a string to find and mark the post tops. Cut the posts to the proper height, then mark each post where you will attach the rails.

Like the picket fence, this sort of privacy screen has a 'show' side and a 'neighbor' side. On the show side, all you see is boards. The neighbors get to see the rails and the other parts of the screen frame. Decide which side of the screen is which before you proceed; the rails must be attached to the show side. When you have decided on the show side, measure down from the tops of the posts and mark where you will nail the rails. The measurements are shown in the *Post Layout* drawings. With a handsaw or circular saw, cut kerfs 1″ deep along the marks. Then chisel out the waste between the kerfs to make a notch or 'dado' for the rails. (See Figure 2.) Use a wide chisel and have a helper brace the posts while you pound on the chisel with a mallet. The corner posts must be notched on two adjacent sides, as shown in Figure 3.

Option: Some people may find it easier to notch the posts *before* they set them in the ground. However, if you notch the posts first, you must dig the post holes to precisely the right depth. This can take a great deal of time: Constantly check the depth you dig. If you dig too deep, partially fill the hole with dirt and tamp down.

4　Notch the posts for the rails.

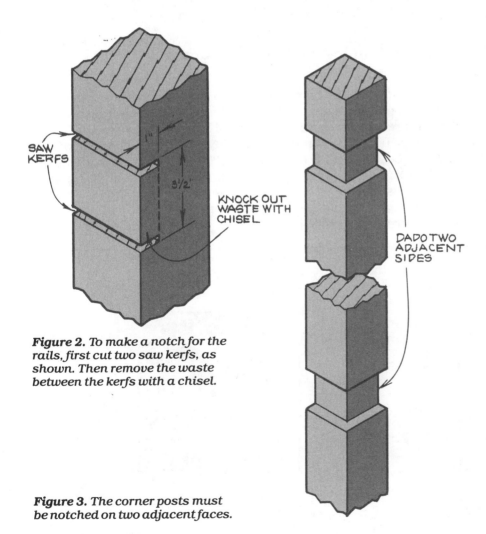

Figure 2. *To make a notch for the rails, first cut two saw kerfs, as shown. Then remove the waste between the kerfs with a chisel.*

Figure 3. *The corner posts must be notched on two adjacent faces.*

SOLID PRIVACY SCREENS

5 Attach the rails to the posts.

Cut the rails to length. Position the rails in the notches and nail them to the posts with 16d nails. Where the rails meet on an intermediate post, butt the ends together so that they meet near the middle of the post. (See Figure 4.) Where they meet on corner posts, lap one rail over the end of the other. (See Figure 5.)

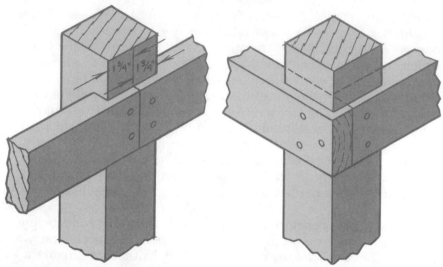

Figure 4. *Where the rails meet on an intermediate post, butt the ends together.*

Figure 5. *Where the rails meet on a corner post, lap one rail over the end of the other.*

Making a Simple Board Privacy Screen

6 Cut the boards.

Shape the tops of the boards, and cut them to length. For a decorative touch, you may want to cut a shape in the tops of the boards. You can also buy pre-shaped boards at your local lumberyard. If you purchase pre-shaped boards, cut the length off the bottom end. If you shape your own boards, cut the boards all 1"-2" longer than needed. Cut the shapes in the tops, using a bandsaw, saber saw, or circular saw. Several possible shapes are shown in the *Board Top Designs* drawing. Once the boards are shaped, set up a 'stop-block' jig on your table saw or radial arm saw to cut all the boards to precisely the same length.

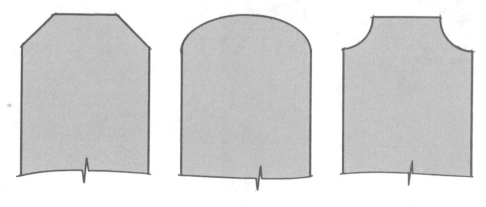

BOARD TOP DESIGNS

Nail the boards to the rails with 6d nails. Position each board so that it's at least 2″ above the ground, and ⅛″-¼″ from the adjacent boards, as shown in the *Simple Board Privacy Screen Assembly* drawing. Even though you're putting up this structure to hide your yard from prying eyes, you don't want to butt the boards up against each other. There must be at least a small space between the boards to allow for expansion and contraction. Wood moves with changes in humidity and temperature. The rule of thumb is that a board will expand or contract *across* the grain up to ¼″ for every 12″ of width. In other words, a board that you rip to 12″ wide in your shop may measure 12¼″ wide after it sits outside in a rainstorm. On a 48′ length of privacy screen (8-12 sections), the total movement of all the boards is 12″. If the boards are butted edge to edge, many of the boards may be pushed loose. So be careful to space your boards according to their width. For boards 6″ or less in width, space them ⅛″ apart; greater than 6″ (up to 12″), space them ¼″ apart.

7 **Attach the boards to the rails.**

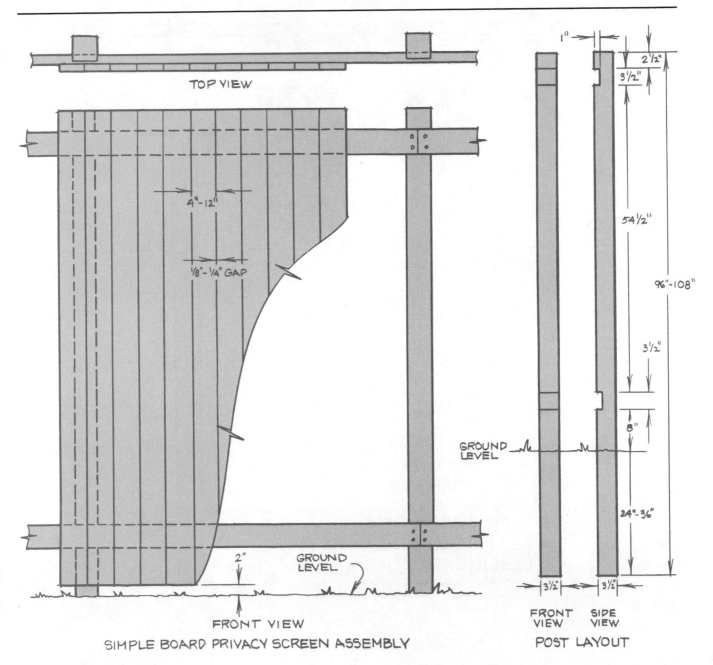

SIMPLE BOARD PRIVACY SCREEN ASSEMBLY

POST LAYOUT

8 **Attach battens over the joints between the boards.**

If you want your privacy screen to be totally private—that is, absolutely no cracks between the boards—you may want to make a 'board-and-batten' privacy screen. To do this, simply attach small boards, 1"-2" wide, over the joints between the larger boards, as shown on the *Board-and-Batten Privacy Screen Assembly, Front View* and *Top View* drawings. These smaller boards are called 'battens'. They can be made from lath, furring strips, or any other small-dimension lumber that you can find at your lumberyard. They should be no thicker than ¾", and will look better if they're ¼"-½" thick. Nail the battens in place with 2d or 4d nails, depending on their thickness.

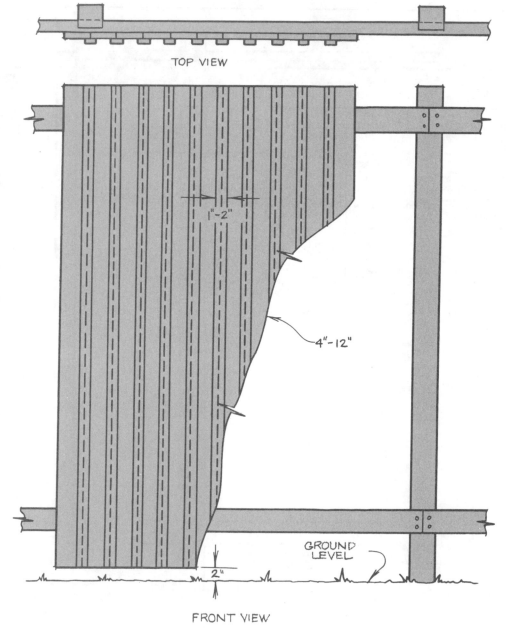

TOP VIEW

1"-2"

4"-12"

GROUND LEVEL

2"

FRONT VIEW

BOARD-AND-BATTEN PRIVACY ASSEMBLY

Finishing the Privacy Screen

9   **Make the gates you need.**

Make as many 40″ wide gates as you need, as shown in the *Privacy Screen Gate Layout* drawing. Use 1 x 4's for the cleats and diagonal brace, and assemble the parts with 4d nails. Be sure the parts are square before you nail them together. If you're building a board-and-batten privacy screen, nail battens to the gate to cover the joints between the boards.

10   **Hang the gates.**

Use two 'T' hinges to hang each gate. Attach the strap of the 'T' to the horizontal cleats with 1¼″ long screws. Then position the gate between the posts so that the gate boards are even with the privacy screen boards. The cleats and the diagonal brace should face the 'neighbor' side—the same side as the rails. When the gate is properly positioned, screw the butt side of the 'T' hinges to one post with 3″ long screws. Attach the gate latch to the other post, according to the manufacturer's directions.

11   **Paint or stain the privacy screen.**

If desired, paint or stain all exposed wood surfaces. (You can paint pressure-treated wood—it will hold paint just as well as untreated lumber.) To help paint between the boards, use a paint sprayer.

TIP   If you're going to paint the privacy screen, it would be easier to paint all the parts *before* you assemble them—even before you set the posts. Then just touch up the paint once the screen is assembled.

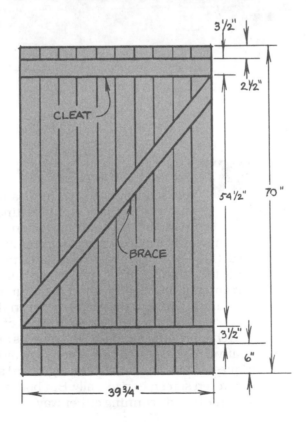

PRIVACY SCREEN
GATE LAYOUT

Ventilated Privacy Screens

While solid privacy screens do an excellent job of screening
your yard from view and cutting down on outside noise,
they restrict the air movement across your yard. On a hot
summer's day, when every little breeze is welcome relief from the
heat, this can be a problem. Ventilated privacy screens don't provide
quite the same degree of privacy and noise reduction as solid
screens, but they don't restrict the air movement, either. The air
flows freely through the spaces between the boards.

We show the two major types of ventilated screens in
this chapter: alternating-board screens and woven screens. Both are
relatively easy to build, and employ many of the same building techniques. As shown, the alternating-board screen is built vertically, and
the woven screen horizontally. But both these screens can be built
with the boards running either way.

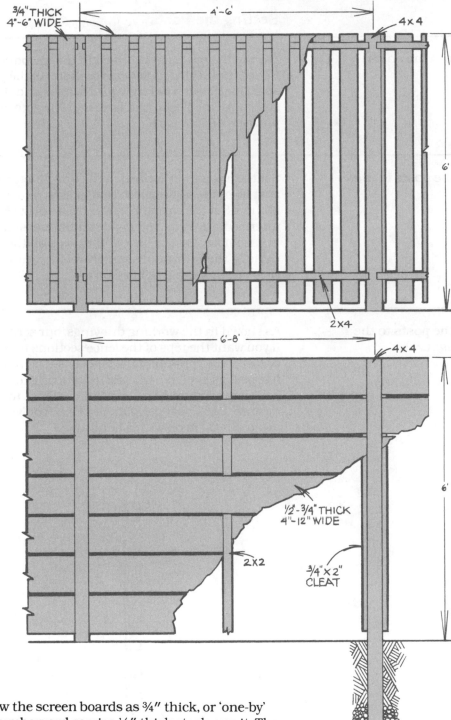

3/4" THICK
4"-6" WIDE

4'-6'

4 x 4

6'

2 x 4

SIDE ELEVATIONS

6'-8'

4 x 4

6'

1/2"-3/4" THICK
4"-12" WIDE

2 x 2

3/4" x 2"
CLEAT

Materials

In both screens, we show the screen boards as ¾″ thick, or 'one-by'
stock. However, if your lumberyard carries ½″ thick stock, use it. The
thinner stock is strong enough to last, but the weight of the completed screen will be lighter and there will be less tendency to sag.
When building a woven screen, thinner lumber is easier to bend.

Even if you can get thin stock for the screen boards, you'll still
need some ¾″ thick stock for the cleats and spreaders in the woven
screen. Make the top and bottom rails in the alternating-board
screen from 2 x 4's. The intermediate, corner, and gate posts should
be made of 4 x 4's. All the wooden parts of the screens should be
made from redwood, cedar, or pressure-treated lumber, so that your
completed screen will withstand moisture and insect damage.

In addition to the wood parts, you'll also need galvanized nails,
'T' hinges, and gate latches.

1 **Lay out the posts.**

Lay out the posts as described in the introductory chapter of the *Fences and Privacy Screens* section. All the posts should be 4'-6' apart for the alternating-board screen, and 6'-8' apart for the woven screen. All the posts, that is, except the gate posts. These should be precisely 40½" apart.

2 **Set the posts in the ground.**

Set the posts as described in the introductory chapter. Dig the post holes deep enough so that at least ⅓ of the posts are below the ground. Set the corner posts and gate posts 6"-12" deeper than the intermediate posts, to provide additional support. Place a large rock in each hole, rest the base of the posts on the rock, brace the post upright, then throw in some gravel for drainage. Finally, fill the hole with dirt and tamp it down firmly.

3 **Cut the posts to the proper height.**

As shown in the working drawings, our screens are 72" high. Decide if you want the tops of the fence sections to be level with each other, or follow the contour of the land. If you want them to be even, mark one post as a reference, then use a string to find the tops of the other posts. (See Figure 1.) If you want the fence tops to follow the contour, measure each post individually. Once you've marked the tops of the posts, cut them off with a handsaw.

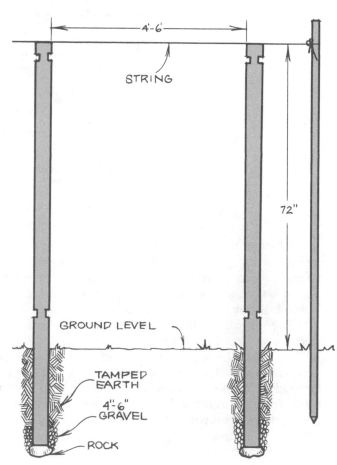

Figure 1. If you want the tops of the fence sections to be level, use a string to find and mark the post tops. Cut the posts to the proper height, then mark each post where you will attach the rails.

Building an Alternating-Board Privacy Screen

The posts must be notched to hold the bottom rails in place. Mark where you will nail the rails; the measurements are given in *Alternating-Board Screen Post Layout* drawings. With a handsaw or circular saw, cut kerfs 1″ deep along the marks. Then chisel out the waste between the kerfs to make a notch or 'dado' for the rails. (See Figure 2.) Use a wide chisel and have a helper brace the posts while you pound on the chisel with a mallet. The corner posts must be notched on two adjacent sides, as shown in Figure 3.

Option: Some people may find it easier to notch the posts *before* they set them in the ground. However, if you notch the posts first, you must dig the post holes to precisely the right depth. This can take a great deal of time: Constantly check the depth you dig. If you dig too deep, partially fill the hole with dirt and tamp down.

4 Notch the posts for the rails.

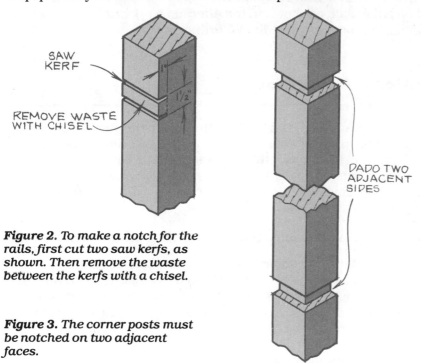

Figure 2. *To make a notch for the rails, first cut two saw kerfs, as shown. Then remove the waste between the kerfs with a chisel.*

Figure 3. *The corner posts must be notched on two adjacent faces.*

Cut the rails to length. Position the bottom rails in the notches and toenail them to the posts with 16d nails. (See Figure 4.) Nail the top rails to the tops of the posts, butting the ends together. Where the rails meet on corner posts, notch the end of one rail to fit around the corner of the other, as shown in the *Fence Frame Assembly* drawing.

5 Attach the rails to the posts.

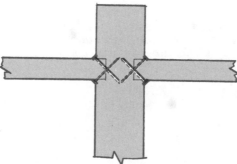

Figure 4. *Toenail the bottom rails to the posts from above and below, as shown. This will hold the rails tight in the posts.*

6 **Attach the boards to one side of the frame.**

Nail the boards to one side of the frame with 6d nails. Carefully space them so the distance between each board is 1½″ less than the width of the board. For example, space 4″ wide boards every 2½″. 5″ boards should be spaced every 3½″, and so on.

TIP To space the boards quickly and precisely, rip a board the width of the space you want, and use this as a spacer when positioning each board.

7 **Attach the boards to the other side of the screen.**

Nail the boards to the other side of the frame, positioning the boards so that they cover the spaces between the boards on the first side. This second set of boards should overlap the position of the first set ¾″ on either side, as shown in the *Alternating-Board Privacy Screen Assembly, Front View* and *Top View* drawings.

Making a Woven Privacy Screen

4 **Attach the cleats to the posts.**

From ¾″ stock, rip cleats 2″ wide. Nail these to the posts with 6d nails, as shown in the *Woven Screen Post Layout, Side View* and *Top View* drawings. Notice that the intermediate posts have cleats attached to opposite sides; the corner posts have cleats attached to adjacent sides, and the gate posts have cleats attached to just one side.

5 **Nail the screen boards to the cleats.**

Carefully measure the distance between the posts, and cut the horizontal screen boards ¼″-½″ longer than this measurement. Working from the bottom, position a board between the posts and nail it to the cleats with 6d nails. (For maximum strength, angle the nails so that they go through the cleats and into the posts, as shown in Figure 5.) The board will bow slightly, because it's slightly too long for the space. Make sure it bows towards the *inside* of the screen. Switch to the other side of the fence and nail another board in place, about ¼″ above the first one. Once again, make sure the board bows towards the inside. Repeat until you have completely filled the spaces between the posts with horizontal boards.

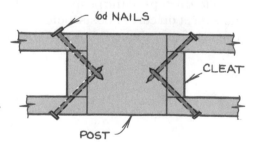

Figure 5. When attaching the boards to the posts, angle the nails so that they go through the cleats and into the posts. This will add strength to the joint.

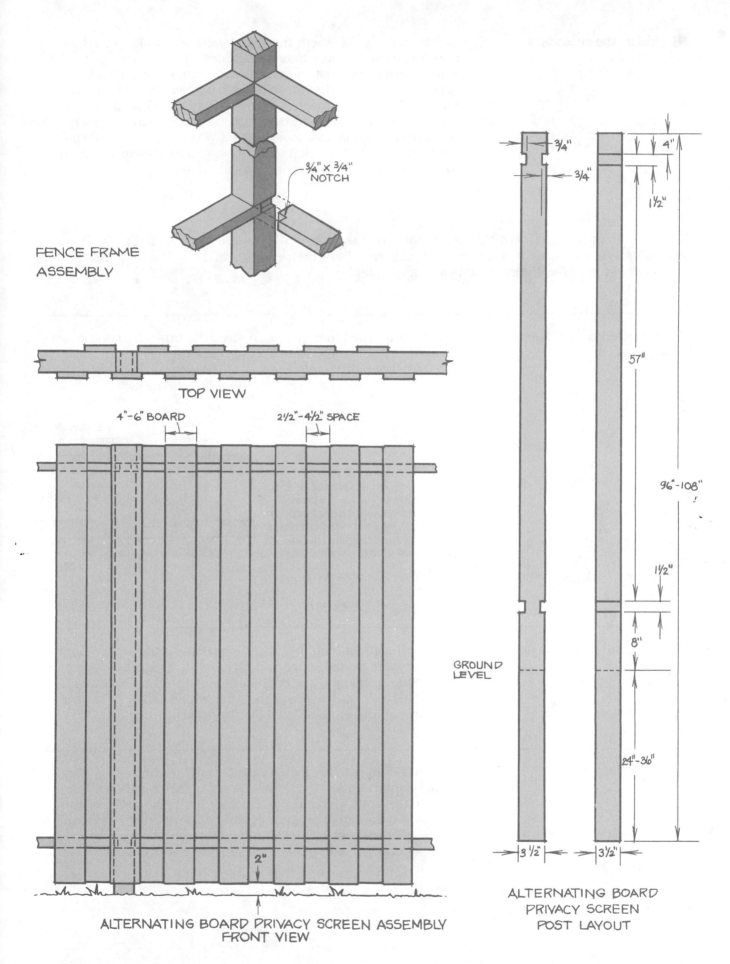

FENCE FRAME
ASSEMBLY

3/4" X 3/4"
NOTCH

TOP VIEW

4"-6" BOARD

2½"-4½" SPACE

2"

ALTERNATING BOARD PRIVACY SCREEN ASSEMBLY
FRONT VIEW

3/4"

3/4"

4"

1½"

57"

96"-108"

1½"

8"

24"-36"

GROUND
LEVEL

3½"

3½"

ALTERNATING BOARD
PRIVACY SCREEN
POST LAYOUT

6 **Insert the spreaders.**

Cut 2 x 2 spreaders exactly the same length as the cleats. In the middle of each section, where the boards bow in towards each other, insert these spreaders to bow the boards even more and create a 'weave', as shown in the *Woven Privacy Screen Assembly, Front View* and *Top View* drawings. Working from the top, slide each spreader down between the boards, pressing or pulling on each board in turn to make way for the 2 x 2. When you've properly positioned each spreader, tack it to the top and bottom screen boards with 4d nails. This will keep it from sliding out.

TIP At the beginning of this step, make just one section at a time. If the boards aren't bowed enough and the spreader is hard to insert, cut the boards on the next section a little longer. If the spreader is too loose, cut the boards a little shorter.

---

7 **If needed, attach keepers to the posts.**

If you're concerned that the woven screen boards might pull loose from the cleats, or if you want to cover the joint between the boards and the posts, nail 2″ wide 'keepers' to the posts on both sides, as shown in Figure 6. These keepers will help hold the boards in place.

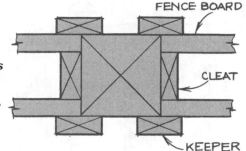

Figure 6. If you wish, nail keepers to the posts to finish the woven screen, as shown. These keepers will help keep the boards in place as well as hide the joint between the boards and the posts.

Finishing the Screen

---

8 **Make the gates you need.**

The gate design for both these privacy screens is the same—simple vertical boards held together by horizontal cleats and a diagonal brace. Make as many 40″ wide gates as you need. Use 1 x 3's for the cleats and brace, and assemble the parts with 4d nails. Be sure the parts are square before you nail them together.

---

9 **Hang the gates.**

Use two 'T' hinges to hang each gate. Attach the strap of the 'T' to the horizontal cleats with 1¼″ long screws. Then position the gate between the posts so that the gate boards are at the level you want them. When the gate is properly positioned, screw the butt side of the 'T' hinges to one post with 3″ long screws. Attach the gate latch to the other post, according to the manufacturer's directions.

If desired, paint or stain all exposed wood surfaces. (You can paint pressure-treated wood—it will hold paint just as well as untreated lumber.) To help paint all the cracks and crevices, use a paint sprayer.

10 **Paint or stain the screen.**

TIP If you're going to paint the screen, it would be easier to paint all the parts *before* you assemble them—even before you set the posts. Then just touch up the paint once the fence is assembled.

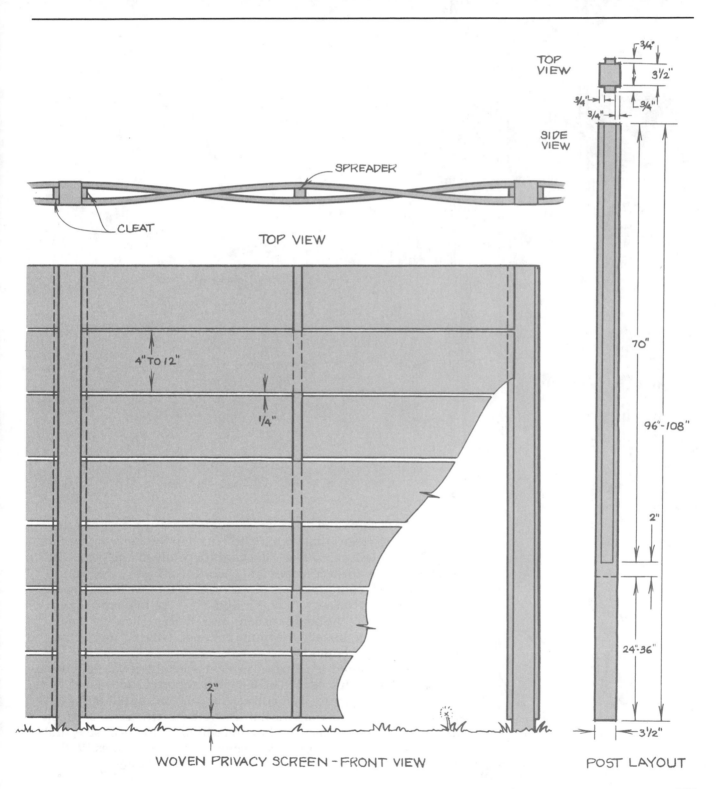

TOP VIEW

SIDE VIEW

SPREADER

CLEAT

TOP VIEW

4" TO 12"

1/4"

2"

WOVEN PRIVACY SCREEN - FRONT VIEW

3/4"
3 1/2"
3/4"
3/4"
3/4"
70"
96"-108"
2"
24"-36"
3 1/2"

POST LAYOUT

Contoured Fences & Privacy Screens

The designs and building techniques for the fences and privacy screens that we have described in the last few chapters work well for relatively flat yards or gently rolling yards. But what of yards with steep slopes and gulleys? For these yards, you need to make a 'contoured' fence or screen. By this, we mean a structure that follows the contour of the land. The bottoms—and sometimes the tops—of the boards of contoured fences follow the slopes of your landscape. Usually, the frame does not. This requires some special building techniques.

A 'contoured' fence can also mean something else—a fence whose top has been scalloped or somehow shaped to add visual interest to the structure. Sometimes, our yards are so flat that we would like some relief from the flatness. A fence with a scalloped top has the same visual effect as a horizon with mountains or steep hills in the distance. This, too, requires some special techniques. We'll discuss how to make both types of contours in this chapter.

Making a 'Contoured-to-the-Land' Fence

Before we go too far, it should be noted that some designs can be made to follow the contour of the land easier than others. Perhaps the easiest are post-and-rail fences with widely spaced horizontal members. The next easiest are structures with vertical members, such as a picket fence or alternating-board privacy screen. The hardest of all is fences and screens with closely-spaced horizontal members. It is all but impossible to put up a woven privacy screen on steeply sloping land. Consider both the time you want to spend on this project and the degree of your expertise when choosing a design for your yard.

1 Pick a design that can be adapted to follow the contour of your landscape.

Every slope has a total 'rise' and a total 'run', just like a stairway. The rise is the vertical distance—how tall the hill is, as measured from the bottom of the gulley. The run is the horizontal distance, as measured from where the hill crests to the lowest point of the gulley. To find the rise and run, first mark the highest point of your fence line —the crest of the hill—with a short stake. 'Eyeball' this location; it doesn't matter if you're off by a few inches. Tie a string around this stake, close to the ground. Next, mark the low point of the fence line— the bottom of the gulley—with another, longer stake. This stake must be long enough to extend above the crest of the hill. Tie the other end of the string to the long stake, near the top. Use a string level to make sure the string is perfectly horizontal. (See Figure 1.) The distance between the ground and the string on the long stake is the total rise of the slope. The distance between the two stakes is the total run.

2 Calculate the total rise and run of the slope.

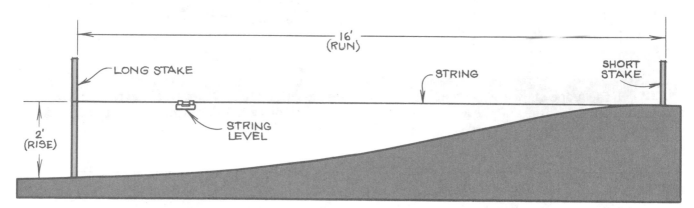

Figure 1. Every slope has a 'rise' and a 'run'. The rise is the vertical distance from the top of the hill to the bottom. The run is the horizontal distance between the top of the hill and the bottom. Find the rise and the run with stakes, string, and a string level.

To keep the spacing of the fence posts even, you may want to adjust the position of the long stake. If the stakes are 15' apart, and your post spacing is 4', move the long stake 1' back from the short stakes, making the run 16'. 16 is evenly divisible by 4. If you move the long stake, figure the rise again at this new position.

3 Adjust the length of the run, if necessary.

4 **Lay out the position of the fence posts along the slope.**

Measure along the string and mark it where you want to locate the fence posts. Use a plumb bob to transfer these marks to the ground, and mark the locations on the slope with nails and scraps of paper. (See Figure 2.)

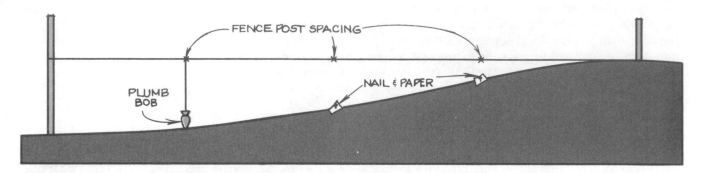

Figure 2. To lay out the locations of the posts, use a plumb bob to locate each post on the hill. Mark the location with nails and paper.

5 **Calculate the rise for each fence section.**

The next step is to figure out how far you want the fence to rise or drop with each section. To do this, simply divide the total rise by the number of fence sections. This will give you the rise per section. For example, if there are four sections in the run, and the rise is 2′, then the rise per section is ½′ or 6″. The equation can be written as follows:

$$\frac{\text{Total Rise}}{\text{Number of Sections}} = \text{Rise Per Section}$$

TIP If the slope of the hill is steep, space the posts closer together. Otherwise, the rise per section will be much too large and the fence or privacy screen will look ridiculous. Try to keep the rise per section under 1′ for structures shorter than 4′, and under 2′ for those taller than 4′.

6 **Set the posts in the ground.**

Set the posts as you would a post in level ground: Use a rock for the base, fill the hole with two shovels full of gravel, fill the rest of the hole with dirt, and tamp down.

7 **Mark the posts where you will attach the rails.**

Don't cut the posts to the proper height just yet; instead, mark the posts where you will attach the rails. To find the locations of the rails on the posts, measure the first post up from the ground and mark where you want the notches. Put a small nail at each mark, and tie strings to these nails. With a string level, find the location on the *uphill* side of the next post down the slope where you will attach the other end of the rails. (See Figure 3.) Then measure up from the ground on the *downhill* side of the second post to find the location of the next set of rails. Mark the location, and repeat the process moving downhill. Use the locations of the rails on the *uphill* side of each post to determine the top of the post.

FENCES AND PRIVACY SCREENS

Figure 3. *Measure up from the ground to find the location of the rails on the* downhill *side of each post. Use string, nails, and a string level to find the locations on the* uphill *sides.*

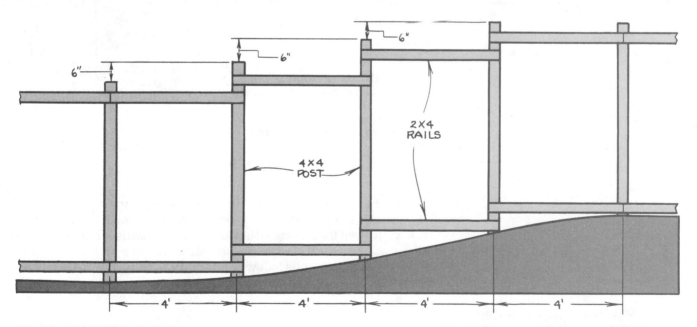

MARK FOR POST TOP
4"
STRING LEVEL
STRING
4"
MARK FOR RAIL
3½"
3½"
59" DOWNHILL SIDE
UPHILL SIDE
59"
68"
3½"
2"
3½"
2"

Notch the posts as described in previous chapters. Cut along each mark with a handsaw or circular saw and clean out the waste with a wide chisel. Then cut the posts to the proper height with a handsaw.

8 **Notch the posts and cut them to the proper height.**

Attach the rails to the posts with 16d nails. Where there is only one rail in a notch, the rail should lap the width of the post, completely filling that notch. (See Figure 4.)

9 **Attach the rails to the posts.**

6"
6"
6"
2 X 4 RAILS
4 X 4 POST
4'
4'
4'
4'

Figure 4. *Nail the rails in the notches on the posts. Where there is only one rail in a notch, the rail should completely lap the width of the post, so that the end of the rail is flush with the side of the post.*

10 **Attach the boards to the rails.**

Nail the boards to the rails with 6d nails. Using a saber saw, cut the bottom of each board so it follows the contour of the land. Position the boards on the fence so that the bottom edge is at least 2″ from the ground. (See Figure 5.)

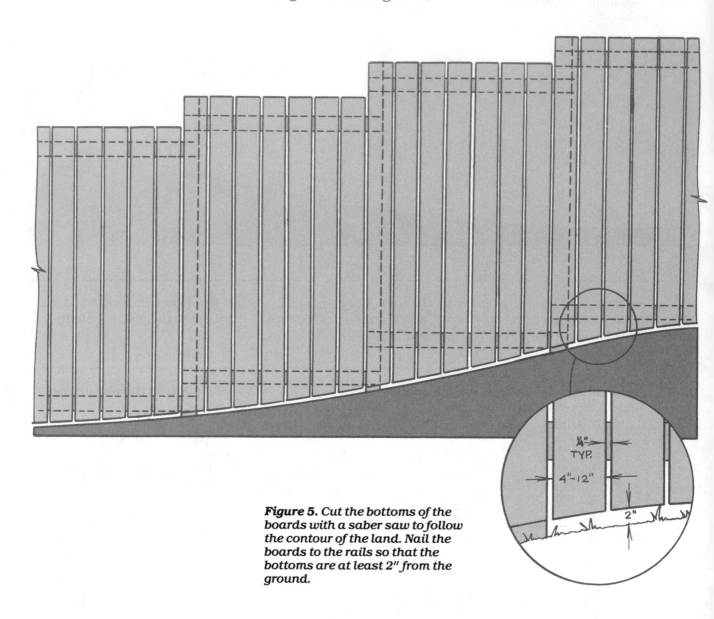

Figure 5. Cut the bottoms of the boards with a saber saw to follow the contour of the land. Nail the boards to the rails so that the bottoms are at least 2″ from the ground.

There are two different treatments that you can use for the top. The top edges can be parallel to the top rails, as shown in Figure 5. This works well for fences with wide, vertical boards such as the solid privacy screen. But if the boards are narrow, as in a picket fence, you can easily make the tops follow the contour of the ground. (See Figure 6.)

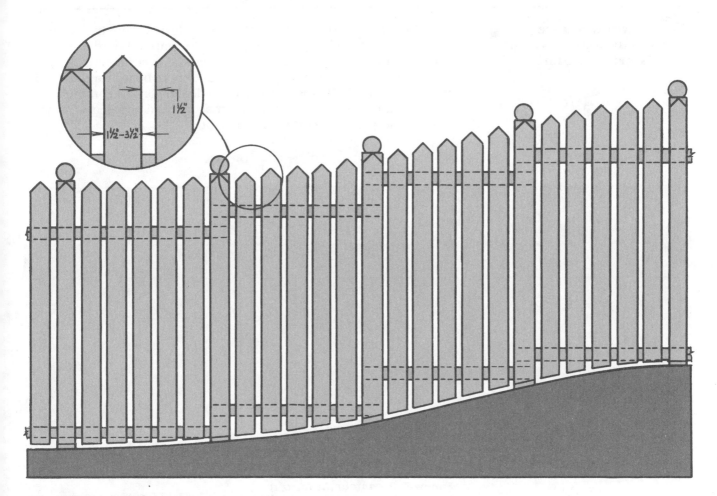

Figure 6. If you wish, you can cut the boards so that the tops also follow the contour of the ground.

Install post tops, gates, and paint or stain the fence as described in previous chapters.

11 **Finish the fence.**

1 Leave extra space between the top edges of the boards and the top rail.

A contoured-top structure works best with designs that incorporate vertical boards, such as picket fences or alternating-board privacy screens. Build the fence or screen as described in earlier chapters, but leave extra space between the top edge of the boards and the top rail. (See Figure 7.) How much space will depend on the contour you wish to cut in the fence boards. This space should be at least 2″ more than the deepest part of the contour.

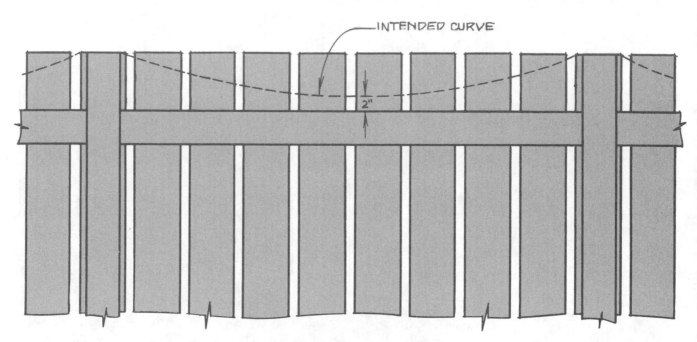

INTENDED CURVE

2″

Figure 7. To contour the top of a fence or privacy screen, build the structure with some extra distance between the tops of the boards and the top rail. This distance should be 2″ more than the depth of the contour.

2 Mark the contour on the fence boards.

Mark the contour in pencil on the 'show' side of the fence boards. The most common contour is a broad scallop. The easiest way to mark this is to use a length of chain. Stretch the chain between the posts and let it go 'slack', so that the middle of the chain sags. Add more slack until the lowest point of the chain is a pre-determined distance away from the top of the fence. The chain will describe a perfectly symmetrical arc between post tops. With a pencil, trace this arc, being careful not to move the chain and destroy the symmetry. (See Figure 8.)

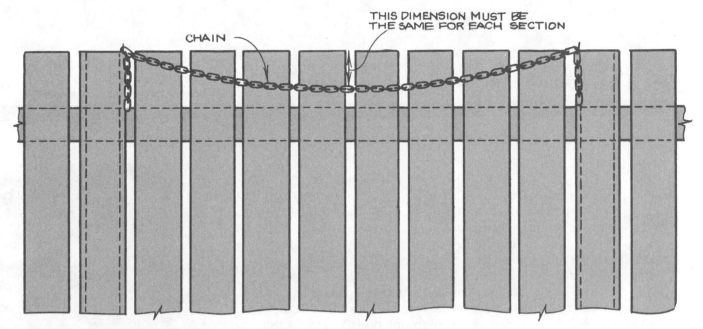

CHAIN

THIS DIMENSION MUST BE THE SAME FOR EACH SECTION

Figure 8. Use a chain to mark a simple scallop. When you hang the chain between the post tops with some slack, the chain will describe a symmetrical arc.

3 Cut the contour.

Carefully cut the contour you've marked with a saber saw or reciprocating saw. Lightly sand the tops of the fence boards with 50# sandpaper to smooth the edges and help eliminate the chance of splinters.

4 Finish the fence.

Install post tops, gates, and paint or stain the fence as described in previous chapters. You may wish to contour the gate boards in the same manner as you contoured the fence boards.

Patios and Decks

The least expensive way to increase your living space is to build a patio or deck. And there's no more popular gathering place in warm weather than your own backyard. So, why not add a patio or a deck to your home and enjoy the great outdoors in style? If you build it yourself, you can tailor the deck to your own tastes *and* your budget.

The following chapters contain step-by-step instructions for building a variety of decks, patios, and coverings for them. There is a plan for a raised deck, as well as an on-ground deck. You can select a solid covering for these decks, or put up an elegant sunshade. The design of your deck and its cover depends on your home and your preferences.

Choosing and Using Wood

No matter what sort of deck or patio you wish to build, keep these pointers in mind:

Types of wood. Most decks are made of Douglas fir, Southern pine, or redwood, all of which are easy to work with and have sufficient strength to support the structure. Other varieties, however, may be less expensive and just as suitable. These include ash, Western red cedar, cypress (rot-proof), larch, gum, hemlock, white fir, soft pines, poplar, and spruce. Check with your local lumberyard for the types of wood that are available in your area.

Use pressure-treated Douglas fir, ash, or Southern pine for your posts because they won't rot like untreated wood. The best decking woods are Douglas fir, Southern pine, redwood, cedar, Western larch, and cypress because they are strong, decay-resistant, and seldom warp.

What size lumber to buy. The size of your patio or deck will determine your materials list, but consider the size of the beams, joists, and planks you will need to use to properly support your structure. On the following page are three charts to help you decide—at a glance—what size is right for your design.

Beam Size For Specific Span

Beam Size	Maximum Span*
4 x 4	4 ft.
4 x 6	6 ft.
4 x 8	8 ft.
4 x 10 (or two 2 x 10's)	10 ft.
4 x 12 (or two 6 x 10's)	12 ft.
6 x 10 (or three 2 x 10's)	12 ft.
6 x 12 (or three 2 x 12's)	14 ft.

*With the beam placed on edge.

Joist Size For Specific Span

Joist	Spacing	Maximum Span*
2 x 6 (minimum)	12" o.c.	8 ft.
	16" o.c.	7 ft.
	24" o.c.	5 ft.
2 x 8	12" o.c.	10 ft.
	16" o.c.	9 ft.
	24" o.c.	7 ft.
2 x 10	12" o.c.	13 ft.
	16" o.c.	12 ft.
	24" o.c.	10 ft.
2 x 12	12" o.c.	16 ft.
	16" o.c.	15 ft.
	24" o.c.	14 ft.

*With the joist placed on edge.

Plank Size For Specific Span

Plank	Maximum Span*
1 x 4	12"
1 x 6	16"
2 x 2	42"
2 x 3	42"
2 x 4	42"
2 x 6	42"

*With the plank placed flat.

Building code requirements vary, so check with your local building inspections office before you buy the lumber and be sure your plans meet specific load and span restrictions. You may also have to obtain a building permit.

Working with the lumber.   Common construction lumber, like that used for framing and decking, is always plain sawn. As such, the annual rings will curve through the width of the board. Look at the end grain of a piece of construction lumber and you'll see what we mean. The outside or convex side of these curves is called the 'bark' side of the board, since it was once nearer the bark of the tree. The inside or concave side of the curve is the 'heart' side, since it was closest to the heartwood. (See Figure 1.)

The heartwood is denser, and more resistant to rot than the softer sapwood near the bark. Also, when the lumber warps, it tends to cup in the *opposite* direction of the annual rings, toward the bark side. For these two reasons, make sure that you install the decking bark side down. The decking will last longer, and if it does cup, the cup will face down. If the cup is up, people may trip over the raised edges of the board.

HEART SIDE

BARK SIDE

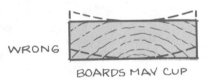

RIGHT

CONTROLS CUPPING

WRONG

BOARDS MAY CUP

Figure 1. Decking stock has a 'bark' side and a 'heart' side. You can tell which is which by looking at the end grain to see which way the annual rings curve. To help control cupping, always install decking with the bark side down.

Open Deck

If you're a nature enthusiast, or if you simply enjoy an outdoor barbecue with a few friends, adding an open deck to your home may be a wise move. Not only will a deck increase your property value, it'll extend your living—and entertaining—space. Open decks are easy to build, require a minimum investment in materials, and most can be constructed in a weekend or two.

The pier-and-post foundation that supports this deck is perfect for all types of terrain—just adjust the length of the posts to compensate for the slope. (And it's easy to replace damaged posts because you don't have to dig posts out of the ground.) Build the deck close to the ground or raise it up to make a second-story balcony. The design you see here is quite versatile; it can be adapted easily to suit your tastes and needs.

266

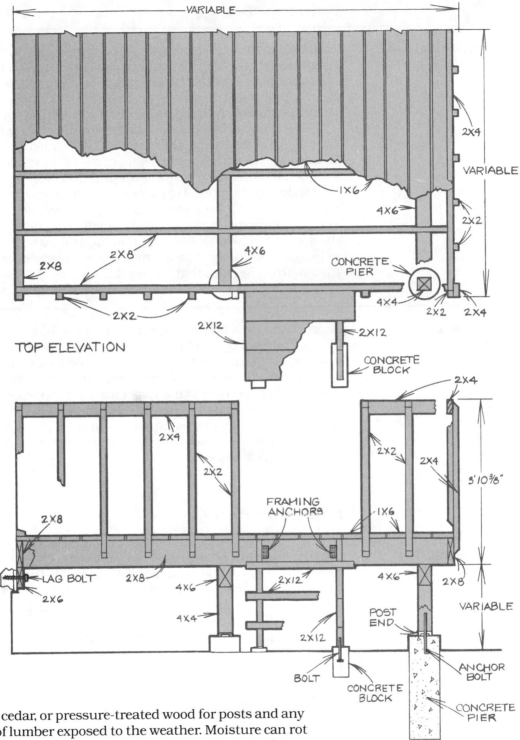

VARIABLE

2X4

VARIABLE

1X6

4X6

2X2

2X8

2X8

4X6

CONCRETE
PIER

2X2

4X4

2X2   2X4

2X12

TOP ELEVATION

2X12

CONCRETE
BLOCK

2X4

2X4

2X2

2X4

5'10⅜"

FRAMING
ANCHORS

1X6

2X8

LAG BOLT

2X6

2X8

4X6

2X12

2X12

4X4

BOLT

CONCRETE
BLOCK

4X6

2X8

VARIABLE

POST
END

ANCHOR
BOLT

CONCRETE
PIER

SIDE ELEVATION

Materials

Use redwood, cedar, or pressure-treated wood for posts and any
other pieces of lumber exposed to the weather. Moisture can rot
other woods. Purchase 4 x 4's for posts, at least 1' longer than you
will actually need, to allow yourself extra stock to 'level' them once
they have been set on the piers. As shown, this deck uses 4 x 6's for
beams and a 2 x 6 ledger. The joists are 2 x 8's and the decking is 1 x 6
'decking' planks. (Decking stock is actually 1¼" thick.) Purchase
2 x 2's for spindles, 2 x 4's for railing, and 2 x 12's for the stair risers.

The posts are set on concrete piers, so you'll need concrete, several lengths of 8" stovepipe or cardboard sleeves to serve as forms,
anchor bolts, and metal post ends to make the foundation. In addition
to these materials, purchase nailing straps, lag screws, and galvanized nails.

1   Adjust the size and find the height of the deck.

As shown in the working drawings, this deck is 10′ wide and 12′ long. However, this may be larger or smaller than you need. To adjust the size, simply alter the dimensions to suit your site and your needs. Be sure that your dimensions are divisible by 2′—otherwise you'll be cutting standard stock to size and that wastes time and money.

This deck can be built close to the ground or raised, depending on where it is attached to the house and how much your land slopes. To calculate the approximate height of the deck, attach strings to your house where the corners of the deck will meet the house. Drive long stakes in the ground to mark the other two deck corners and stretch the strings between the house and the stakes. Use a string level to make sure the strings are perfectly horizontal. Then measure the distance between the strings and the ground where you intend to run beams. Use these measurements to calculate the number and the length of the posts that you need.

2   Check the building codes.

Because this deck is attached to a permanent foundation, it will likely be affected by building codes. Check your local codes before you begin construction to be sure your design—and materials list—meets code requirements. If you have increased the dimensions of the deck, you may have to use larger lumber for the posts, beams, and joists. Likewise, if you have scaled down the size of the deck, you may be able to save money by using smaller lumber. Check the span and load requirements for your area. If needed, secure a building permit before you begin work.

Pouring the Piers

3   Lay out the piers.

Use the stakes and strings that you've already set in place to lay out the locations of your piers. (See Figure 1.) Locate the center of each pier no more than 6′ apart. As we show in the *Pier Layout* drawing, the maximum distance between piers on our deck is 70¼″. For this 10′ x 12′ deck, we need six piers. Mark the locations of the piers with stakes, then remove the string while you dig the holes.

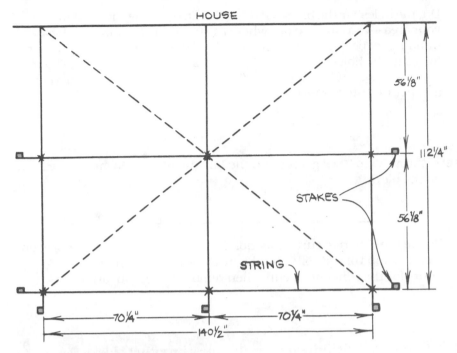

Figure 1. *Use stakes and string to lay out your deck, then find the locations of the piers by measuring along the strings. Check that your layout is square by measuring diagonally from corner to corner. AD must equal BC.*

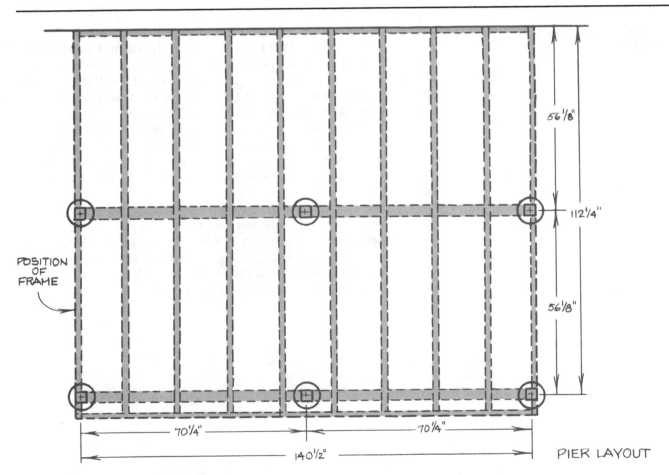

PIER LAYOUT

4 Pour the piers.

Dig the holes for the piers 24″ to 36″ deep or below the frost line for your area—you can find out where the frost line is from your local building inspections department. Set 8″ diameter round cardboard forms in the holes. (You can also use 8″ stovepipe for forms.) The tops of the forms should be at least 4″ above ground level. Mix concrete and pour it into the forms.

TIP Before you pour the concrete, throw 2″ to 3″ of gravel into the bottom of each form. This will help to drain ground water away from the piers.

5 Set the anchor bolts.

Before the concrete cures, position anchor bolts in the center of each pier. The tops of the bolts should protrude 6″-8″. Wait at least 24 hours for the concrete to cure, then remove the cardboard forms.

TIP To hold the anchor bolts at the proper height, drive two small stakes on either side of the piers. Wrap a wire around the end of the bolt, then wrap the ends of the wire around the stakes so that the bolt is suspended in the wet concrete.

Building the Deck Frame

6 Attach the ledger strip to the house.

Calculate where you want to attach the ledger strip to your house. You'll want the surface of the deck just below the sill of the door that will lead from the house to the deck. Therefore, you'll want the ledger somewhere below that sill. Add up the thickness of the decking (1¼″) and the width of the joists (7¼″). On our deck, the result is 8½″. Measure down along the wall of the house 8½″ from the bottom of the door sill. This will give you the position of the *top edge* of the ledger strip. Stretch a chalk line along the wall of the house where you want to attach the ledger, using a string level to make sure the chalk line is perfectly horizontal. Remove the string level and 'snap a line' on the house.

Cut a 2 x 6 ledger strip exactly as long as the deck will be. Place the ledger even with the chalk mark and hold it in place. Use a carpenter's level to be sure the ledger is level, then attach it permanently to the house. To attach the ledger to a wood frame wall, use lag screws spaced to bite into the frame studs, as shown in Figure 2. To attach the ledger to a masonry wall, use expansive lead anchors every 2′, as shown in Figure 3.

270

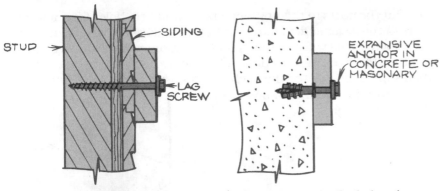

Figure 2. Secure the ledger to a wood frame wall by using lag screws. Make sure each screw sinks into a stud, otherwise the ledger won't support the weight of the deck.

Figure 3. Secure the ledger in concrete or masonry walls by using expansive lead anchors.

With strings and a string level, carefully calculate the length of the post on each pier. The posts must support the beams so that the top edge of the beams is level with the top edge of the ledger, as shown in Figure 4. Attach a string to the top edge of the ledger, and stretch it out over a pier. Use a string level to make sure that the string is perfectly horizontal. Measure the distance from the string to the pier. From this distance subtract the width of the beam (5½", for a 4 x 6), and the thickness of the metal post end (variable with the manufacturer). This will give you the length of the post for that pier. Repeat this process with other piers to find the length of all the posts.

7 **Set the posts on the piers.**

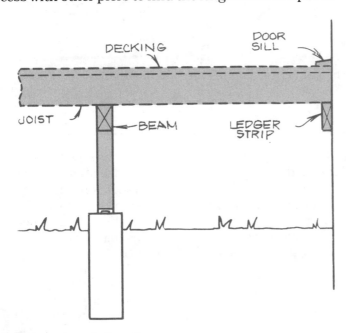

Figure 4. Cut the posts to length so that they will support the beams at the same level as the ledger strip. The ledger strip must be attached to the wall several inches under the door sill, so that when you assemble the deck, the top surface of the decking will be just under the sill.

Cut the posts to their proper length, and drill holes for anchor bolts in the lower end, making the holes $\frac{1}{16}''$ larger than the diameter of the bolts. This will give the wood some room to swell during wet weather. Place metal post ends on the piers, then slip the posts over the anchor bolts. (See Figure 5.) The anchor bolts will hold the posts upright.

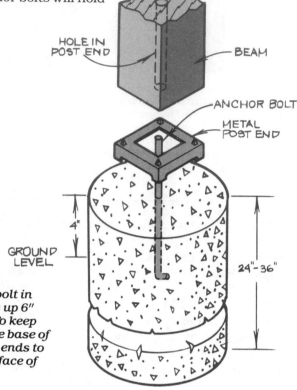

HOLE IN POST END

BEAM

ANCHOR BOLT

METAL POST END

4"

GROUND LEVEL

24"-36"

Figure 5. Set an anchor bolt in each pier so that it sticks up 6" into the end of the post. To keep water from rotting out the base of the posts, use metal post ends to keep the posts off the surface of the pier.

TIP   To bore deep holes in the ends of the posts, use an aircraft drill or a drill bit 'extender' and an auger bit.

8   **Attach the beams to the posts.**

Cut the 4 x 6 beams 144" long for this deck, as shown in the *Post and Beam Layout* drawing, or to stretch the width of your deck if you have altered the dimensions. If you can't get a 4 x 6 long enough, join two shorter beams with a lap joint. Plan this lap joint so that it occurs right over a post. Tack the beams in place, then use a carpenter's level to be sure they are level. Attach the beams to the posts with no-rust metal nailing straps, as shown in Figure 6.

NO-RUST METAL NAILING STRAP

Figure 6. Attach the beams to the posts using no-rust metal nailing strips.

Cut the joists from 2 x 8 stock. Place the joists on edge over the ledger strip and beams, as shown in the *Deck Frame, Side View* drawing. Be sure the joists are spaced 16″ on center, as shown in the *Deck Frame, Top View* drawing. Toenail the joists to the ledger strip and beams using 16d nails, or secure the joists in place using metal joist hangers. Cut a 2 x 8 facing strip the same length as the ledger strip. Place the facing strip over the ends of the joists at the front of the deck and nail to the joists using 12d nails.

9 Attach the joists and facing board to the beams.

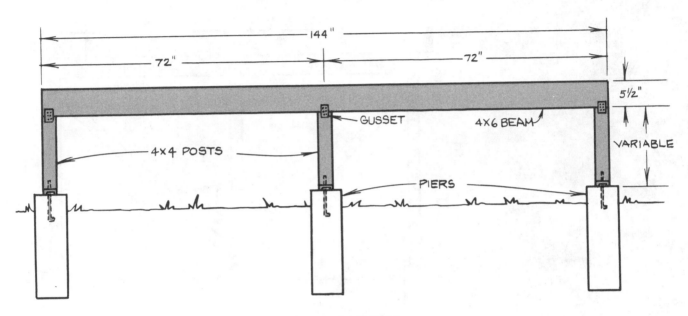

POST & BEAM LAYOUT

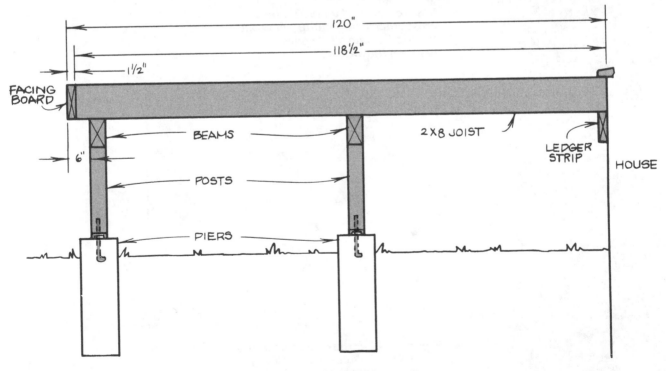

DECK FRAME, SIDE VIEW

10 **Add bracework, if necessary.**

Check your local building codes to determine whether you must install bracework. (A rule of thumb is that if a post is over 2' long, it's a good idea to brace it.) If necessary, cut 2 x 4's for 2 x 6's to fit diagonally from the bottom of one post to the top of another, as shown in Figure 7. Miter the ends of the bracework so that the end grain faces to one side. If the end grain faces up, water may collect, soak into the wood and rot it. (See Figure 8.) Attach the bracework to the posts, leaving a ¼" space between the braces, as shown, so that water doesn't collect there. Use ⅜" bolts or lag screws to attach the bracework to the posts.

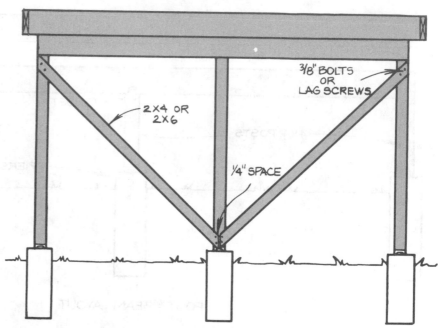

Figure 7. *Use 2 x 4's or 2 x 6's to brace the supporting posts, if the posts are over 2' long. Nail these crossbraces diagonally from post to post, leaving a ¼" space between the braces.*

Figure 8. *Miter the braces so that the end grain faces to one side. If it faces up, rainwater will soak into the brace and rot the wood.*

274

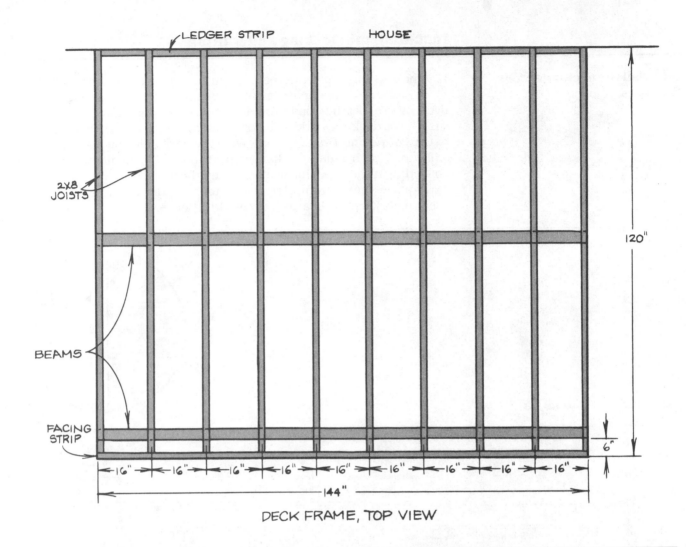

LEDGER STRIP   HOUSE

2x8 JOISTS

BEAMS

FACING STRIP

120"

6"

16" 16" 16" 16" 16" 16" 16" 16" 16"

144"

DECK FRAME, TOP VIEW

Cut spacer blocks from 2 x 8 lumber to provide a nailing surface for the first row of decking. Fit the spacer blocks between the joists and place them flat against the house. The spacer blocks must be flush with the top of the joists, as shown in Figure 9. Attach the spacer blocks to the joist and the house with 16d nails. If you're attaching the deck to a masonry wall, use masonry nails.

11 **Attach spacer blocks to the house and joists.**

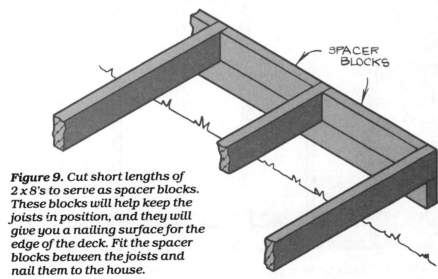

SPACER BLOCKS

Figure 9. *Cut short lengths of 2 x 8's to serve as spacer blocks. These blocks will help keep the joists in position, and they will give you a nailing surface for the edge of the deck. Fit the spacer blocks between the joists and nail them to the house.*

Installing the Decking and Railing

12 Nail the decking across the joists.

Lay the decking in place perpendicular to the joists, as shown in the *Deck Assembly, Top View* drawing. Leave a ½″ space between the decking to allow the wood to swell without warping in wet weather. Attach the decking 'bark' side down, as explained in the introduction to the *Patios and Decks* section. With the bark side down, the wood will cup in such a manner that the rainwater will run off instead of collecting in the cup. This will help keep the wood from rotting. Use 12d square-shanked spiral nails to attach the decking to the joists—these nails keep the boards from working loose.

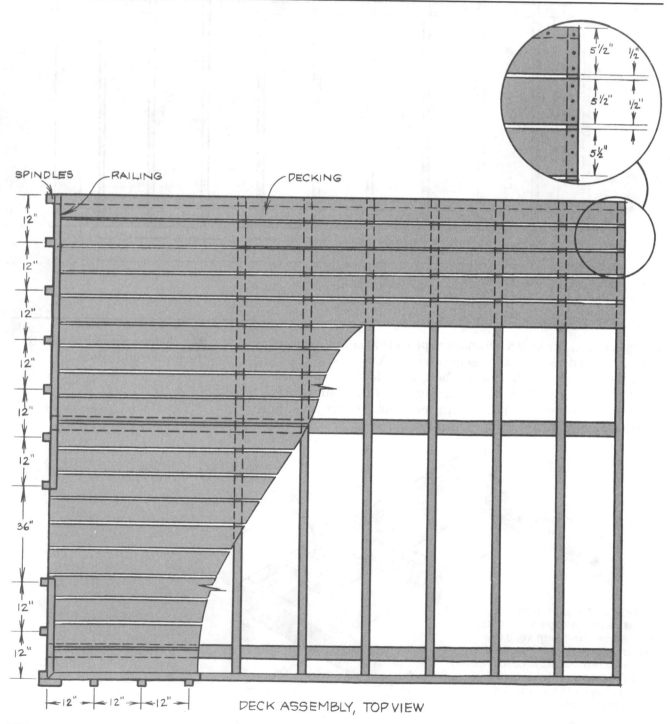

DECK ASSEMBLY, TOP VIEW

Cut the decking to length *after* you've nailed it in place. While you're installing it, let the boards stick out 1"–3". Then snap a chalk line even with the outside edge of the end joists, and cut along the mark with a circular saw. (See Figure 10.)

13   Cut the decking flush with the joists and facing.

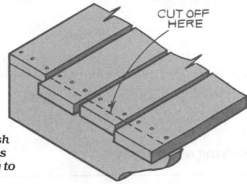

Figure 10. Cut the decking flush with the joists and facing strips after you've nailed the decking to the frame.

Cut the spindles from 2 x 2 stock and miter the top and bottom ends at 45°, so that water will run off the wood, as shown in the *Railing Detail* drawing. To make the railing turn a corner, make an L-shaped spindle from a 2 x 4 and a 2 x 2, as shown in Figure 11. Double-miter the ends of the corner spindles, so that they slope in two directions. Then, cut 2 x 4 stock for rails, mitering the ends where the rails come together in the corners.

14   Cut the spindles and rails.

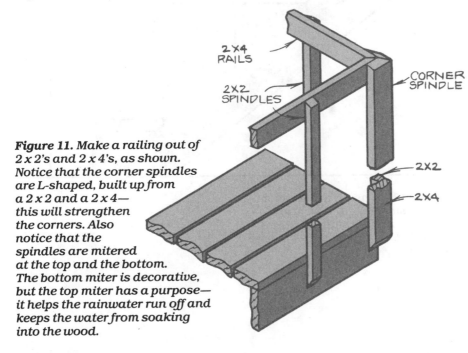

Figure 11. Make a railing out of 2 x 2's and 2 x 4's, as shown. Notice that the corner spindles are L-shaped, built up from a 2 x 2 and a 2 x 4— this will strengthen the corners. Also notice that the spindles are mitered at the top and the bottom. The bottom miter is decorative, but the top miter has a purpose— it helps the rainwater run off and keeps the water from soaking into the wood.

15 Attach the spindles and rails.

Lay out the spindles for one side of the deck, inside surface up, on the deck. Lay the railing over the spindle, making sure the top edge of the rail is flush with the top ends of the spindles. Nail the rail to the spindles with 12d nails. With a helper, lift this railing assembly into place, and attach the bottom of the spindles to the outside joists or facing strips. If you need to, drill through the spindles to prevent the wood from splitting. Repeat for the other open sides of the deck. Be sure to leave an opening for an entrance onto the deck. When all the railing assemblies are in place, attach the railing to the house with framing anchors.

TIP If you want to really finish off the deck and hide the end grains of the decking, nail a 1 x 4 molding strip in between the spindles all along the edge of the deck.

Making a Stairway

Unless your deck sits very close to the ground, you'll need one or more stairways to get from the deck to the yard. To calculate the rise and run of a stairway, first measure the total rise (the distance from the ground to the deck), as shown in Figure 12.

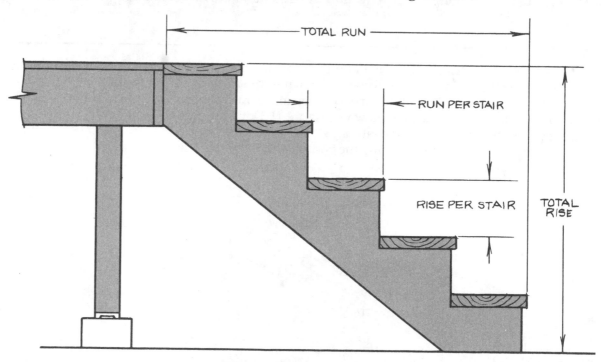

Figure 12. Measure the total rise of your stairway, and arbitrarily choose a run per stair. Then use these figures to calculate the rise per stair.

278

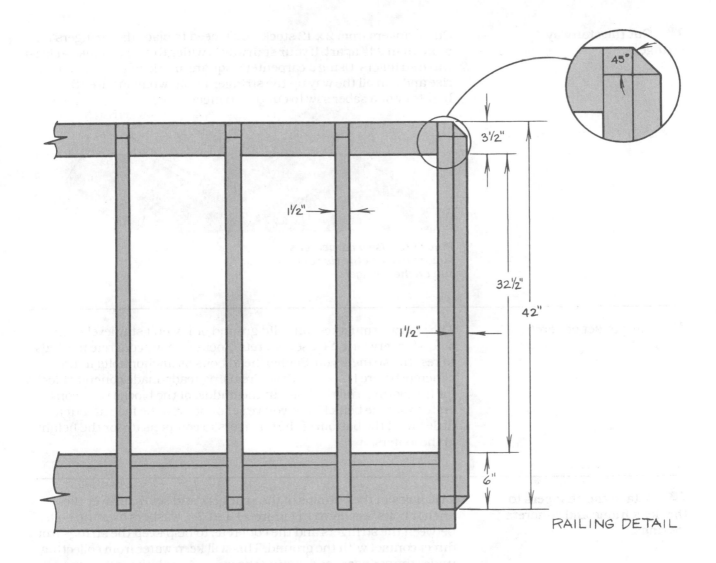

RAILING DETAIL

Then pick a measurement for the run (the horizontal distance of each step—for outdoors it should be no less than 10½"). To arrive at the number of steps you will need, divide the total rise by 7 and round it off to get an integer. This integer is the number of steps in your staircase. Divide the number of steps into the total rise again to get the rise of each step. The run of the stairs should be in proportion to the rise. The general rule is that the rise times the run in inches should equal 70 to 75. Here are the formulas for the math you just performed:

$$\frac{\text{Total Rise}}{7} = X.y$$

If y < .5, round down so that X = Number of Stairs
If y > or = .5, round up so X + 1 = Number of Stairs

$$\frac{\text{Total Rise}}{\text{Number of Stairs}} = \text{Rise Per Stair}$$

Rise per Stair x Run per Stair > or = 70
Rise per Stair x Run per Stair < or = 75

Check your local building code for restrictions that may apply to stairways in your area.

16 Cut the stairway stringers.

Cut stringers from 2 x 12 stock. You'll need to place the stringers no more than 24" apart. If your stairway is wider than 24", make at least three stringers. Using a carpenter's square, mark off the cuts for the rise and run all the way up the stringer, as shown in Figure 13. Use a handsaw or a saber saw to cut the stringers.

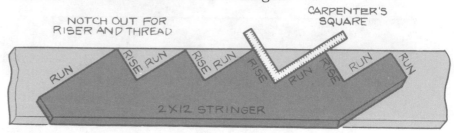

Figure 13. *Use a carpenter's square to mark the rise and the run on the stringers.*

17 Pour or set concrete blocks.

The stairway must rest on solid ground or it won't stay level—and safe—for very long. So use concrete blocks or pour concrete footings to rest the stringers on. Set hex head bolts as anchor bolts in the concrete before it sets up. (If you're using ready-made concrete blocks for the footings, fill the holes in the middle of the blocks with concrete to set the bolts.) Once you've set or poured the footing, cut a little bit off the bottom of the stringers to compensate for the height of the footers.

18 Attach the stringers to the deck frame and concrete footings.

Drill holes in the bottoms of the stringers and set them over the anchor bolts, as shown in Figure 14. Put 3-4 washers over the bolts, between the stringers and the concrete, to help keep the strings from direct contact with the ground. This will keep water from collecting under the stringers and rotting the wood. Attach the stringers to the deck frame using metal framing anchors, as shown in Figure 15.

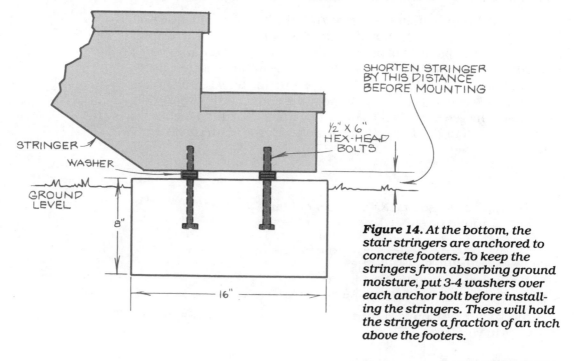

Figure 14. *At the bottom, the stair stringers are anchored to concrete footers. To keep the stringers from absorbing ground moisture, put 3-4 washers over each anchor bolt before installing the stringers. These will hold the stringers a fraction of an inch above the footers.*

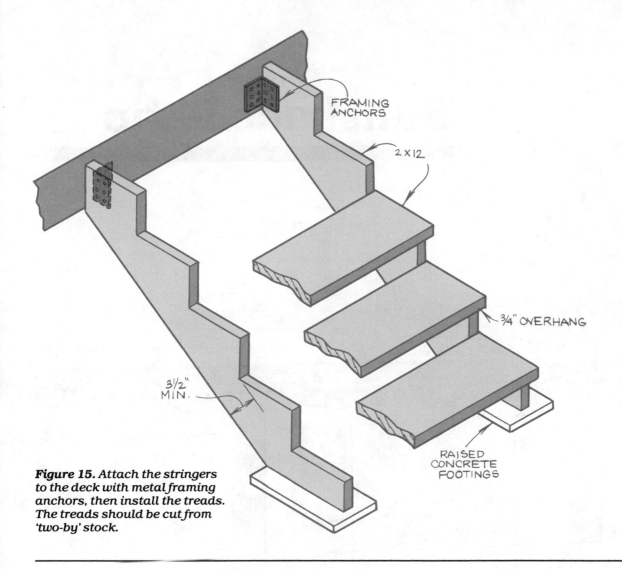

FRAMING
ANCHORS

2 × 12

3/4" OVERHANG

3½"
MIN.

RAISED
CONCRETE
FOOTINGS

Figure 15. *Attach the stringers to the deck with metal framing anchors, then install the treads. The treads should be cut from 'two-by' stock.*

Cut treads from 2 x 10 or 2 x 12 stock (depending on the run per stair) and lay them across the stringers. Attach the treads to the stringers using 16d nails.

19　Nail the treads to the stringers.

Sunshade Patio

Delicate patterns of filtered light create a new dimension in outdoor living when you sit under the sunshade roof of this patio. It not only provides shade and ventilation, but this roof is a perfect trellis for vining plants. By adding vines as a natural cover, you can create a cozy hideaway for yourself, your family, and friends. If you wish, add trellises to the sides to close in the hideaway.

As shown, this patio cover is attached to the house, but you can build it to stand by itself by building two freestanding walls, as outlined in the *Stand-Alone Carport* chapter. Attach this cover to an existing wood deck or a concrete slab, or build it over one of the deck plans given in this section. Adjust the spacing between the slats to allow more (or less) light to filter through—whatever suits your preferences.

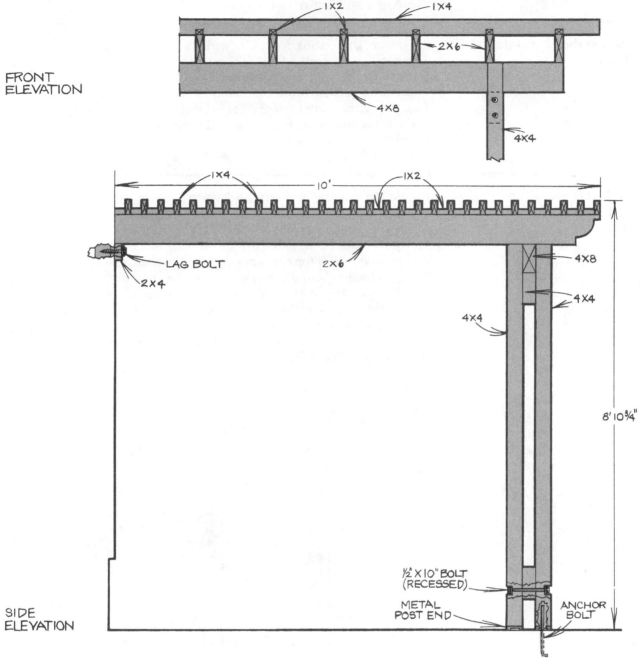

Materials

To build the sunshade patio as shown, use 4 x 4's for posts, a 4 x 8 for the beam, 2 x 6's for the joists and spacers, and a 2 x 4 for the ledger strip. Purchase 1 x 4 slats for the roof. Use pressure-treated lumber for all these parts, since they are exposed to the weather.

In addition to these materials, purchase galvanized nails and lag bolts. If you're attaching the ledger strip to a masonry wall, you'll also need expansion bolts. Select metal post anchors if you will be attaching the posts to a wood deck, or post ends if setting the posts on a concrete pad or footers.

Note: If you will be using your sunshade roof as an arbor, make sure your lumber hasn't been treated with chemicals that can kill vines. Check with your local lumberyard dealer for appropriate woods available in your area, or use heart cedar, redwood, or other "naturally" treated lumbers.

1 **Adjust the size of the sunshade.**

As shown, this patio cover is 10' wide, 8' high, and stretches 12' long. However, you can adjust any of these dimensions to cover an existing patio or to suit your needs. Simply alter the length of the beam, joists, or posts as needed. But if you extend the span of the joists or the beam, you may have to use larger lumber. The maximum span for a 4 x 8 beam (under load) is 12', and the maximum span of a 2 x 6 (on edge) is 10'.

2 **Adjust the spacing between the slats for your latitude.**

Adjust the spacing between the slats to provide the maximum amount of shade on hot summer days. To do this, you'll have to do a little sketching on a piece of paper. First, find the latitude for your area on a map. Then sketch a 1 x 4 on edge, sitting on a flat surface. Draw a line from a top corner of the 1 x 4 to the flat surface. The angle between the line and the surface should be the same as your latitude. (See Figure 1.) If your latitude is 45°, then draw the line at 45° to the flat surface. Measure the distance between the 1 x 4 and where the line meets the surface. (For 45° of latitude, this distance will be 3½" —the same as the width of the 1 x 4.) This is the maximum spacing that you should use for the slats.

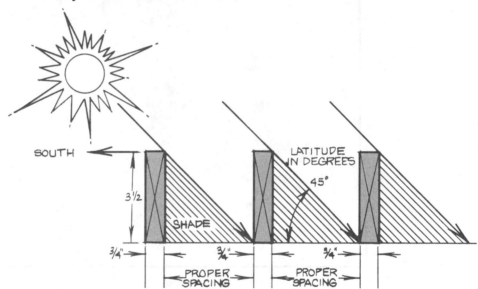

Figure 1. To calculate the proper spacing of the slats for your area, draw a picture of a 1 x 4 on edge, resting on a flat surface. Then draw a line from the upper corner of the 1 x 4 to the surface. The angle between the line and the surface should equal your latitude, in degrees. The distance from the 1 x 4 to the point where the line meets the surface is the maximum spacing between the slats for your area.

Building the Freestanding Wall

Use stakes and string to lay out the locations of your posts. Locate each post as shown in the *Post Layout* drawing. The posts must support the beam so that it is parallel with the side of the house, so measure carefully. Check to be sure the layout is square by measuring diagonally from corner to corner. AD should be equal to BC, as shown in Figure 2.

3 Lay out the supporting posts.

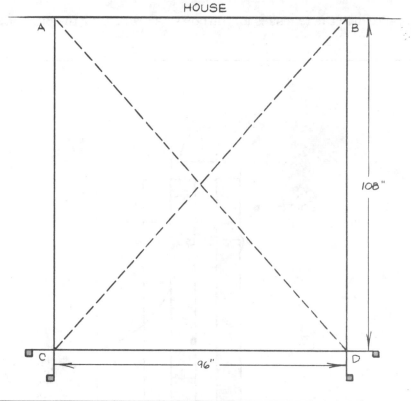

Figure 2. Lay out the location of the support posts using stakes and string. Check to be sure the lines are in square by measuring diagonally from corner to corner. AD should equal BC.

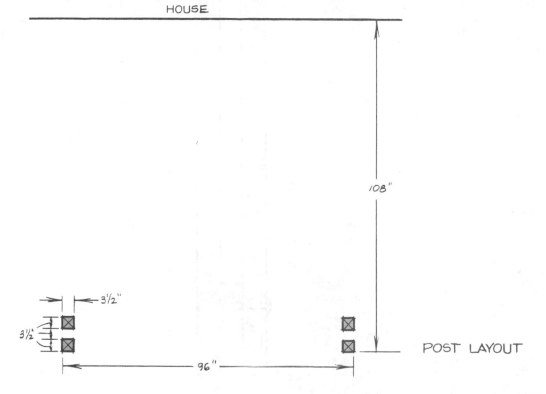

4 **Build the posts.**

To support the load of the beam and the sunshade roof, make the posts from two 4 x 4's, as shown in the *Post, Side View* drawing. Put two spacers in between each 4 x 4, and bolt the parts together with ½" x 10" bolts. Recess the head and the nut on each bolt in the wood, as shown in the working drawings.

TIP   It's easiest to drill the recesses for the bolts *before* drilling the pilot holes.

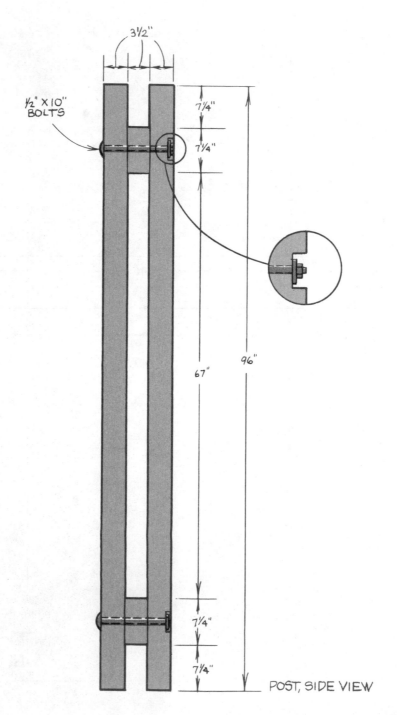

3½"

½" X 10"
BOLTS

7¼"

7¼"

96"

67"

7¼"

7¼"

POST, SIDE VIEW

To attach the posts to an existing wood deck: Set metal post anchors on the deck beam where the posts will be anchored—you'll need two per post. Make sure the anchors are in the proper position, then attach them to the deck with lag bolts and fender washers, as shown in Figure 3.

To attach the posts to concrete: Set anchor bolts in the wet concrete. Once again, you'll need two per post. Drill holes for the anchor bolts in the bottom of the posts, making sure the holes are 1/16″ larger than the diameter of the bolts. This will give the wood some room to swell during wet weather. Place metal post ends over the anchor bolts, then place the posts on the post ends. (See Figure 4.)

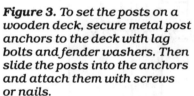

Figure 3. To set the posts on a wooden deck, secure metal post anchors to the deck with lag bolts and fender washers. Then slide the posts into the anchors and attach them with screws or nails.

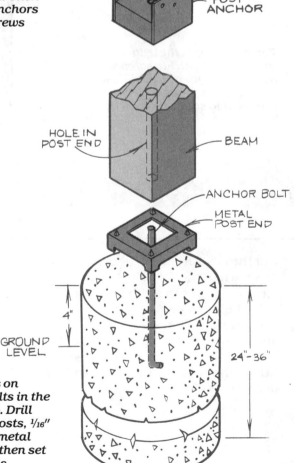

Figure 4. To set the posts on concrete, cast anchor bolts in the concrete before it sets up. Drill holes in the ends of the posts, 1/16″ larger than the bolt. Put metal post ends over the bolts, then set the posts on the post ends.

6 Set the beam in the posts.

Cut a 4 x 8 beam 126″ long, and place it between the 4 x 4's that make up each post. The beam will rest on the upper spacer. Adjust the beam so that it overhangs each post by 15″, as shown in the *Wall Frame, Front View* drawing. Use a carpenter's level to make sure the posts are plumb, then brace them upright. Bolt the beam in place with ½″ x 10″ bolts. Recess the bolt heads and the nuts as you did when you assembled the posts.

Building the Sunshade Roof

7 Attach the ledger strip to the house.

Cut a 2 x 4 ledger strip exactly as long as the beam. Snap a chalk-line across the house to mark the position of the ledger, then lift the ledger in place flat against the house, as shown in the *Sunshade Frame, Side View* drawing. Attach the ledger to the house using lag screws. (See Figure 5.) These screws must bite into the frame studs of the house so that the ledger strip can support the weight of the roof.

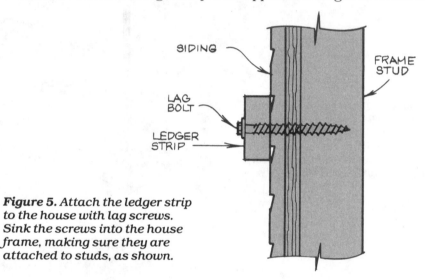

Figure 5. Attach the ledger strip to the house with lag screws. Sink the screws into the house frame, making sure they are attached to studs, as shown.

TIP If you are attaching your ledger to a masonry wall, use ½″ lead expansion shields and bolts. Space the expansion shields and bolts 24″ on center.

8 Attach the joists and spacer blocks.

Cut the joists from 2 x 6 stock, and shape the ends with a saber saw, as shown in the *Joist End Detail* drawing. Set the joists on the ledger strip and the beam, with the shaped end overhanging the beam by 12″. The joists should be spaced 18″ on center, as shown in the *Sunshade Frame, Top View* drawing. Toenail the joists to the ledger strip and beam with 12d nails. Next, cut 2 x 6 'spacers' and nail them between the joists above the ledger strip, as shown in Figure 6. Attach the blocks to the house with nails. These blocks will help to hold the 2 x 6 on edge.

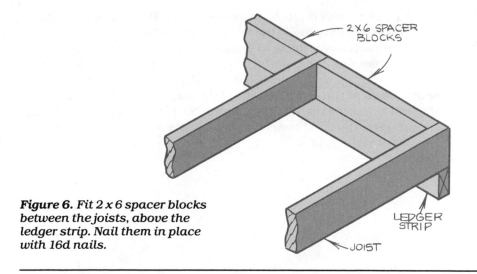

Figure 6. Fit 2 x 6 spacer blocks between the joists, above the ledger strip. Nail them in place with 16d nails.

2X6 SPACER BLOCKS

LEDGER STRIP

JOIST

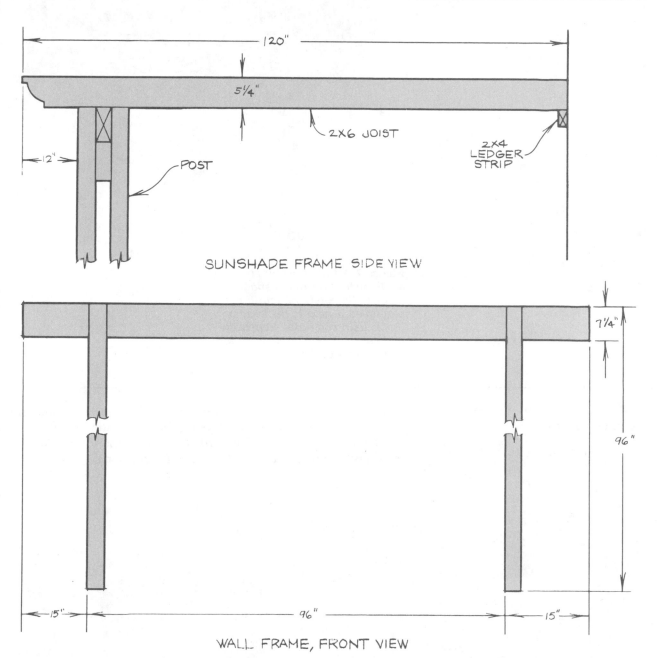

120"

5¼"

12"

2X6 JOIST

POST

2X4 LEDGER STRIP

SUNSHADE FRAME SIDE VIEW

7¼"

96"

15"

96"

15"

WALL FRAME, FRONT VIEW

9 **Attach the slats and small spacer blocks.**

Cut 1 x 4 slats to fit on edge across the joists. The slats should be 144″ long—they overhang the joists by 9″ each side. Next, snap chalklines across the joists to mark the exact location of the slats. Cut the spacer blocks from 1 x 2 stock—they should be as long as the distance you figured between slats. To attach the slats, first nail a spacer block to the joist, then stand the slat on edge and toenail it to the joist with 6d nails, as shown in the *Slat Detail* drawing. Repeat the process until you have covered the joists with slats and spacers. (See Figure 7.)

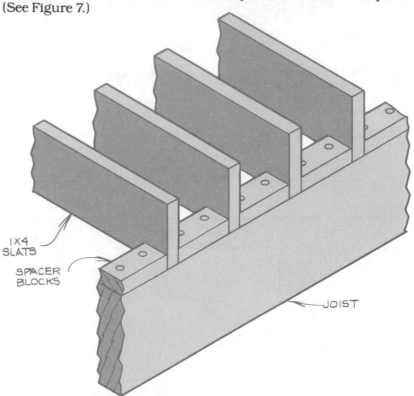

1X4 SLATS

SPACER BLOCKS

JOIST

Figure 7. Attach the slats to the joists by first nailing a spacer block to the joist, then nailing the slat on edge next to the spacer block, using 6d nails. Repeat the process until you have attached all the slats and spacers.

---

10 **Paint or stain all exposed wood surfaces.**

Paint or stain all exposed wood surfaces, if desired, to protect them from the weather. If you will be using the sunshade roof as an arbor, check with your local lumberyard dealer or nurseryman to be sure the paint or stain you are using is not harmful to plants.

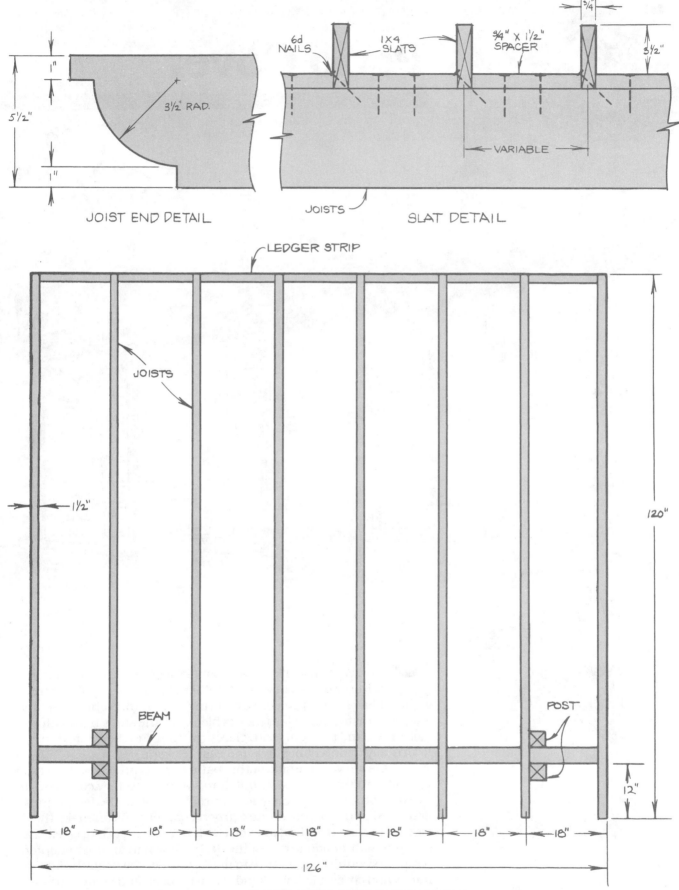

JOIST END DETAIL

1"

5½"

1"

3½" RAD.

6d NAILS

1X4 SLATS

¾" X 1½" SPACER

¾"

3½"

JOISTS

VARIABLE

SLAT DETAIL

LEDGER STRIP

JOISTS

1½"

120"

BEAM

POST

12"

18"   18"   18"   18"   18"   18"   18"

126"

SUNSHADE FRAME, TOP VIEW

Patio Cover

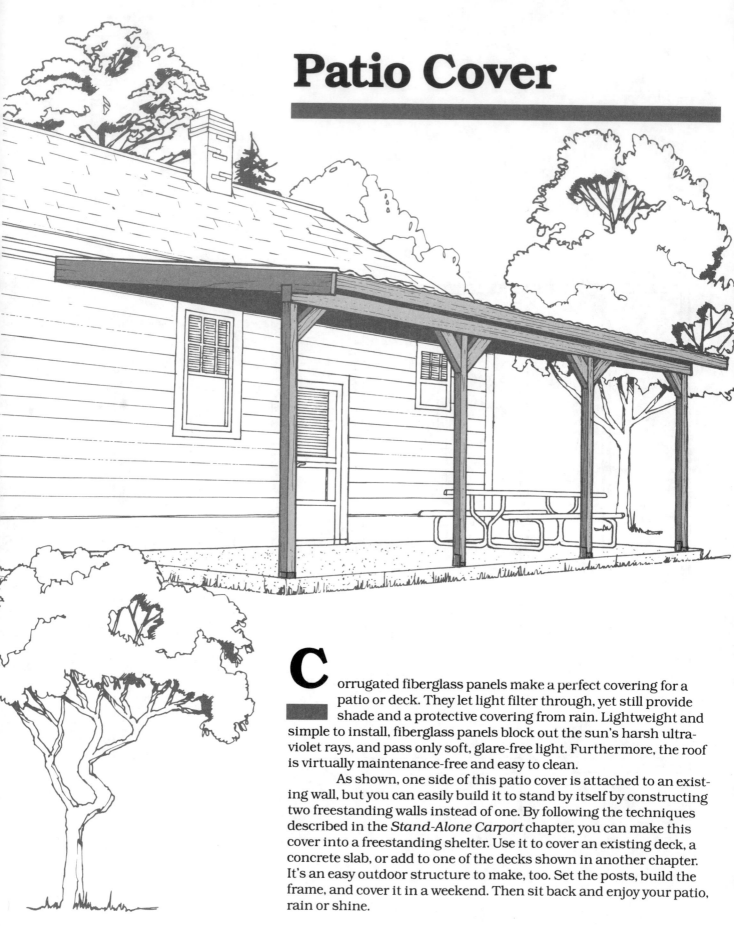

Corrugated fiberglass panels make a perfect covering for a patio or deck. They let light filter through, yet still provide shade and a protective covering from rain. Lightweight and simple to install, fiberglass panels block out the sun's harsh ultraviolet rays, and pass only soft, glare-free light. Furthermore, the roof is virtually maintenance-free and easy to clean.

As shown, one side of this patio cover is attached to an existing wall, but you can easily build it to stand by itself by constructing two freestanding walls instead of one. By following the techniques described in the *Stand-Alone Carport* chapter, you can make this cover into a freestanding shelter. Use it to cover an existing deck, a concrete slab, or add to one of the decks shown in another chapter. It's an easy outdoor structure to make, too. Set the posts, build the frame, and cover it in a weekend. Then sit back and enjoy your patio, rain or shine.

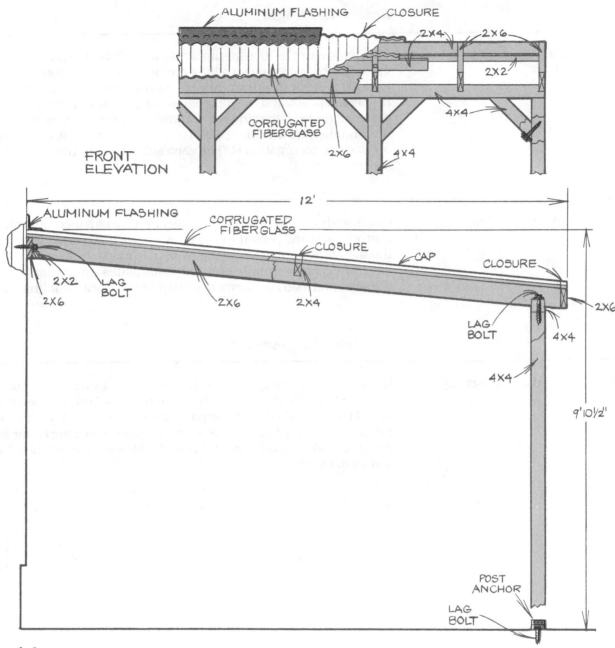

FRONT ELEVATION

SIDE ELEVATION

Materials

To build the patio cover as shown, use 4 x 4's for the posts, the top plate, and the braces, and 2 x 6's for the rafters and the header. The ledger strip is a 2 x 2, and the facing strip is cut from 1 x 6 stock. You can use pressure-treated wood for the posts if you want, but it's not really necessary—the posts do not contact the ground. Use untreated lumber for all the other parts; this will save you money.

Use corrugated fiberglass paneling (standard width is 26″), corrugated redwood closure strips and caps, and corrugated flashing for the roof. In addition to these materials, you'll also have to purchase clear silicone sealant, aluminum nails with neoprene washers, and galvanized nails. Select metal post anchors if you will be attaching the posts to a wood deck, or post ends if setting the posts on footers or piers. Use lag screws to attach the header and ledger strip to wood frame walls or expansion bolts to attach the assembly to concrete or masonry.

1 **Adjust the size of your patio cover.**	This patio cover is 10' wide, approximately 8' high, and stretches 12' long; however, you can adjust any of these dimensions to cover an existing patio or to suit your needs. Simply alter the position and the number of posts, and the dimensions of the frame. If you change the dimensions, keep in mind the standard size for fiberglass panels is 26" and they are available in 8', 10' or 12' lengths. The panels should overlap one corrugation (1") on each side, so space the rafters 24" on center.
2 **Check the building codes.**	Check with your local building inspections department for information on codes and permits, before you begin. If you have decreased the size of the patio cover, you may be able to use smaller (and therefore less expensive) lumber for the posts, plates, and rafters. Check the span and load requirements listed in your local building code.

Building the Freestanding Wall

3 **Lay out the supporting posts.**	Use stakes and string to lay out the locations of your posts. Locate each post exactly 4' apart, as shown in the *Post Layout* drawing. This line of posts must be precisely parallel with the side of the house, so measure carefully. Check to be sure the layout is square by measuring diagonally from corner to corner. AD should be equal to BC, as shown in Figure 1.

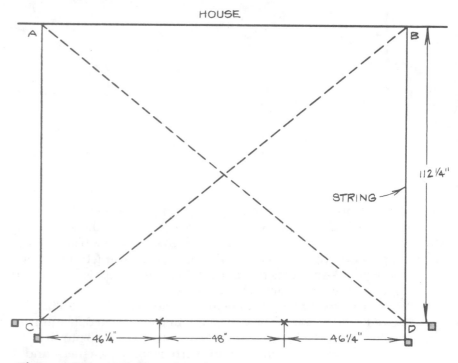

Figure 1. Lay out the location of the support posts using stakes and string. Check to be sure the layout is square by measuring diagonally from corner to corner. AD should equal BC.

To attach the posts to an existing wood deck: Set metal post anchors on the deck beam where the posts will be anchored. Make sure the metal post anchor is in the proper position, then attach it to the deck using a lag bolt and fender washer, as shown in Figure 2.

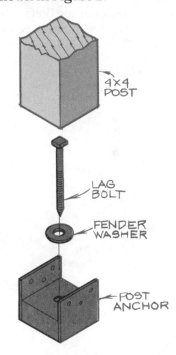

Figure 2. To set the posts on a wooden deck, attach metal post anchors to the deck with lag screws and fender washers. Then slide the posts into the anchors and attach the posts with screws or 16d nails.

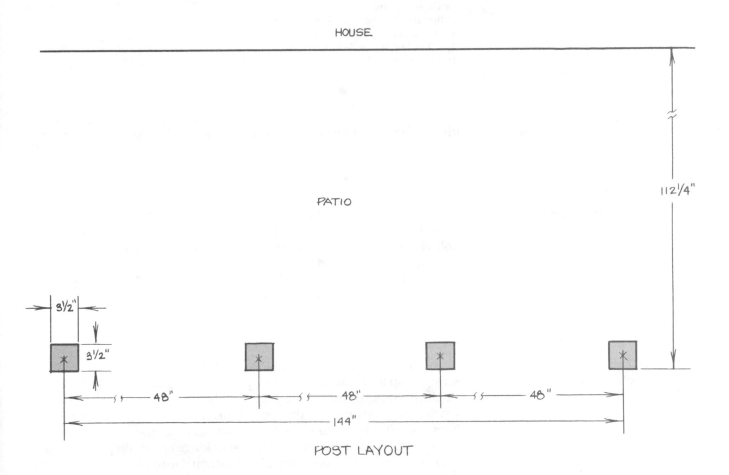

To attach the posts to concrete footers: Set anchor bolts in the footers, before the concrete cures. These anchor bolts should protrude at least 6″ from the footers. Drill holes for the anchor bolts in the bottom of the posts, making sure the holes are 1/16″ larger than the diameter of the bolts. This will give the wood some room to swell during wet weather. Place metal post ends on the footers, as shown in Figure 3.

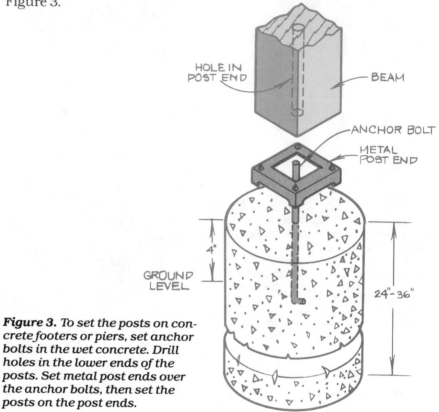

Figure 3. To set the posts on concrete footers or piers, set anchor bolts in the wet concrete. Drill holes in the lower ends of the posts. Set metal post ends over the anchor bolts, then set the posts on the post ends.

TIP To bore deep holes in the ends of the posts, use an aircraft drill, or a drill bit 'extender' and an auger bit.

---

5 **Set and brace the posts.**

On a wood deck, slide the posts into the metal post anchors and attach with screws or 16d nails. On concrete footers, fit the anchor bolts into the holes in the ends of the posts, and set the posts in place over the metal post ends. The post anchors or anchor bolts will hold the posts upright. Use a carpenter's level to be sure the posts are plumb, then brace them with stakes and temporary braces. Use two braces for each post.

---

6 **Cut the tops of the posts level.**

With a level, check the posts again to be sure they are perfectly straight up and down. If necessary, adjust the braces that hold them in place. Mark the top of one post 96″ above the ground. Using this mark as a reference, find the tops of the other posts with a string level. (See Figure 4.) Take the posts down and cut them off with a handsaw. Be sure to mark which post belongs on which post anchor or bolt! Put the posts back up and re-attach the temporary braces.

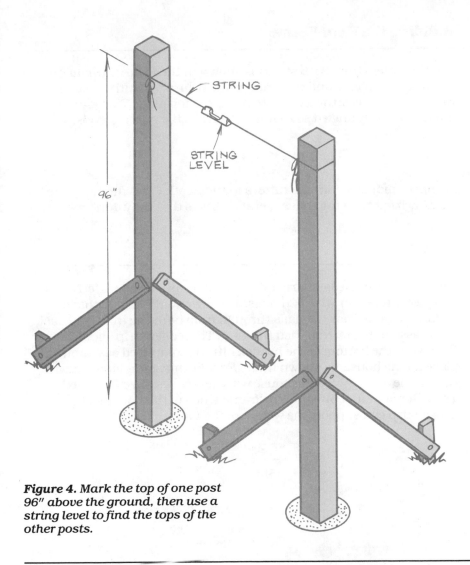

Figure 4. *Mark the top of one post 96" above the ground, then use a string level to find the tops of the other posts.*

Cut a top plate to the proper length. If you can't get a 4 x 4 long enough, you will have to join two shorter boards with a lap joint. Plan this lap joint so that it occurs right over a post. (See Figure 5.) Put the top plate in place, as shown in the *Wall Frame Layout* drawing. Attach the top plate to the posts with lag screws. Cut the braces from 4 x 4 stock, mitering the ends at 45°. Attach the braces to the posts and the top plate with lag screws.

Figure 5. *If you can't buy a 4 x 4 long enough for the top plate, join two shorter boards with a lap joint. Plan the joint so that it will be directly over a post.*

8 **Cut the rafters.**

Cut the rafters from 2 x 6 stock, as shown in the *Rafter Layout* drawing. Miter each end, and cut a notch and a 'bird's mouth' in each rafter so it fits over the ledger strip and the top plate. This joinery must be slightly angled so that it's square to the mitered ends.

TIP Instead of cutting notches and bird's mouths, join the rafters to the header and the top plate with joist hangers and rafter ties. The drawback to using these metal plates is that they don't look as good as plain wood.

9 **Attach the header and ledger strip to the house.**

Cut a header and ledger strip exactly as long as the top plate. Nail the two pieces together with 12d nails, as shown in the *Ledger Strip Detail* drawing. Pass the nails through from both the front and back of the assembly to strengthen the joint. The ledger strip should be flush with the bottom of the header. Lift the completed assembly in place on the house, as shown in the *Roof Frame, Side View* drawing. Attach the assembly to the house with lag screws. (See Figure 6.) These screws must bite into the frame studs of the house so that the ledger strip can support the weight of the roof.

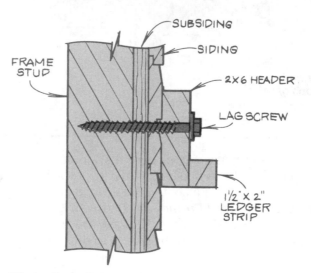

Figure 6. Attach the header and ledger strip assembly to the house with lag screws. Sink the screws into the frame studs, as shown.

TIP If you are attaching your cover to a masonry wall, use ½″ lead expansion shields and bolts. Space the expansion shields and bolts 24″ on center.

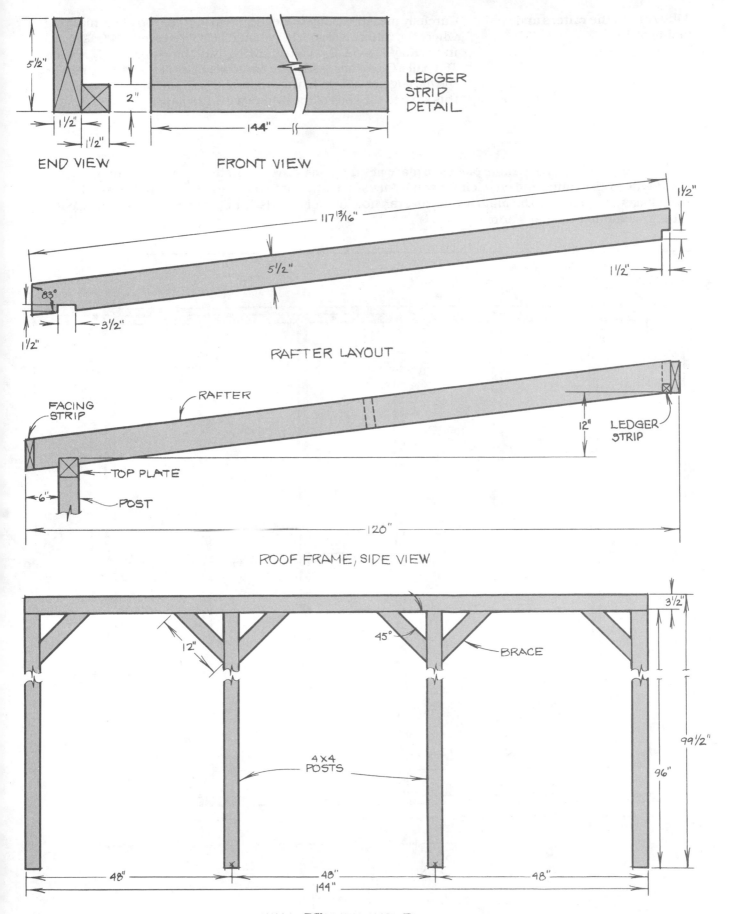

5½"

2"

1½"

1½"

END VIEW

144"

FRONT VIEW

LEDGER
STRIP
DETAIL

117¹³⁄₁₆"

5½"

1½"

1½"

83°

3½"

1½"

RAFTER LAYOUT

FACING
STRIP

RAFTER

12"

LEDGER
STRIP

TOP PLATE

6"

POST

120"

ROOF FRAME, SIDE VIEW

3½"

12"

45°

BRACE

4 X 4
POSTS

99½"

96"

48"

48"

48"

144"

WALL FRAME LAYOUT

10 **Attach the rafters and facing strip.**

Carefully measure along the ledger strip and the top plate, marking where the rafters will go. They should be spaced every 24″, as shown in the *Roof Frame, Top View* drawing. Nail the rafters in place with 16d nails. Cut a facing strip for the lower ends of the rafters, beveling the top edge, as shown in the *Facing Strip Detail* drawing. Attach this strip to the ends of the rafters with 12d nails.

TIP Just tack the roof frame pieces in place at first; don't drive the nails all the way into the wood. Build the entire frame and check it for squareness. If a rafter or two has to be realigned, it will be much easier to remove the nails if you can grab hold of their heads. When you are sure the frame is square, hammer the nails home.

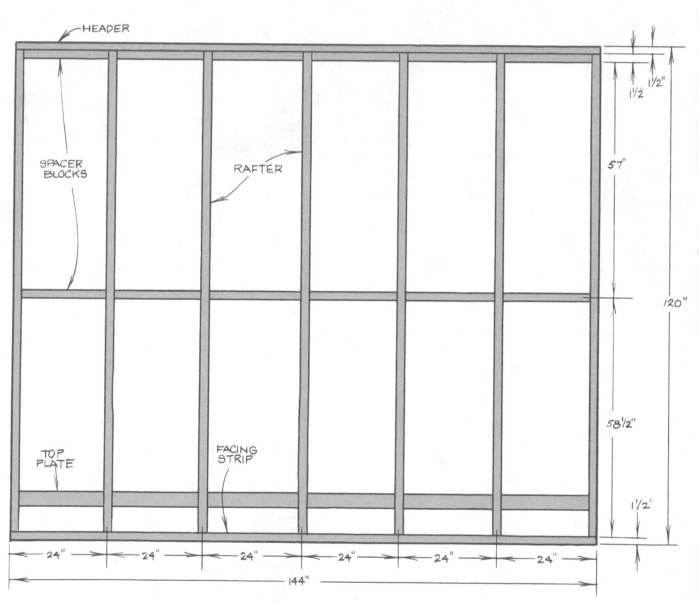

ROOF FRAME, TOP VIEW

Cut 2 x 4 'spacers' and nail them between the rafters with 16d nails. Fit the topmost row of spacers over the ledger strips and nail them to the header, as shown in Figure 7. Attach the other row of spacers between the rafters, about halfway along their length. The top edge of the spacers must be flush with the top edge of the rafters.

11 **Attach the spacer blocks.**

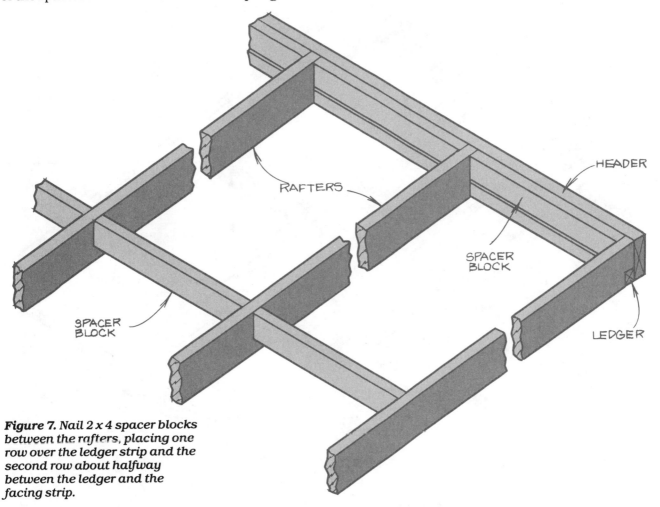

Figure 7. *Nail 2 x 4 spacer blocks between the rafters, placing one row over the ledger strip and the second row about halfway between the ledger and the facing strip.*

TIP To fit the spacers over the heads of the lag screws, drill 1″ holes, ½″ deep in the faces of the spacers.

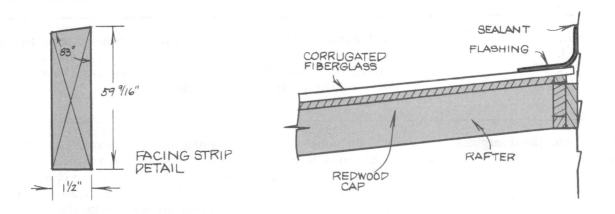

12 **Attach the closure strips and caps.**

Attach 1½" wide corrugated redwood closure strips to the spacers and the facing strip, as shown in Figure 8. Nail half-round redwood molding caps to the rafters.

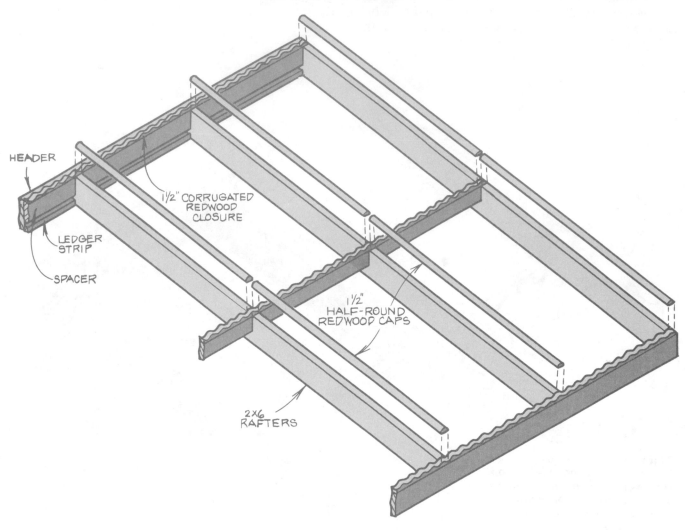

HEADER

1½" CORRUGATED REDWOOD CLOSURE

LEDGER STRIP

SPACER

1½" HALF-ROUND REDWOOD CAPS

2X6 RAFTERS

Figure 8. Attach corrugated redwood closure strips and caps to the rafters and spacer blocks to properly support the fiberglass panels on the roof frame.

Finishing Up

13 **Paint all wood surfaces, if desired.**

Paint or stain all exposed wood surfaces, if desired, to protect them from the weather. It's easier to paint them now, before you install the fiberglass panels, than to wait until afterwards and have to worry about getting paint on the fiberglass.

14 **Install the fiberglass panels.**

Lift the panels into place, one at a time. Position each panel so that it overhangs the rafters 1" on either side. The corrugation of the panel must match up exactly with the corrugations of the closure. Nail the panel in place with aluminum nails and neoprene washers, but *don't* put nails where the next panel will overlap the panel you're working on. When you're finished nailing, put a bead of clear silicone along the edge of the panel where the next panel will overlap. Put the next panel in place and put a few nails through *both* panels, where they overlap. Repeat the process until you've covered the entire roof. (See Figure 9.)

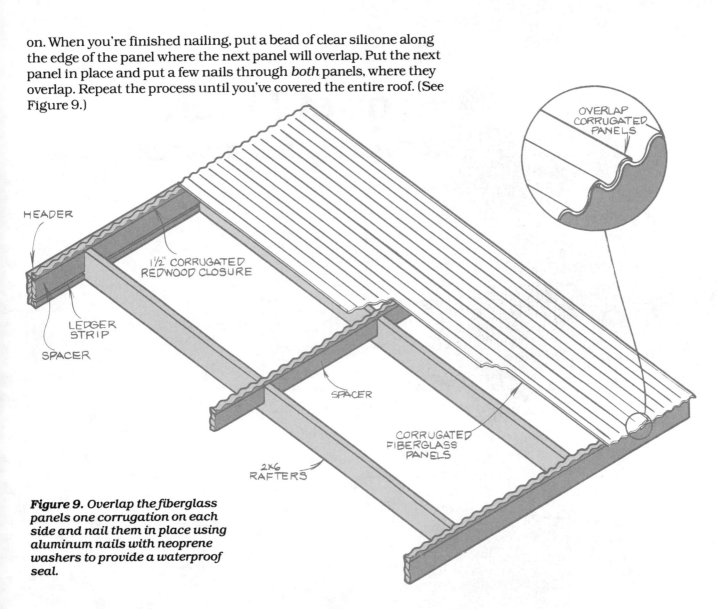

OVERLAP CORRUGATED PANELS

HEADER

1½" CORRUGATED REDWOOD CLOSURE

LEDGER STRIP

SPACER

SPACER

CORRUGATED FIBERGLASS PANELS

2×6 RAFTERS

Figure 9. Overlap the fiberglass panels one corrugation on each side and nail them in place using aluminum nails with neoprene washers to provide a waterproof seal.

TIP To cut the panels cleanly, use a plywood blade on your circular saw and *reverse* the blade, so that the teeth are facing in the opposite direction of rotation.

15 **Install the flashing.**

Install corrugated flashing over the fiberglass sheets, where the roof meets the wall of the house. Seal the upper edge of the flashing with mastic or sealant to keep out the rainwater, as shown in the *Roofing Detail* drawing.

Parquet Deck

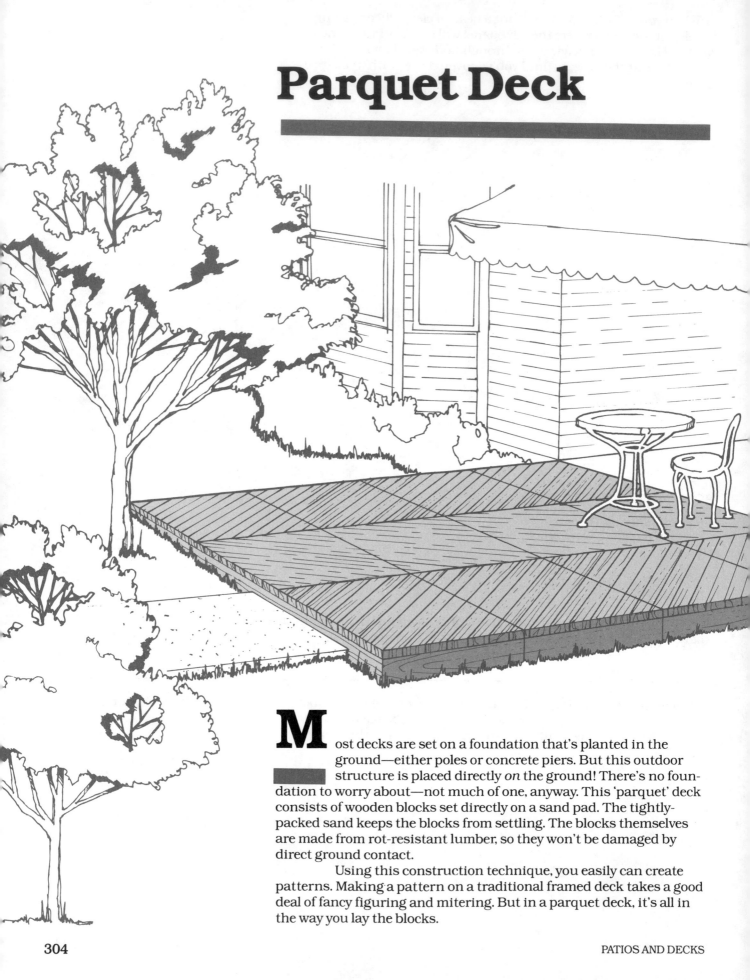

Most decks are set on a foundation that's planted in the ground—either poles or concrete piers. But this outdoor structure is placed directly *on* the ground! There's no foundation to worry about—not much of one, anyway. This 'parquet' deck consists of wooden blocks set directly on a sand pad. The tightly-packed sand keeps the blocks from settling. The blocks themselves are made from rot-resistant lumber, so they won't be damaged by direct ground contact.

Using this construction technique, you easily can create patterns. Making a pattern on a traditional framed deck takes a good deal of fancy figuring and mitering. But in a parquet deck, it's all in the way you lay the blocks.

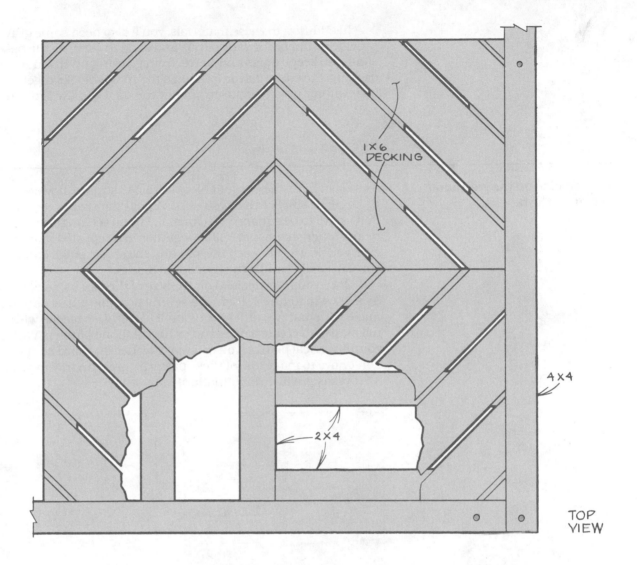

1×6 DECKING

4×4

2×4

TOP VIEW

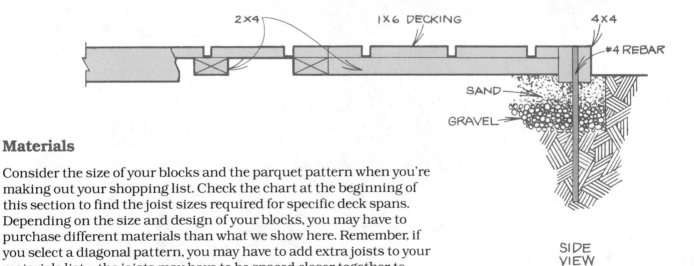

2×4   1X6 DECKING   4X4

#4 REBAR

SAND

GRAVEL

SIDE VIEW

Materials

Consider the size of your blocks and the parquet pattern when you're making out your shopping list. Check the chart at the beginning of this section to find the joist sizes required for specific deck spans. Depending on the size and design of your blocks, you may have to purchase different materials than what we show here. Remember, if you select a diagonal pattern, you may have to add extra joists to your materials list—the joists may have to be spaced closer together to compensate for the longer span of the planks.

As shown, the parquet blocks are framed with 2 x 4 joists, then covered with 1 x 6 decking planks. The blocks are laid out and surrounded by 4 x 4 'keepers' to prevent them from shifting. Use redwood, cedar, or pressure-treated lumber for all wooden parts.

In addition to these materials, you'll also need some gravel (to provide drainage for your patio), and sand to serve as the base. The sand also keeps grass and weeds from growing up through your decking. Purchase galvanized nails to put the blocks together, and #4 reinforcing rod (rebar) to anchor the 4 x 4's to the ground.

Before You Begin

1 Decide on the size and design of the blocks.

As shown here, the parquet blocks are 24″ square; however, this may be larger or smaller than you need. Adjust the size as required. They don't have to be square; you can make them rectangular, triangular, or any other shape that will fit together in a repeated pattern. However, you do want to keep the size and the shape small enough that you can easily handle the individual blocks.

Once you have decided on the size of the blocks, select a pattern. As shown in Figure 1, there are several possible patterns you can make with just two different types of blocks. Use blocks with diagonal planks to create a diamond or herringbone pattern. With horizontal/vertical planks, you can make a checkerboard or stripes. You don't have to make one of these patterns; you can invent your own. All it takes is a little imagination!

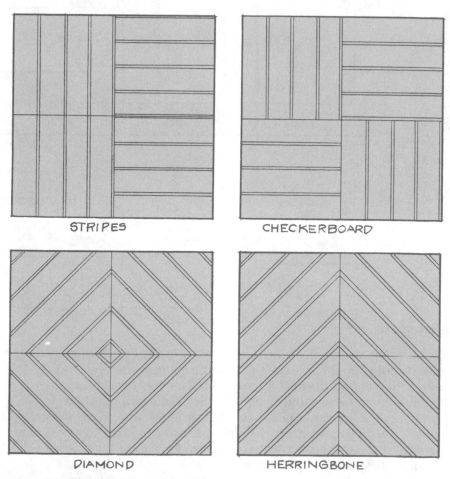

STRIPES   CHECKERBOARD

DIAMOND   HERRINGBONE

Figure 1. Create decorative 'parquet' patterns in your deck by arranging the blocks in one of these designs.

Prepare the Pad

Use stakes and string to lay out the pad, as shown in Figure 2. Drive the stakes outside the pad—the points where the strings intersect will mark the corners of your deck. Check to make sure the pad is square by measuring diagonally from corner to corner. In your layout, be sure to leave room for the blocks *and* the 4 x 4 keepers.

2   Lay out the pad.

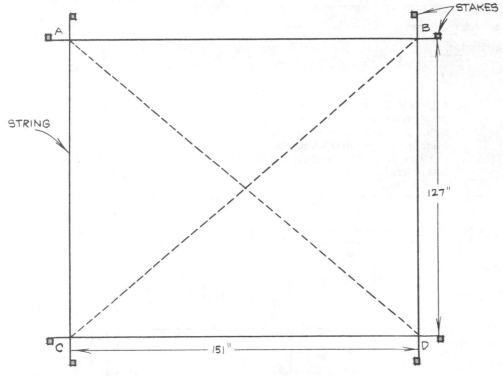

Figure 2. Use stakes and string to lay out the location of your deck. Check to be sure the deck will be square by measuring diagonally from corner to corner. AD should equal BC.

Remove all sod, rocks, and debris from the site. Then excavate 6″ below ground level, and haul away the dirt. Add tile drainage, if needed. Check to be sure the area is level by placing a 2 x 4 on edge in the bed and setting a carpenter's level on it. Fill in low areas and rake high areas until the bed is level. Then tamp the earth down and check once more.

3   Prepare the area.

4 **Fill with gravel and sand, then level.**

Shovel pea-size gravel into the bed until you have a 3″ layer. This will provide drainage for your deck. Level the gravel with a hand rake. Then cover the gravel with a 3″ layer of sand. Level the sand and tamp. Check to be sure the area is level by placing a 2 x 4 on edge in the bed and setting a carpenter's level on it, as shown in Figure 3. Fill in low areas and rake high areas until the bed is level.

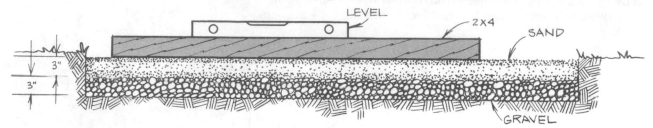

Figure 3. To determine whether your gravel and sand bed is level, place a 2 x 4 on edge in the bed, then put a carpenter's level on the 2 x 4. Rake the sand until the bed is level.

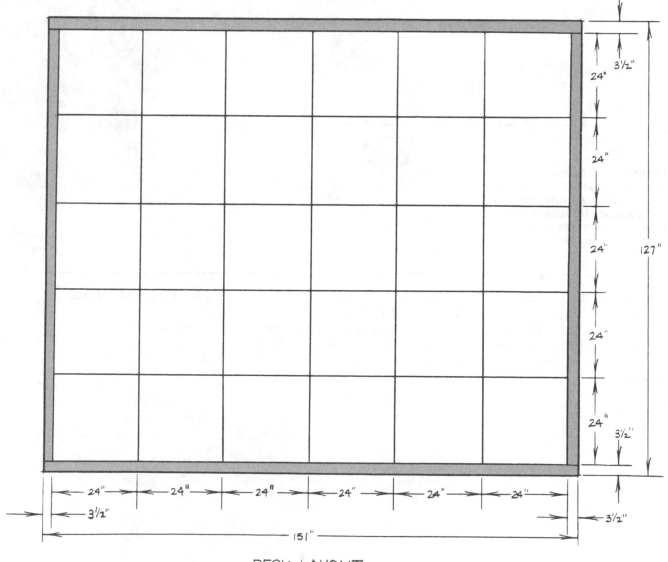

DECK LAYOUT

Lay the Pad

To keep the blocks from shifting in the sand, you'll need to 'frame' the pad with 4 x 4 'keepers', as shown in the *Deck Layout* drawing. Cut the keepers to the proper length, then set them in the sand. To set a keeper, put it in place, then work it back and forth a few inches at a time, until it settles ⅞″ into the sand, as shown in Figure 4. With very long keepers, you may need a second person to help you work the 4 x 4 back and forth.

5 **Enclose the pad with 4 x 4 keepers.**

Figure 4. Set the 4 x 4 keepers in place on the sand, then work them back and forth until they settle ⅞″ into the sand.

Using scrap lumber, make a nailing jig for the blocks. The inside dimensions of the jig must be the same as the outside dimensions of your blocks. (See Figure 5.) To make the blocks, first precut the parts. Set the 2 x 4 joists in the jig, and place the decking on top of them. Nail the decking to the joists with 8d nails. Leave a ½″ space between the planks to allow the wood to swell in wet weather. Two possible designs for blocks are shown in the *Block, Side View, Diagonal Block Layout*, and *Horizontal/Vertical Block Layout* drawings.

6 **Make the blocks.**

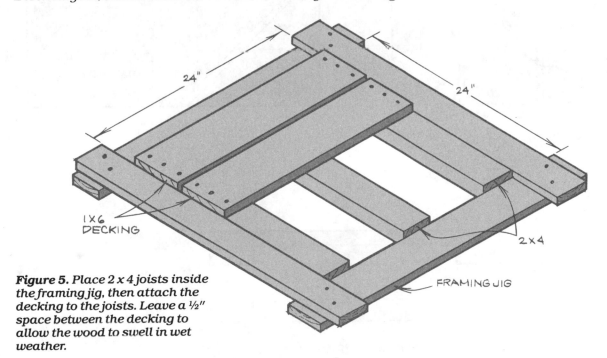

Figure 5. Place 2 x 4 joists inside the framing jig, then attach the decking to the joists. Leave a ½″ space between the decking to allow the wood to swell in wet weather.

TIP Use serrated square-shanked nails to attach the decking to the joists. They provide more holding power than regular nails and reduce the tendency of the wood to split.

7 **Lay the blocks.**

Set the first block in position on the sand, butted up against the 4 x 4 keepers. Fit the remaining blocks in place on the sand bed. As you work, walk on the blocks, not on the sand. The blocks should fit tightly together inside the keepers without rocking. If any block wobbles or rocks excessively, pick it up and use some extra sand to shim up the high corners. A little wobble is okay; this will cease as soon as the deck has settled.

Option: Leave out a block or two and use the empty squares in the deck as planters or as a sandbox for the kids.

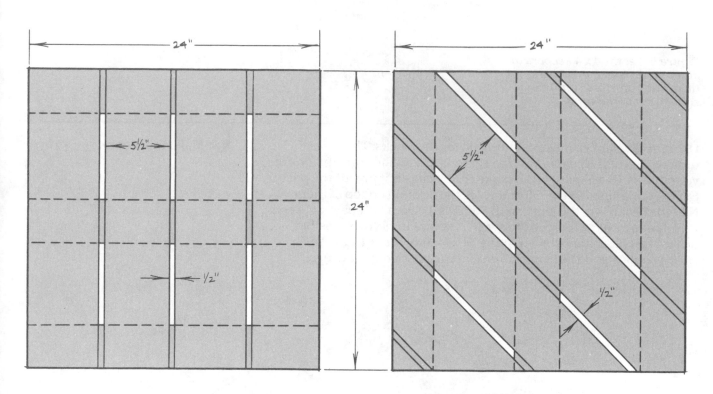

HORIZONTAL/VERTICAL
BLOCK LAYOUT

DIAGONAL
BLOCK LAYOUT

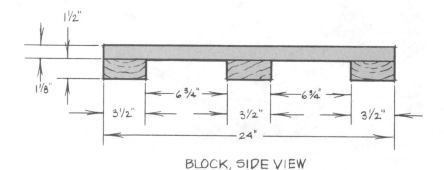

BLOCK, SIDE VIEW

After you have placed all the blocks inside the keepers, anchor the keepers to the ground. To do this, drill ⅝″ holes through the 4 x 4's, as shown in Figure 6. Drill the holes at the corners and about every 4′, as shown. Cut #4 reinforcing bars in 18″ lengths. Then drive the rebar rods down through the 4 x 4's and into the ground, as shown. When anchored, the keepers will prevent the blocks from shifting on the sand.

8   **Stake the 4 x 4's down with rebar.**

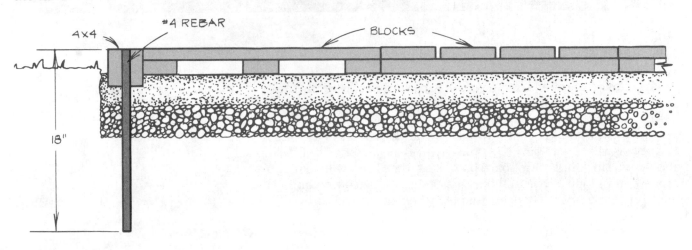

Figure 6. Anchor the keepers to the ground with 18″ lengths of #4 rebar. This will keep the blocks from shifting in the sand.

Playhouses and Play Centers

A playhouse or a jungle gym will turn your backyard into an outdoor activity center for the kids. And it doesn't have to break your back or your budget. Playhouses and jungle gyms are easy to construct and you can go as plain or fancy as your pocketbook will allow. Some playhouses are just a few sheets of inexpensive exterior-grade chipboard bolted together. They will hold up in the weather for a few years—about as long as it takes your children to outgrow them. Then you just tear them down and throw them away.

Of course, you can go the other way. You can look on a playhouse as an investment, a *permanent* structure that will enhance the value of your real estate. Put it on a permanent foundation; build it to last; and when the kids outgrow it, turn it into a storage shed or guest house.

But whether you treat these structures as transient or permanent, remember the real reason for building them—so your kids can have fun. If you build the playhouse you always wanted to have when you were a kid, chances are it'll be pretty close to the playhouse your kid wants to have, too.

Before You Begin

Adjust the design of the playhouse or jungle gym to suit yourself and your kids. We have several examples in this section: a traditional playhouse, a tower playhouse, and a jungle gym with several different activities. Mix and match as much as you like. Many of the play center projects that we've seen have combined playhouses, raised platforms, and jungle gym equipment all in the same grand design.

Besides trying to pack as much fun as you can into this project, there are some other factors you should carefully consider:

Choosing a site. Select a site that's visible from the house so you can keep an eye on the children's activities. Don't give in to the temptation to locate them way out in the backyard, where the noise won't bother the adults. If you will be adding other components to the play center later on, select a large, open space. This will give both you and the kids room to expand. And be sure to consider the traffic flow to and from the playhouse or jungle gym—you don't want children tromping through your garden on their way to play.

Selecting materials. Don't skimp on materials, even though you may be planning to tear this project down when the kids outgrow it. Substandard materials may not be dangerous, exactly; but they will be a nuisance. You will be forever picking splinters out of little hands and feet. If you work with pressure-treated lumber, make sure you know what chemical was used to treat it and avoid anything that

312

may endanger your children's health. Creosote and pentachlorophenol are poisonous; opt for wood preserved with chromated copper arsenate (CCA). This wood is easy to spot in the lumberyard—the CCA turns the lumber a light green.

Considering kits.   Because of the popularity of backyard activity centers, a wide variety of playhouse and jungle gym kits are available at lumberyards and department stores. A good kit will come with good materials; this will save you the trouble of searching through the lumberyard for splinter-free 2 x 4's. Also, kits may be less expensive than building from scratch. But if you're looking for something a little different, you may want to work from your own plans. Generally, you'll use the same building skills to construct a kit as you would to build from scratch, so let your budget—and your vision of the ideal playhouse—be your guide.

Checking local codes.   Whether you are building your project from scratch or from a kit, check with your local building inspections office for any restrictions that may apply to your playhouse or jungle gym. Unless you obtain a variance, you may have to locate your project several yards back from your property line. If needed, secure a building permit before you start work.

Tower Playhouse

Most kids like a bird's-eye view of things. They climb trees and build treehouses as if they were born to it. And it isn't just trees—they'll climb anything that gets them off the ground. So when it comes time to build your kids a playhouse, why not put it up on stilts? With the proper railings and stairs, a 'tower playhouse' is safer than a tree, and it still satisfies the youngsters' fascination with heights.

This playhouse sits on a pole foundation, and the walls are covered with sheet siding. The gable roof is extended to cover a 'front porch'—which your kids will probably pretend is the rampart of a fort. The design is compatible with the slide, swings, and other jungle gym structures in the following chapter, so you can turn this playhouse into a complete play center. In fact, you can hang swings and a glider from the underside of the playhouse, since the platform framework is almost the same as the platform in the *Jungle Gym* chapter.

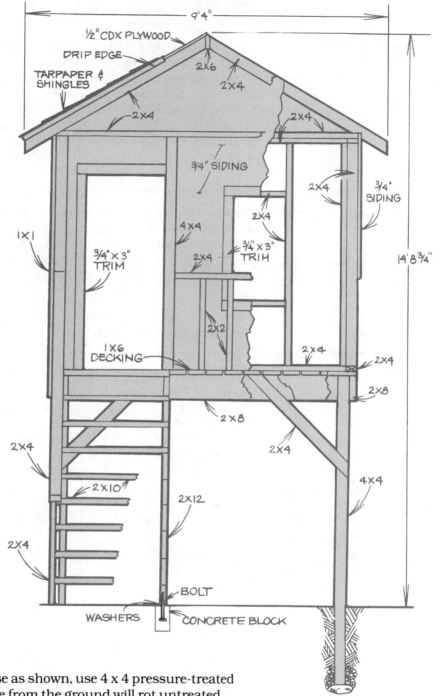

9'4"

½" CDX PLYWOOD
DRIP EDGE
TARPAPER &
SHINGLES
2×6
2×4
2×4
2×4
¾" SIDING
2×4
¾" SIDING
1×1
¾"×3"
TRIM
4×4
2×4
¾"×3"
TRIM
2×4
14'8¾"
2×2
1×6
DECKING
2×4
2×4
2×8
2×8
2×4
2×4
2×10
2×12
4×4
2×4
BOLT
WASHERS
CONCRETE BLOCK

FRONT ELEVATION

Materials

To build the tower playhouse as shown, use 4 x 4 pressure-treated posts for the poles. Moisture from the ground will rot untreated wood. Select 2 x 8's for the floor joists and beams, 2 x 12's for the stairs, and 2 x 4's for the wall frame, roof frame, and railings. Use 1 x 6 decking planks for the floor. The walls are covered with ¾" thick sheet siding and the roof sheathing is ½" CDX (exterior) plywood. Purchase 'one-by' (¾" thick) stock for the trim and moldings and 2 x 2's for the spindles.

The posts are set in the ground, so you'll need gravel for drainage at the base of each post hole. In addition to these materials, purchase galvanized nails, roofing nails, tarpaper, shingles, and metal drip edge.

TIP Select posts that are at least 1' longer than you'll need so you can cut them off to the proper level after you've set them.

Before You Begin

1 **Adjust the dimensions and the design.**

Consider the age of your children before settling on a design for their playhouse. As shown, the playhouse is 6' above the ground; however, if you have very young children you may want to lower it. Your children's ages and relative sizes will dictate many safety concerns—not just the height, but the positioning of the windows in the wall, the rise and run of the stairs, and the spacing between the spindles on the railing. Consider each feature of the playhouse carefully before settling on a final design.

2 **Check the building codes.**

Because this playhouse is set on a permanent foundation, it will likely be covered by local building codes. Before you begin, check with your local building inspections office to be sure your playhouse meets code requirements. If needed, obtain a building permit before you begin work.

Building the Platform

3 **Lay out the posts.**

Use stakes and string to mark the locations of your posts. (See Figure 1.) Use the *Post Layout* drawing to determine the exact placement of the posts. Mark the location of the posts with stakes, then dig the holes 24″ to 36″ deep. The holes must be deeper than the frost line. Check with your local building inspections officer to determine the exact frost line for your area. The holes must also be twice as wide as your posts to allow space for packing gravel and dirt around them.

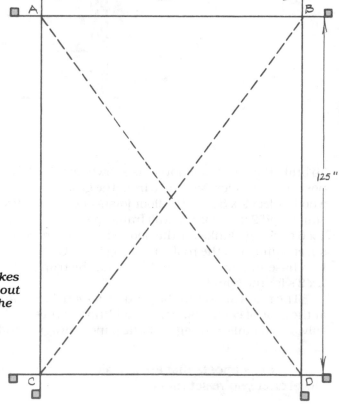

Figure 1. Use stakes and string to lay out the positions of the posts. Measure diagonally from corner to corner to be sure your layout is square. AD should equal BC.

Place a large rock in the base of each hole to keep the posts from settling. These rocks should be twice the diameter of the posts—at least 8″ in diameter. (See Figure 2.) Set the posts in place, then shovel gravel in the hole to a depth of 12″. This will help drain the ground water away from the posts. Finally, fill the rest of the hole with dirt and tamp lightly around the post. Brace the posts, if necessary, with scrap lumber and stakes to hold them upright. Then use a carpenter's level to be sure they are plumb.

4 **Set the posts.**

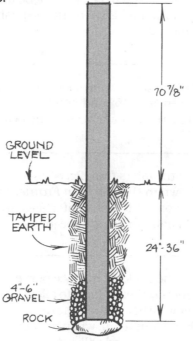

Figure 2. Set the posts at least 24″ deep in the ground, or below the frost line for your area. Plant the posts on a large rock, and throw some gravel in around the bases of the posts to provide drainage.

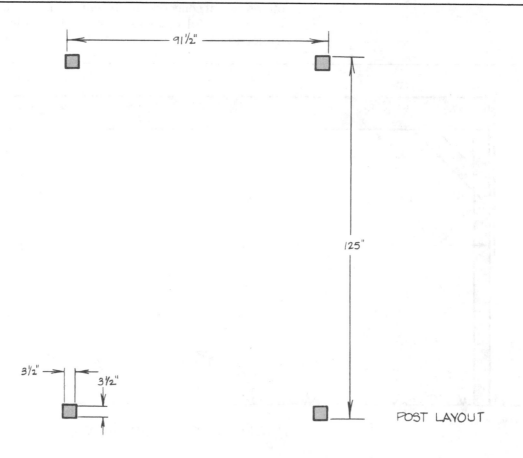

5 **Nail the beams to the posts.**

Measure one front post and mark it 70⅞″ above the ground. Using this mark as a reference, find the tops of the other posts with a string level. However, do *not* cut the posts to their proper height just yet.

Once you've marked the posts, cut the 2 x 8 beams that frame the platform. Tack these beams to the outsides of the posts, as shown in the *Platform Frame, Side View* drawing. Make sure the tops of the beams are flush with the marks for the tops of the posts. Use a carpenter's level to be sure the beams are level, and secure them in place with 16d nails. Then cut the posts off to the proper level, using a handsaw.

6 **Brace the platform.**

To keep the platform from wobbling, cut braces from 2 x 4's. Miter the ends at 45° and notch them as shown in the *Brace Detail* drawing. Attach the braces to the beams and the posts with lag screws.

7 **Nail the joists to the beams.**

Cut the joists from 2 x 8 stock. Place the joists between the beams, as shown in the *Platform Frame, Top View* drawing. Be sure the joists are spaced 16″ on center, as shown. Nail the joists to the beams using 16d nails, or secure the joists in place using metal joist hangers.

8 **Cover the floor frame with decking.**

Lay the decking in place perpendicular to the joists. Leave a ½″ space between the decking to allow the wood to swell without warping in wet weather. Attach the decking 'bark' side down—the annual rings should curve up, when you look at the end grain. With the bark side down, the wood will cup in such a manner that the rainwater will run off instead of collecting in the cup. This will help keep the wood from rotting. Use 12d square-shanked spiral nails to attach the decking to the joists—these nails keep the boards from working loose.

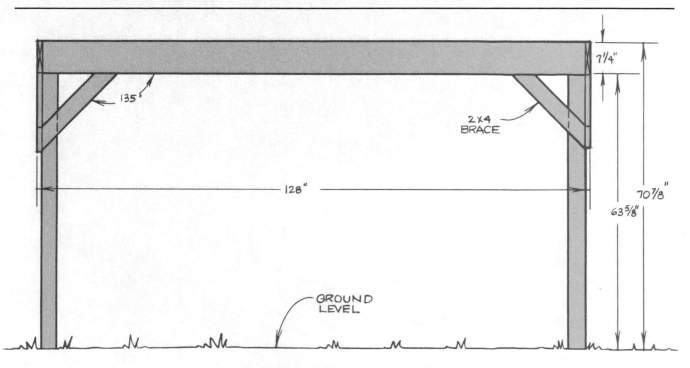

PLATFORM FRAME, SIDE VIEW

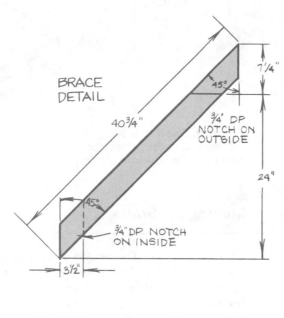

BRACE
DETAIL

40¾″

45°

7¼″

¾″ DP
NOTCH ON
OUTSIDE

24″

45°

¾″ DP NOTCH
ON INSIDE

3½″

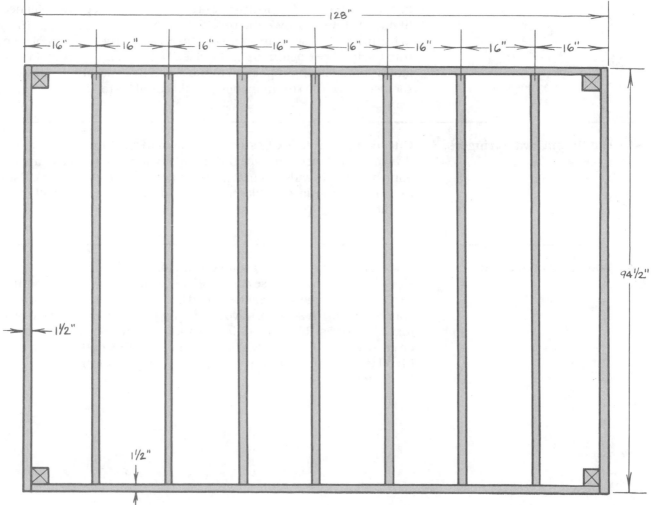

128″

16″ 16″ 16″ 16″ 16″ 16″ 16″ 16″

94½″

1½″

1½″

PLATFORM FRAME, TOP VIEW

9 Cut the decking flush with the joists and beams.

Cut the decking to length *after* you've nailed it in place. While you're installing it, let the boards stick out 1-3". Then snap a chalk line even with the outside edge of the beams, and cut along the mark with a circular saw. (See Figure 3.)

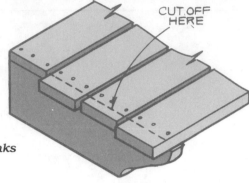

CUT OFF HERE

Figure 3. Cut the decking planks off flush with the beam after you've nailed them in place.

Making the Stairs

10 Calculate the rise and run of the stairway.

If you've changed the height of the platform, you'll need to calculate the rise and run of the stairway. First, measure the total rise (the distance from the ground to the platform. Then pick a measurement for the run (the horizontal distance of each step—for outdoors it should be no less than 10½"). To arrive at the number of steps you will need, divide the total rise by 7 and round it off to get an integer. This integer is the number of steps in your staircase. Divide the number of steps into the total rise again to get the rise of each step. The run of the stairs should be in proportion to the rise. The general rule is that the rise times the run in inches should equal 70 to 75.

11 Cut the stairway stringers.

Cut stringers from 2 x 12 stock, as shown in the *Stair Stringer Detail* drawing. You'll need to place the stringers no more than 24" apart. If your stairway is wider than 24", make at least three stringers. Using a carpenter's square, mark off the cuts for the rise and run all the way up the stringer. Use a handsaw or a saber saw to cut the stringers.

12 Pour or set concrete blocks.

The stairway must rest on solid ground or it won't stay level—and safe—for very long. So use concrete blocks or pour concrete footings to rest the stringers on. Set hex head bolts as anchor bolts in the concrete before it sets up. (If you're using ready-made concrete blocks for the footings, fill the holes in the middle of the blocks with concrete to set the bolts.) Once you've set or pour the footing, cut a little bit off the bottom of the stringers to compensate for the height of the footers.

Drill holes in the bottoms of the stringers and set them over the anchor bolts, as shown in Figure 4. Put 3-4 washers over the bolts, between the stringers and the concrete, to help keep the stringers from direct contact with the ground. This will keep water from collecting under the stringers and rotting the wood.

13 Attach the stringers to the deck frame and concrete footings.

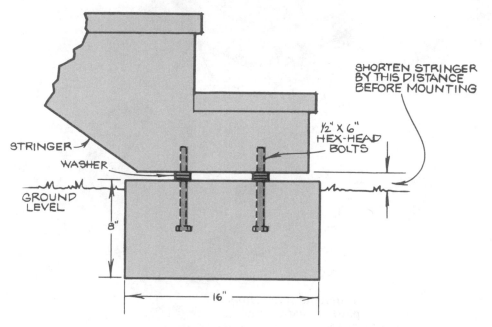

Figure 4. Place 3-4 washers over each anchor bolt before setting the stringers on the footers. This will keep the stringers up off the concrete and help prevent rot.

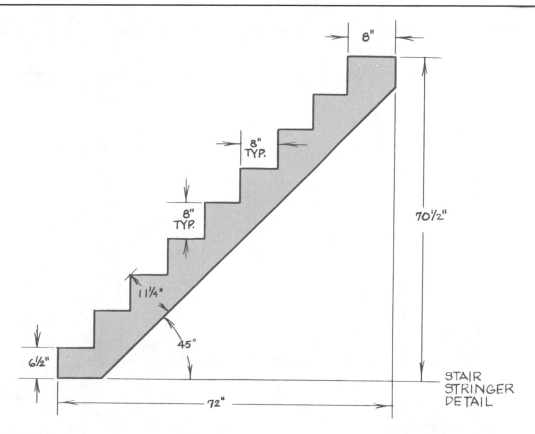

Attach the stringers to the deck frame using metal framing anchors, as shown in Figure 5.

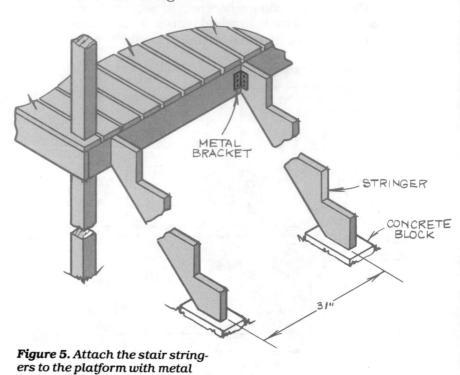

Figure 5. *Attach the stair stringers to the platform with metal framing anchors.*

14 **Nail the treads to the stringers.**

Cut treads from 2 x 10 or 2 x 12 stock (depending on the run per stair) and lay them across the stringers. Attach the treads to the stringers using 16d nails.

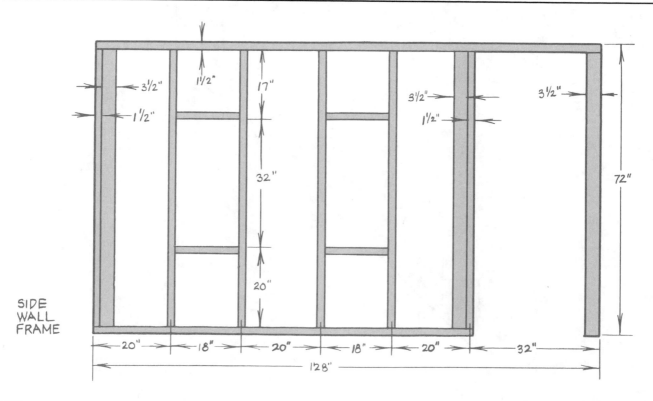

SIDE WALL FRAME

Building the Frame

Use 2 x 4's to frame the walls. Cut the studs, sole plates and top plates and assemble them as shown in the *Side Wall Frame*, *Back Wall Frame*, and *Front Wall Frame* drawings. Frame the windows and door, as shown, by nailing 2 x 4 headers and sills in place between the studs. Assemble the wall frames using 16d nails.

15 **Build the wall frames.**

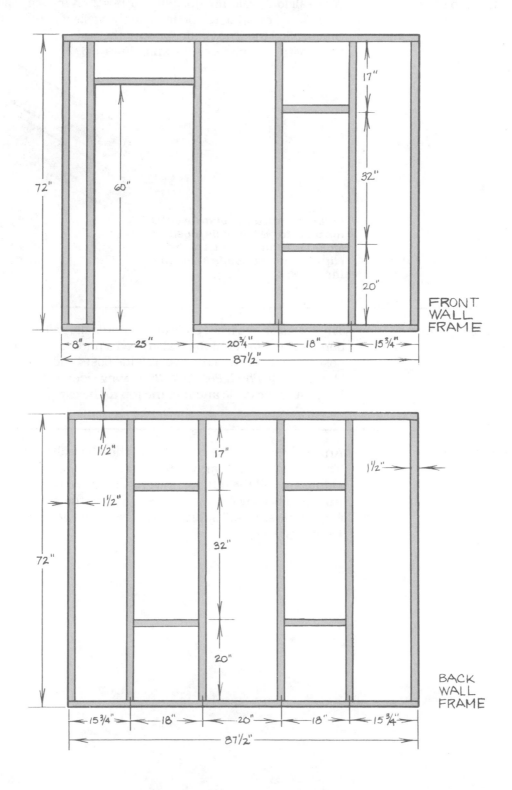

16 Raise the wall frames.

With the help of a friend, raise the walls in place, as shown in the *Stud Layout* drawing. Temporarily brace the walls upright with scrap lumber. Nail the sole plates to the floor frame with 16d nails spaced every 16″ to connect the sole plates to the joists. Connect the walls to each other at the corners with 16d nails spaced every 24″ on center. Check with a level to be sure the walls are plumb.

17 Attach the cap plate.

When all four walls are up, nail another 2 x 4 top (cap) plate flat against the existing top plate to tie the walls together. This is sometimes called the "cap" plate. Be sure the ends of the cap plate overlap those of the top plate, as shown in Figure 6.

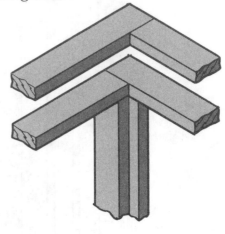

Figure 6. Nail a cap plate over the top plate to tie the walls together. Make sure that the ends of the cap plate lap the joints between the top plates.

18 Put the middle porch post in place.

The 4 x 4's on the front ends of the side walls will serve as the outside porch posts. However, you'll also need a middle porch post. Cut this post 1½″ longer than the two outside posts and put it in place where shown in the *Porch Post Detail* drawing. Toenail the bottom end of the post to the deck, and nail the top to the cap plate.

19 Put siding on the side walls.

Attach ¾″ thick plywood siding to the side walls only, using 6d nails. This will save you some work later on—you won't have to notch the siding around the rafters. However, before you can put siding on the front and back walls, you'll have to put up the roof frame. The bottom edge of the side wall siding should be flush with the bottom edge of the floor frame.

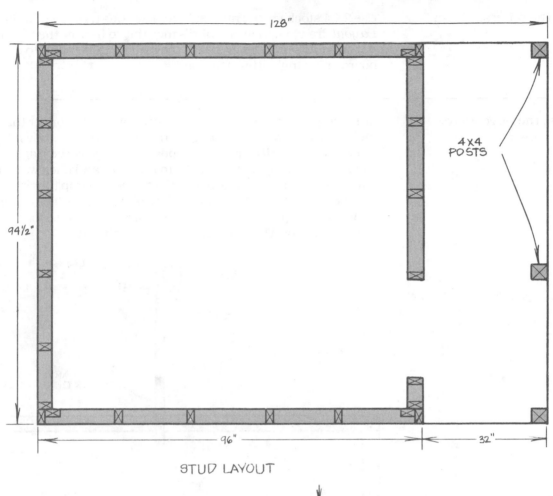

128"

94½"

96"

32"

4×4
POSTS

STUD LAYOUT

1½"

3½"

73½"

35¼"

59¼"

94½"

PORCH POST DETAIL

20 Cut the rafters.

Use 2 x 4 stock for the rafters. Miter the ends as shown in the *Rafter Layout* drawing, then cut 'bird's mouths' to fit over the top plates. Do not cut bird's mouths in four of the rafters; these will serve as outboard or 'facing' rafters later on.

---

21 Raise the ridgeboard.

Cut a ridgeboard from 2 x 6 stock, 140″ long—12″ longer than the platform. Bevel the top edge of the ridgeboard as shown in the *Ridgeboard Detail* drawing. Temporarily support the ridgeboard above the frame by nailing 2 x 4's to the front and back wall frames so that they stick up about 3′ above the frames. Clamp (don't nail) the ridgeboard to these supports, as shown in Figure 7. The clamps will allow you to adjust the height of the board, if you need to. The ridgeboard should overhang the frame 6″ on each end.

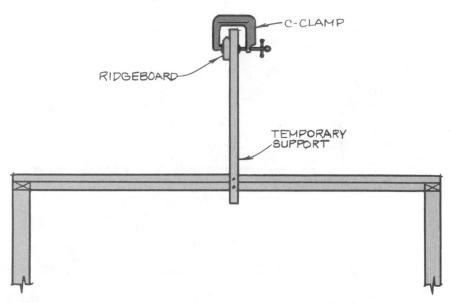

Figure 7. *Use temporary 2 x 4 supports and clamps to hold the ridgeboard in place while you attach the rafters. This arrangement will let you easily shift the position of the ridgeboard, if you need to.*

---

22 Raise the rafters.

After the ridgeboard is clamped in place, lift the rafters in place. Lay them against the ridgeboard and fit the bird's mouths over the cap plate, as shown in the *Roof Frame, Front View* drawing. Nail the rafters to the ridgeboard and the top plate, every 16″ on center, as shown in the *Roof Frame, Side View.* Do not attach the facing rafters yet. After the rafters are in place, remove the temporary ridgeboard supports.

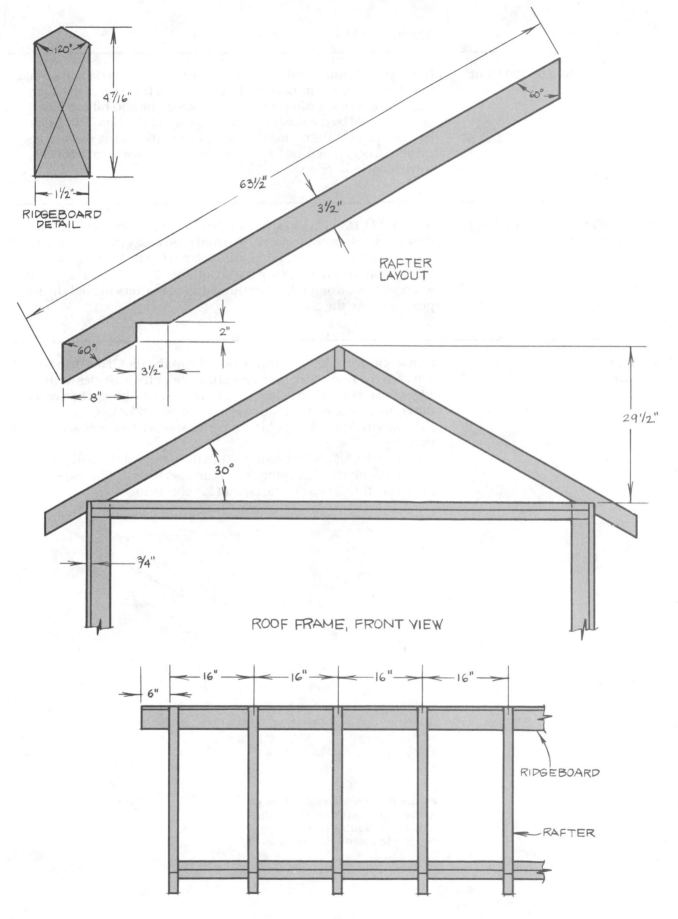

120°

4 7/16"

1 1/2"

RIDGEBOARD
DETAIL

63 1/2"

3 1/2"

60°

RAFTER
LAYOUT

60°

2"

3 1/2"

8"

30°

29 1/2"

3/4"

ROOF FRAME, FRONT VIEW

16"   16"   16"   16"

6"

RIDGEBOARD

RAFTER

ROOF FRAME, SIDE VIEW

23 **Install siding on the front and back walls.**

Nail plywood siding to the front and back walls using 6d nails. Make sure the siding on the back wall goes down to the base of the floor frame. Attach the siding at the gable ends to the end rafters. The siding should be cut flush with the top edge of these rafters. Don't lap the siding at the corners. Instead, install the siding so that the inside corners touch, but don't lap. This will leave room to install corner blocks later on.

24 **Install the roof sheathing and the facing rafters.**

Attach ½″ CDX plywood sheathing to the roof frame, making sure that it extends over the porch area and overhangs the front and the back of the house by 6″. Temporarily clamp the facing rafters in place, flush with the edges of the sheathing. Drive #12 x 1½″ flathead wood screws through the sheathing into the rafters to hold them in place. Remove the clamps.

25 **Install the roofing materials.**

Run metal drip edge around all the sides of the roof. This drip edge will keep the rain water from collecting under the shingles at the edge of the roof and possibly rotting out the sheathing. Then cover the entire roof with a double layer of tarpaper and a layer of shingles, as shown in Figure 8. 'Double-wrap' the peak with shingles, to form a ridge cap.

Note: Use ⅝″ roofing nails so that the points of the nails don't come through the sheathing. If you can't get short nails, use ¾″ thick sheathing, or a double layer of ½″ sheathing.

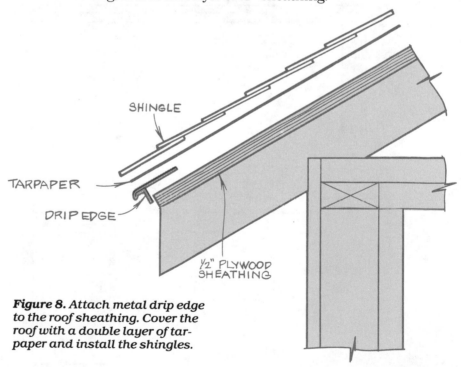

SHINGLE

TARPAPER

DRIP EDGE

½″ PLYWOOD SHEATHING

Figure 8. Attach metal drip edge to the roof sheathing. Cover the roof with a double layer of tarpaper and install the shingles.

Rip ¾" thick stock to trim the window and door openings, as shown in the *Trim Detail* drawing. Place the trim flat on the siding and position it so that the inside edge is flush with the edge of the opening. Secure the trim in place using #12 x 1¼" flathead wood screws. Drive these screws through the siding and into the trim from *inside* the playhouse, for a finished look. Or, simply attach the trim to the siding using 6d nails.

26 **Install the window and door trim.**

Cut 2 x 4 supports for both sides of the stairway, as shown in the *Stair Railing, Side View* drawing. Attach the supports to the outside of the stringers using lag screws. Then cut 2 x 4 railing for each side and miter the ends to match the slope of the stairs. Screw the railing to the supports and to the porch posts using #12 x 3" flathead wood screws.

Cut the porch railings from 2 x 4 stock, and the spindles from 2 x 2 stock. Toenail the porch railings between the porch posts, as shown in the *Porch Railing, Front View* drawing. Put the spindles in place, 9" on center along the front and 9¼" on center along the sides. Drill pilot holes to prevent the spindles from splitting, then nail them to the railings and the platform with 12d nails.

27 **Install the railings.**

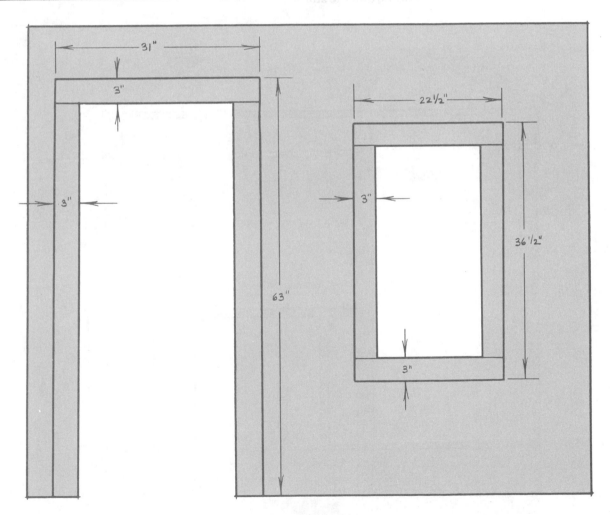

TRIM DETAIL

28 Install the corner blocks.

Rip 1 x 1 corner blocks from ¾″ stock. Run a generous bead of caulk along the corners where the siding comes together. This will seal the corners and keep moisture from creeping under the siding. Press the corner blocks in place, spreading out the caulk. (See Figure 9.) Then nail them to the frame with 6d nails.

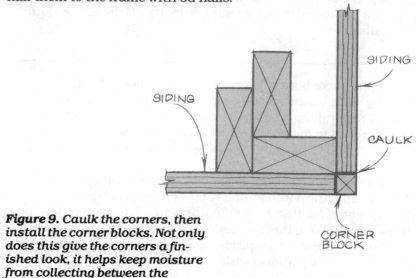

Figure 9. Caulk the corners, then install the corner blocks. Not only does this give the corners a finished look, it helps keep moisture from collecting between the siding and the studs.

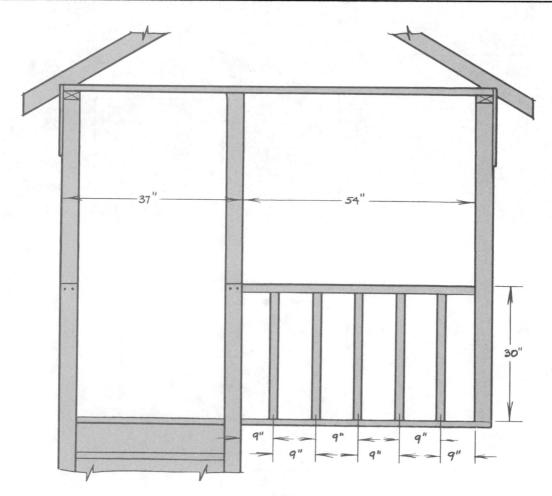

PORCH RAILINGS, FRONT VIEW

Paint or stain the playhouse to match—or compliment—your home. Prime the raw wood first, then cover it with two coats of paint.

29 **Paint or stain all exposed wood surfaces.**

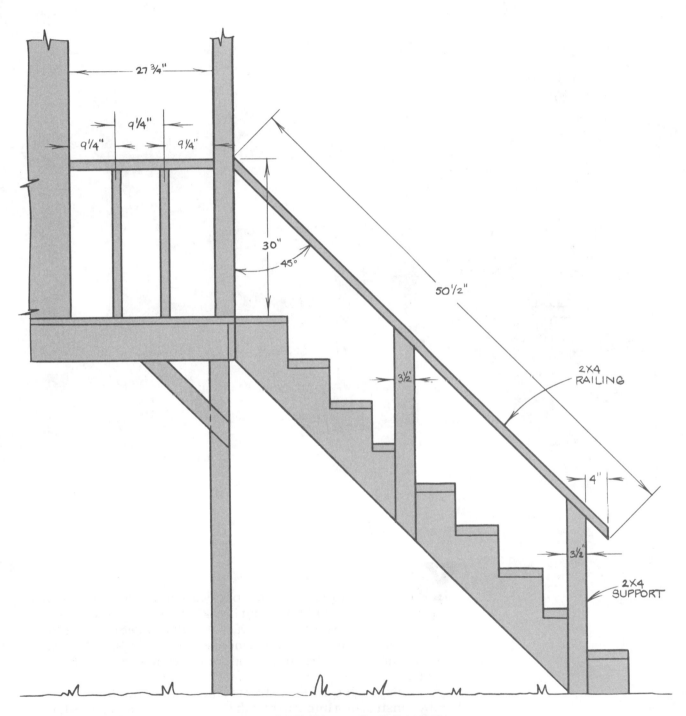

STAIR RAILING, SIDE VIEW

Jungle Gym

You can turn a corner of your backyard into an outdoor activity center for your kids without breaking the budget—or your back—in the process. This easy-to-build 'jungle gym' has four popular components: a rope climb, monkey bars, swings, and a slide. Build all four, or just start out with one or two components and add to them as your family grows.

This jungle gym is constructed mostly from two-by and four-by construction lumber, though several of the components have metal, rope, or hardwood parts. For example, the bars in the monkey bars are iron pipe, and the glider in the swing set is made from oak or rock maple. All the components are mounted on 4 x 4 posts set in the ground, and the major structural members are joined with lag screws.

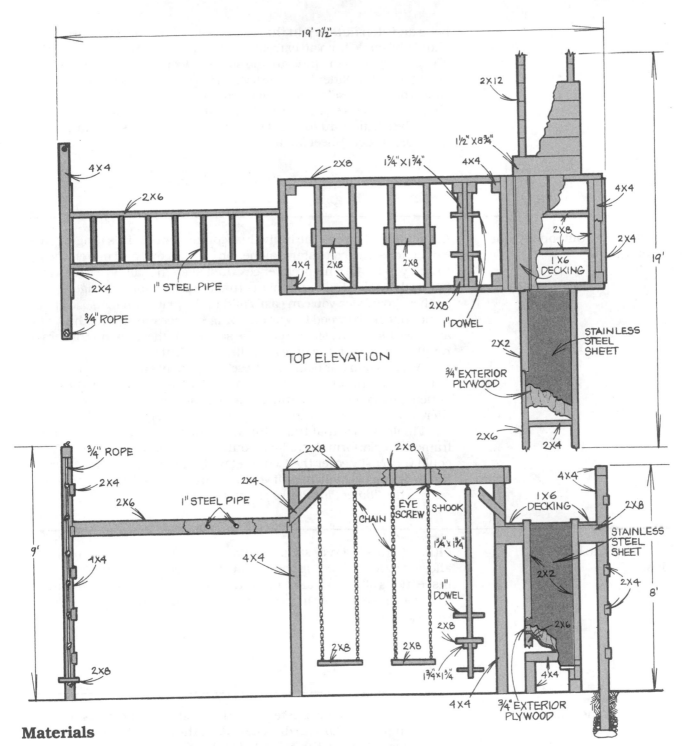

19' 7½"

2X12

1½"X8¾"

4X4

4X4

2X8

4X4

2X6

1¾"X1¾"

4X4

19'

2X8

2X4

4X4

2X8

2X8

1X6
DECKING

2X4

2X4

1" STEEL PIPE

2X8

¾" ROPE

1" DOWEL

STAINLESS
STEEL
SHEET

2X2

TOP ELEVATION

¾" EXTERIOR
PLYWOOD

2X6

2X4

¾" ROPE

2X8

2X8

4X4

2X4

2X4

2X6

1" STEEL PIPE

EYE
SCREW

S-HOOK

1X6
DECKING

2X8

STAINLESS
STEEL
SHEET

9'

CHAIN

1¾" x 1¾"

2X2

2X4

4X4

4X4

1"
DOWEL

8'

2X8

2X8

2X8

2X8

2X6

1¾"x1¾"

4X4

2X8

4X4

¾" EXTERIOR
PLYWOOD

Materials

FRONT ELEVATION

For the posts, beams, joists, stringers, treads, and other structural members, use redwood, cedar, or lumber that has been pressure-treated to resist moisture and insect damage.

To build the jungle gym as shown, use 4 x 4's for all of the posts and for the beam supporting the rope climb. Purchase 2 x 8's to frame the swing set and slide platform and 2 x 6's for the sides of the slide. Use 2 x 4's for the ladders at each end of the jungle gym, and 2 x 12's for the slide stairway. Join all these structural members with ⅜" lag screws—nails may pull out.

In addition to these materials, you'll need cotton rope for the rope climb, and cold-rolled iron pipe for the monkey bars. Purchase chains or ropes to support the swings on the swing sets, and ½" bolts and stop nuts to join the parts of the glider. The glider itself should be made from an extremely hard wood such as oak or maple. (Cedar, redwood, and pine are too soft and won't hold up.) Hang the swings and the glider from eye screws and S-hooks. To make the slide, you'll need ¾″ exterior plywood and a 12-, 14-, or 16-gauge sheet of stainless steel. This is the one item on the materials list that might be a little hard to find. Look in the Yellow Pages of your phone directory under "Sheet Metal".

Before You Begin

1 **Plan the components for your children.**

As mentioned before, the individual components of this jungle gym can be constructed to stand alone. Or, you can mix and match them to suit your children's preferences. Just keep the safety of your kids in mind by selecting components that are suited for their age. Then, as they grow older, you can easily add to the play center. Also keep in mind that all the wood for the climbing surfaces, as well as the seats, and the sides of the slide should be sanded so there are no corners or rough edges to scratch or leave splinters in little hands.

As shown in our plans, all these components are laid out in a long line. But this, too, is variable. You may wish to arrange the components in some other manner to save space. Just make sure that there is ample room for the swings and the glider to swing freely.

Finally, notice that the swing set is constructed as if it were the frame for a platform. Actually, it could be a frame for a platform. You could very easily cover the swing set with decking and build a playhouse on top of the swings. If you want to add this fifth component, refer to the *Tower Playhouse* chapter to see how it's done.

2 **Check the building codes.**

Although this is a very simple structure to build, it may still be affected by local building codes. Check with your local building inspections office to make sure your play center complies with any local code restrictions. If necessary, secure a building permit before you begin work.

Setting the Posts

3 **Lay out the posts.**

With strings and stakes, determine the position of the posts, as shown in the *Post Layout* drawing. Mark the location of the posts with stakes, then dig the holes 24″ to 36″ deep, or below the frost line. Check with your local building inspections office to determine the exact frost line for your area. The holes must also be twice as wide as your posts to allow space for packing gravel and dirt around them.

Place a large rock in the base of each hole to keep the posts from settling. These rocks should be twice the diameter of the posts—at least 8″ in diameter. (See Figure 1.) Set the posts in place, then shovel gravel in the hole to a depth of 12″. This will help drain the ground water away from the posts. Finally, fill the rest of the hole with dirt and tamp lightly around the post. Brace the posts, if necessary, with scrap lumber and stakes. With a carpenter's level, be sure the posts are plumb.

4 **Set the posts.**

Figure 1. Set posts at least 2′ into
the ground or below the frost line
in your area. Plant the posts on
a large rock and surround them
with gravel to provide drainage.

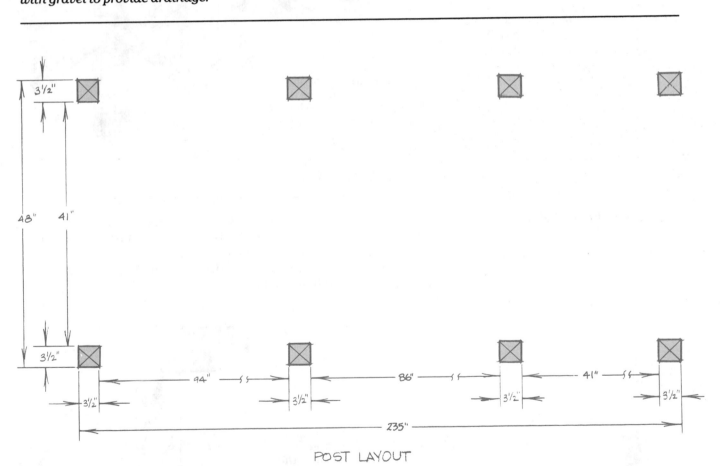

POST LAYOUT

5 Cut the posts to the proper height.

Once the posts are plumb, cut them to the proper height. There are two different heights among the components—the slide and the swing set posts are 96″ high, and the rope climb posts are 104½″. (The monkey bars are suspended between two other components; it needs no supporting posts of its own.) Start by cutting the shortest posts first. Measure one 96″ post and mark to the desired height above the ground. With strings and a string level, use this mark as a reference and find the tops of all the other 96″ posts. (See Figure 2.) Mark all the higher posts, too, at 96″. Then measure from the 96″ to find where you want to cut the higher posts. On the rope climb posts, measure 8½″ above the 96″ marks. Remove the strings and cut the posts off with a handsaw. Check again that the posts are still plumb, and tamp the dirt down as tight as you can.

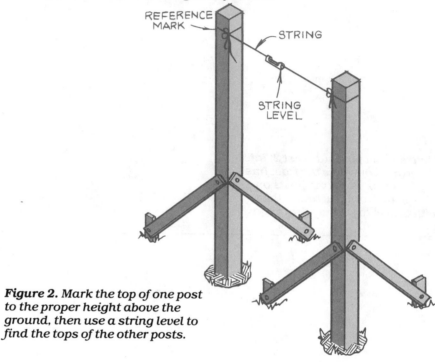

Figure 2. Mark the top of one post to the proper height above the ground, then use a string level to find the tops of the other posts.

Making a Rope Climb

6 Attach the top beam.

Cut a 4 x 4 beam 84″ long and drill two ¾″ holes, one near each end. Place the beam on top of the rope climb posts. Make sure the ends of the beam overhang the posts 18″ on either side, as shown in the *Rope Climb, Side View* drawings. Secure the beam in place with a ½″ x 8″ lag screw, as shown in Figure 3.

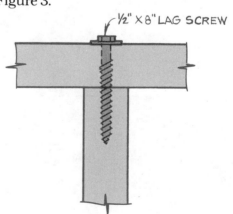

Figure 3. Drill a pilot hole through the beam and into the post, then secure the beam in place with a ½″ x 8″ lag screw.

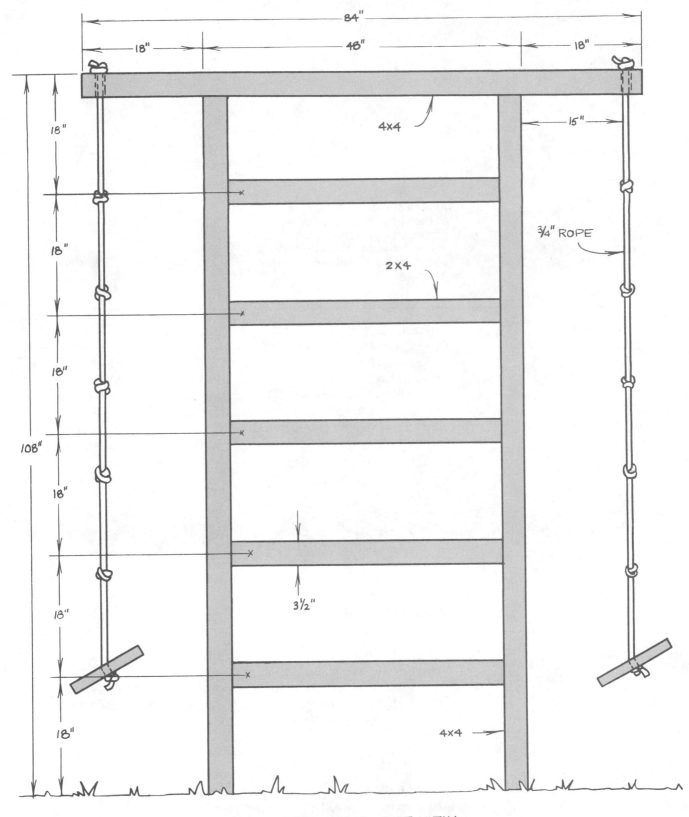

84"

18"   48"   18"

18"

18"   4×4

18"   15"

18"   ¾" ROPE

2×4

108"   18"

18"   3½"

18"

18"   4×4

ROPE CLIMB—SIDE VIEW

7 Attach the rungs.

Use 2 x 4's to make the rungs of the rope climb. Cut the 2 x 4's 48″ long, so that they lap the posts. Mark where the rungs will be attached on the posts and, with a handsaw, cut a 1″ deep kerf along each mark. (See Figure 4.) Remove the waste between the kerfs to form a dado for the rungs. Secure the rungs in place with ¼″ x 3½″ lag screws.

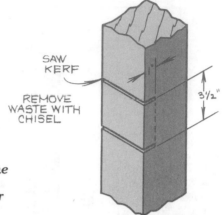

Figure 4. To attach the rungs to the posts, first saw 1″ deep kerfs in the post, then remove the waste between the kerfs with a chisel. This will make a dado for the rungs.

8 Attach the ropes and seats.

Cut seats for the rope climb from 2 x 8 stock, as shown in the *Rope Climb Seat Detail* drawing. Drill a ¾″ hole through the center and round the edges of each seat. Cut two lengths of ¾″ rope 9′ long. Tie knots in the ropes every 12″-18″ to give the kids something to hang on to while they're climbing. Thread one end of the ropes through the holes in the seats and tie a knot to keep the seats from slipping off the ends of the ropes. Then thread the other ends of the ropes through the holes in the beam, and tie knots to keep the ropes from slipping out of the beams.

TIP Use *cotton* or *nylon* rope instead of hemp because the softer roping won't hurt your children's hands.

Making the Monkey Bars

9 Mark and drill the stringers.

Cut the stringers from 2 x 6 stock. Cut notches in the ends of the stringers, 1½″ deep and 3½″ long, as shown in the *Monkey Bars, Front View* drawing. Then mark the locations of the bars, spacing the bars 12″ apart, as shown. Drill 1″ holes 1″ deep in the inside faces of the stringers at each mark.

10 Attach the steel bars.

Slide 1″ diameter steel pipe into the 1″ holes in the stringers, as shown in the *Monkey Bars, Top View* drawing. Temporarily keep the stringers from spreading apart and the bars from dropping out of their holes by tying the stringers together with some scrap wood braces and nails. You can remove these braces after the monkey bars are mounted on the supporting rungs.

In our design, the monkey bars stretch between the rope climb and the swing set. One end of the assembly is mounted on a rope climb rung, 72″ above the ground. You'll need to attach another rung to the swing set posts, also 72″ above the ground. Depending on where you mount your monkey bars, you may have to install one or two supporting rungs.

11 Attach supporting rungs to the posts.

TIP Depending on the ages of your kids, 72″ may be too high—or too low—for the monkey bars. Mount the assembly so that it gets the kids off the ground, but not so high that they may injure themselves if they slip. You can always raise the bars as your children grow.

Fit the notches in the ends of the stringers over the rungs. Then secure the stringers to the rungs with ¼″ x 3½″ lag screws.

12 Attach the stringers to the supporting rungs.

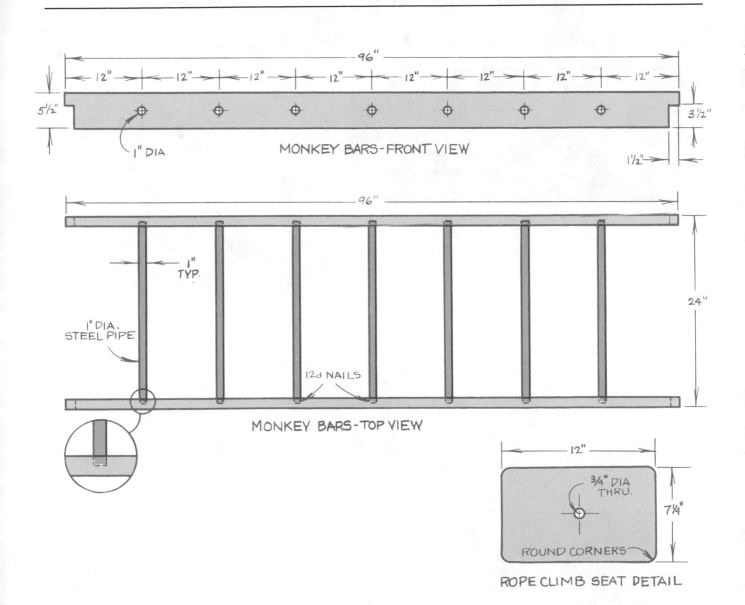

MONKEY BARS-FRONT VIEW

MONKEY BARS-TOP VIEW

ROPE CLIMB SEAT DETAIL

Making the Swing Set

13 Attach the beams and joists.

Cut the beams and joists from 2 x 8 stock. Attach the beams to the outside edges of the posts with ⅜" x 4" lag screws. The tops of the beams should be flush with the tops of the posts, as shown in the *Swing Set, Front View* drawing. Attach the joists between the beams, spacing them every 16" on center, as shown in the *Swing Set, Top View* drawing. Secure them to the beams with ¼" x 3½" lag screws. Cut 2 x 4 braces at each corner of the beams. Miter and notch the braces as shown in the *Brace Detail* drawing. Attach the braces to the beams and the posts with ¼" x 3½" lag screws.

14 Make the swings.

Cut the swing seats from 2 x 8 stock, as shown in the *Swing Seat Detail* drawing. Round and sand the corners. Then drill pilot holes for eye screws, where shown in the working drawings.

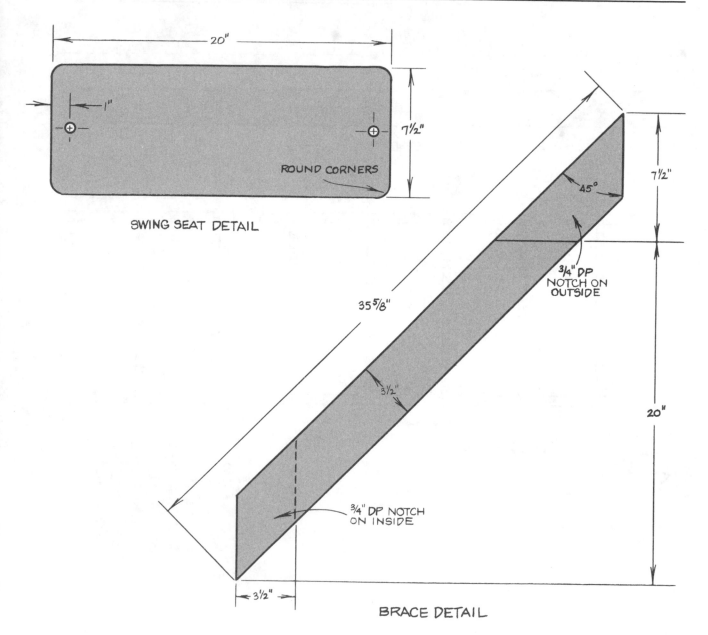

SWING SEAT DETAIL

BRACE DETAIL

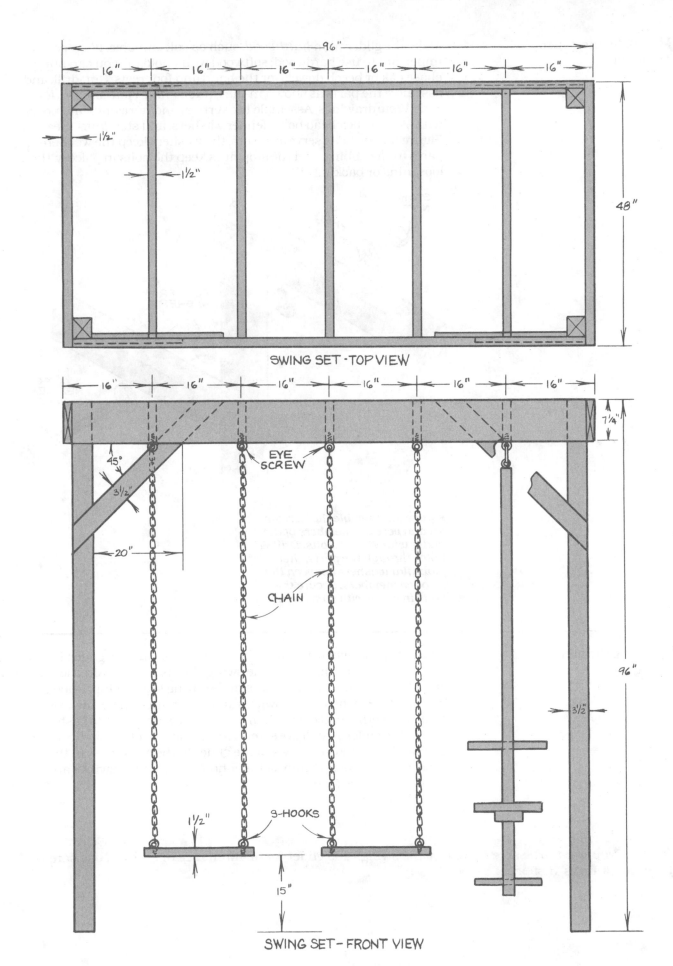

96"

16"   16"   16"   16"   16"   16"

1½"

1½"

48"

SWING SET - TOP VIEW

16"   16"   16"   16"   16"   16"

7¼"

45°

3½"

EYE
SCREW

20"

CHAIN

96"

3½

S-HOOKS

1½"

15"

3½

SWING SET – FRONT VIEW

15 Make the glider, if desired.

Make the glider from *hardwood*, such as oak or maple. It's very important to use hardwood; softwood won't hold up. You can purchase oak or beech dowels for the hand and foot rests. Cut, drill, and assemble the parts as shown in the *Glider, Front View* and *Glider, Side View* drawings. Assemble the vertical and horizontal members with ½" x 6" hex head bolts, fender washers, and stop nuts. (See Figure 5.) The bolts serve as pivots, the washers keep the wooden parts from rubbing, and the stop nuts keep the bolts in place without loosening or backing off.

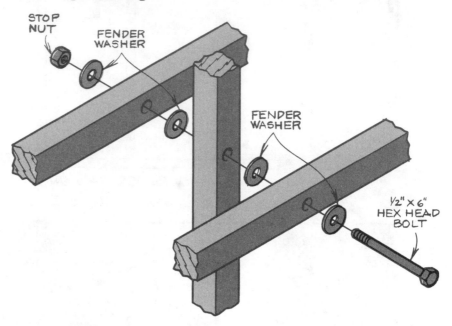

Figure 5. Assemble the horizontal and vertical members of the glider with ½" x 6" bolts. Drill ½" holes through the parts, then place flat washers between the wooden members. Secure the bolts in place with stop nuts.

16 Hang the glider and swings.

Mark the joists where you will hang the swings and the glider. Install eye screws at each mark. Cut the swing chains to the desired length and attach one end of the chains to the swing seats using S-hooks. 'Crimp' the S-hooks so the swings and the chain won't come loose. Attach the other end of the chains to the eye screws in the joists. To attach the glider, install eye screws in the top end of the vertical members, and link the eye screws in the glider to the eye screws in the joists with S-hooks. Crimp all the S-hooks so that the swings and glider can't come loose.

TIP If you want to be able to remove the swings and glider for maintenance or storage, use screw-type 'quick-links' instead of S-hooks.

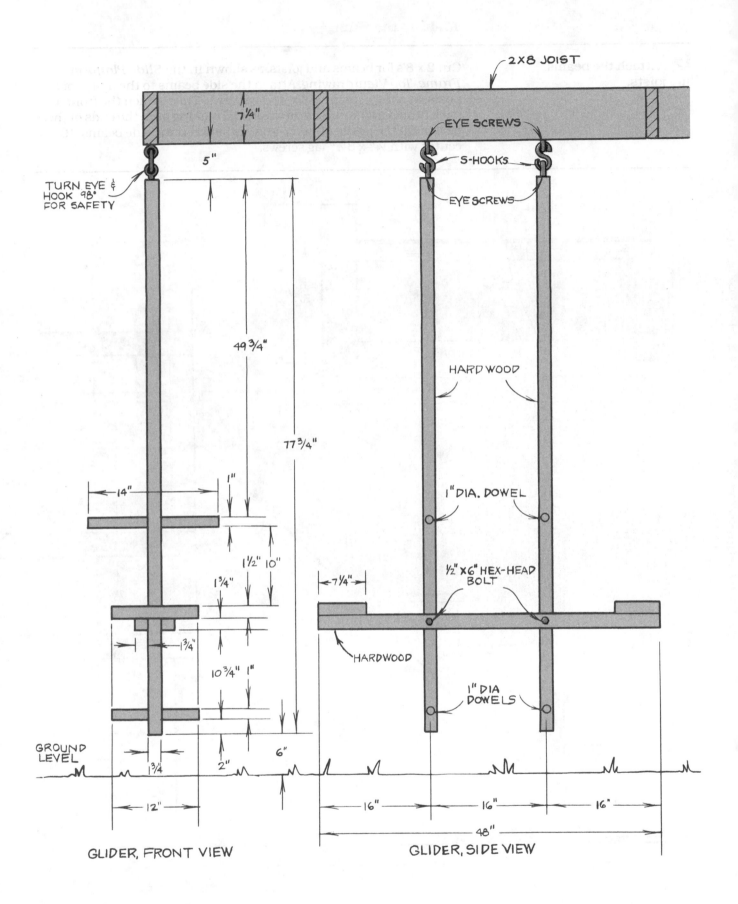

2×8 JOIST

7¼"

EYE SCREWS

S-HOOKS

EYE SCREWS

TURN EYE &
HOOK 90°
FOR SAFETY

5"

49¾"

77¾"

HARD WOOD

1" DIA. DOWEL

½"×6" HEX-HEAD
BOLT

14"

1"

1½" 10"

1¾"

1¾"

10¾" 1"

7¼"

HARDWOOD

1" DIA
DOWELS

GROUND
LEVEL

1¾"

2"

6"

12"

16"

16"

16"

48"

GLIDER, FRONT VIEW

GLIDER, SIDE VIEW

Making the Slide

17 **Attach the beams and joists.**

Cut 2 x 8's for beams and joists, as shown in the *Slide Platform Frame, Top View* drawing. Attach the side beams to the inside of the posts, as shown, using ⅜″ x 4″ lag screws. Then attach the front and back beams to the outside of the posts, making sure the ends of these beams lap the posts. Attach the joists between the side beams, 16″ on center, with ¼″ x 3½″ lag screws.

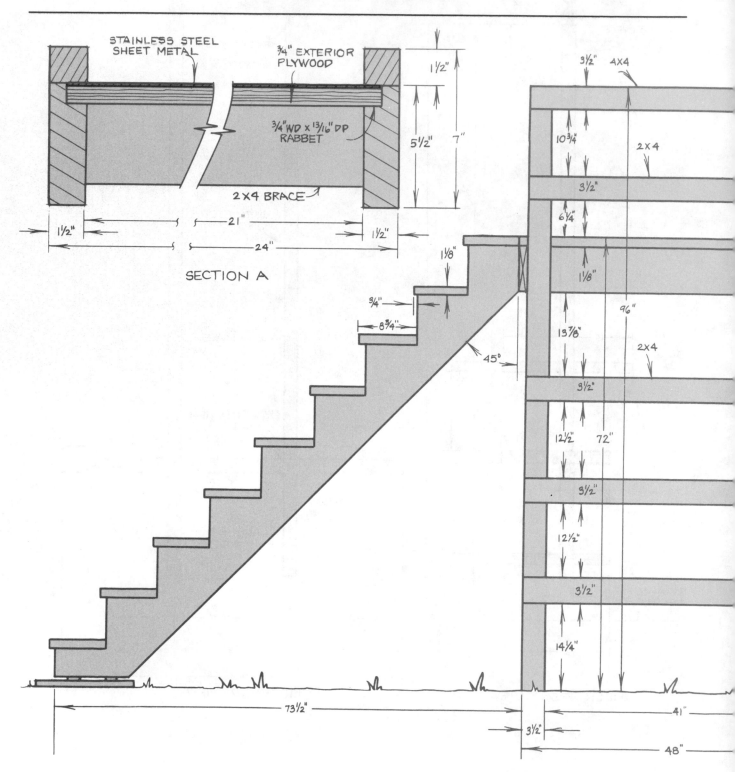

STAINLESS STEEL SHEET METAL

¾″ EXTERIOR PLYWOOD

1½″

5½″   7″

¾″ WD x ¹³⁄₁₆″ DP RABBET

2 X 4 BRACE

1½″   21″   1½″

24″

SECTION A

3½″   4X4

10¾″

2 x 4

3½″

6¼″

1⅛″

1⅛″

96″

13⅞″

45°

¾″

8¾″

2 x 4

3½″

12½″   72″

3½″

12½″

3½″

14¼″

73½″   41″

3½″

48″

Cut the rungs from 2 x 4 stock, then cut dadoes for the rungs in the posts, as you did when installing the rungs on the Rope Climb. Refer to the *Slide, Side View* drawing to see where the rungs are placed. Attach the rungs to the posts with ¼″ x 3½″ lag screws.

18 **Install the rungs.**

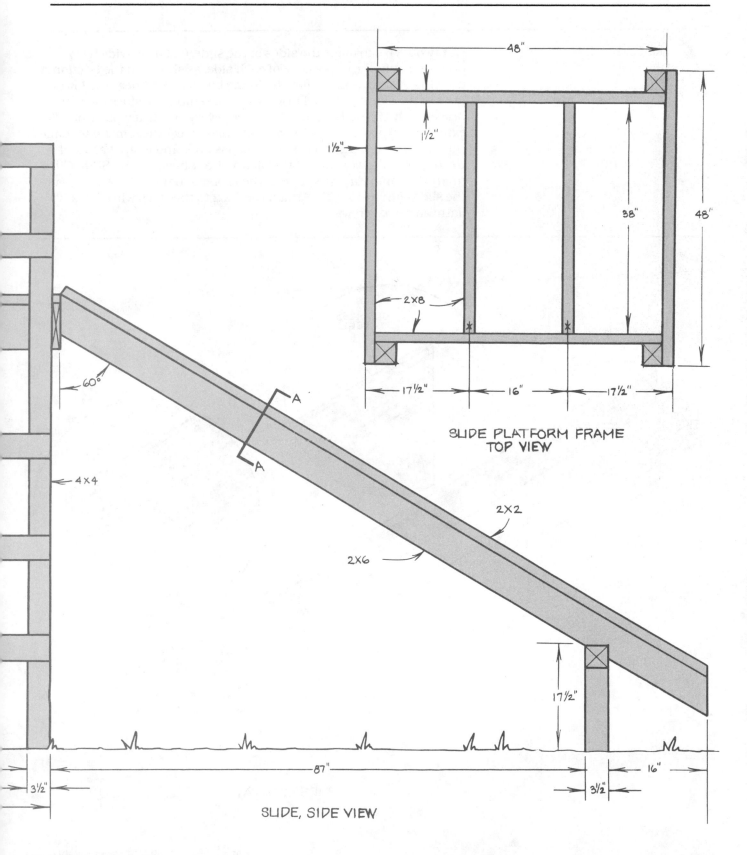

SLIDE PLATFORM FRAME
TOP VIEW

SLIDE, SIDE VIEW

19 Attach the decking.

Cut 1 x 6 decking planks to size, then lay them perpendicular to the joists. Leave a ½″ space between the planks to allow the wood to swell without warping in wet weather. Attach the planks to the joists and beams with square-shanked spiral 12d nails, then trim the planks flush with the beams.

20 Make the slide.

Cut two 2 x 6's to make the sides of the slide. Cut a ¾″ wide by ¹³/₁₆″ deep rabbet in the upper edge of each side, as shown in the *Section A* drawing. Also, cut bird's mouths in the bottom edge, near the lower ends of the sides. Cut 2 x 4 braces, and assemble the sides and the braces with 16d nails. Space the braces every 16″. Cut a piece of ¾″ exterior plywood and cover it with stainless steel sheet metal to make the slide. Stretch the steel as tight as possible, and wrap it around the upper and lower ends of the plywood, as shown in the *Slide Cutaway* drawing. Place the slide in the rabbets in the sides, then cover the slide with two 2 x 2's. Attach the 2 x 2 to the sides with #10 x 3″ flathead wood screws.

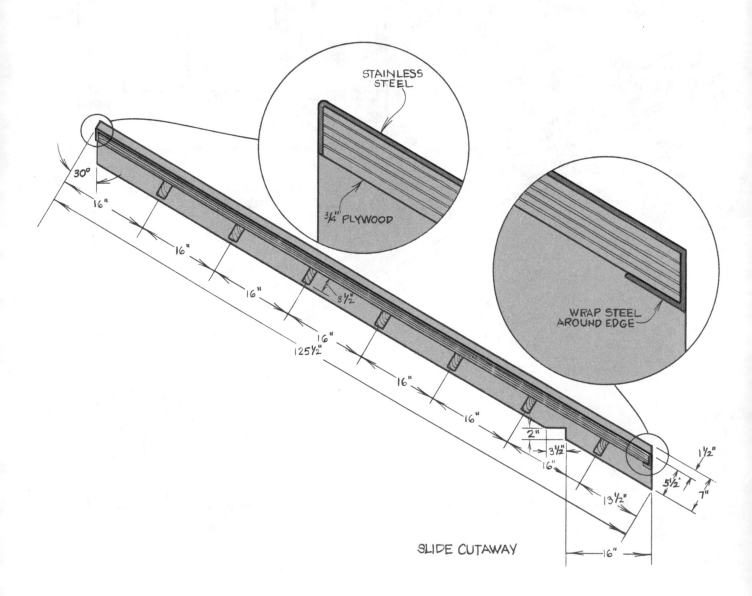

STAINLESS STEEL

¾″ PLYWOOD

WRAP STEEL AROUND EDGE

30°

16″

16″

16″

16″

3½″

16″

125½″

16″

16″

2″

3½″

16″

1½″

5½″

7″

13½″

16″

SLIDE CUTAWAY

Build a slide support from 4 x 4's, as shown in the *Slide Support Detail* drawing. Put the slide in place, with the upper end against the platform and the bird's mouths in the sides fitted over the support. Attach the slide to the support and the platform with framing anchors and nails. If you want, reinforce the slide-to-platform connection with ½" x 8" lag screws. Pass these screws through the front beam and into the side of the slide from inside the platform frame.

21 Install the slide.

Cut two stringers from 2 x 12 stock. Using a carpenter's square, mark off the rise and run for each stair all the way up the stringer, as shown in the *Stair Stringer Detail* drawing. Use a handsaw to cut the stringers.

22 Cut the stairway stringers.

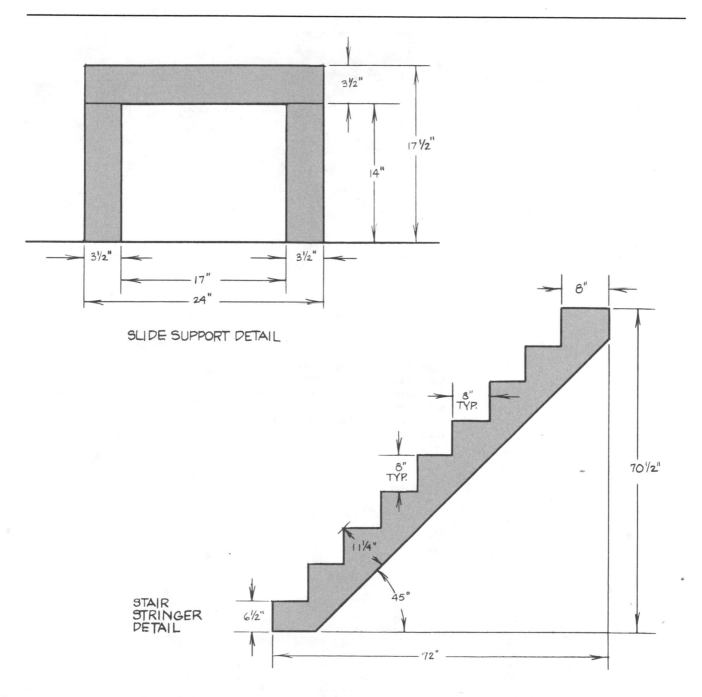

SLIDE SUPPORT DETAIL

STAIR STRINGER DETAIL

23 Set the blocks and anchor bolts.

The lower end of the stairway must rest on solid ground or it won't stay level—and safe—for very long. Use concrete blocks as footers, planting them in the ground so that about 1″ sticks up above the surface. Fill the insides of the blocks with concrete, and set ½″ x 6″ bolts in this concrete to serve as anchor bolts.

24 Attach the stringers to the frame and concrete blocks.

Drill holes in the bottoms of the stringers to fit over the anchor bolts. Place 3-4 washers over each bolt, then put the stringers in place on the footers. (See Figure 6.) The washers will hold the stringers off the footers slightly, and help prevent water damage. Attach the upper end of the stringers to the platform using metal framing anchors, as shown in Figure 7.

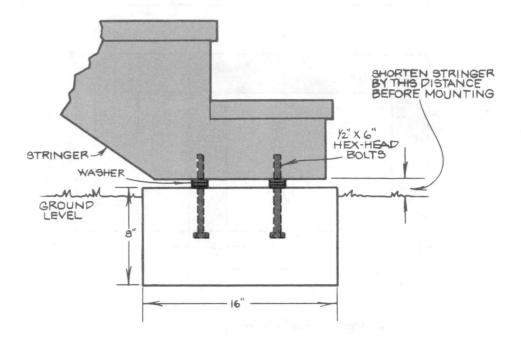

Figure 6. Place washers on the stair anchor bolts to hold the stringers up off the footers. This will keep the wood away from the concrete so moisture won't rot the stringers.

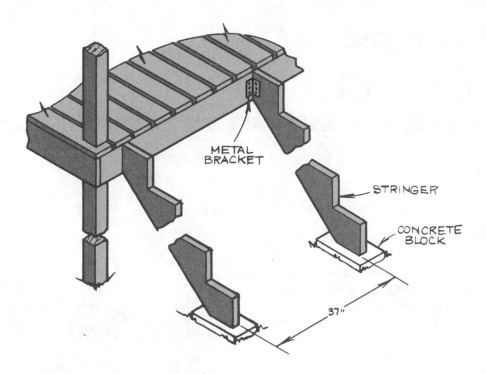

METAL
BRACKET

STRINGER

CONCRETE
BLOCK

37"

*Figure 7. Attach the stair string-
ers to the platform with metal
framing anchors.*

Cut treads from 2 x 10 stock and lay them across the stringers. Attach the treads to the stringers using 16d nails.

25 Nail the treads to the stringers.

Miscellaneous Outdoor Structures

Not all outdoor structures fit into neat little categories like storage buildings, gazebos, and fences. What about all those hard-to-classify projects that make the great outdoors more comfortable for man—and beast?

We've lumped all those other popular projects into this last section. Here you'll find wishing wells and raised garden beds to add a charming touch to any landscape. Many a pedestrian would welcome a footbridge over rocky terrain, so we've got one of those, too. Likewise, there's a trash can enclosure. It may not be as romantic a structure as a wishing well or a footbridge, but the enclosure has a rustic appeal and a very practical purpose. And, not to forget the beasts, we've included a sturdy dog house and several types of birdhouses, plus simple instructions for altering these structures to fit your dog, or the types of birds you hope to attract. So, if you've got a little time and a little scrap wood left over from the garage or the fence you just built, get out the hammer and saw.

Before You Begin

As with all the designs in this book, these projects can be easily altered to suit your own tastes and needs. You can mix and match features, designs, and construction techniques as you feel appropriate. For example, you can put a wood floor in your trash can enclosure by following the steps for framing a floor outlined in many other chapters in this book. Or, put the trash can enclosure on a slab foundation, if you want something that's easy to clean up. Just flip the pages and you'll find the techniques you need.

Once you've adjusted the design, consider the materials. The specific materials for each project are listed in the following chapters, but generally use pressure-treated wood only for the trash can enclosure, the footbridge, and the wishing well. The chemicals used to treat wood can be poisonous to animals, so use naturally rot-resistant wood such as redwood or cedar for the dog house and bird houses.

Also, consider your neighbors. Although these are little projects, they may still be restricted by local building codes. Visit your local building inspections office before you begin work. If needed, obtain a building permit.

Dog House

A dog's home is his castle, so if you're planning to build a new abode for your pooch you'll want to make it as comfortable for him as possible. But that doesn't mean you have to spend the month's mortgage on top-quality lumber and fancy trim. With a little planning, some basic carpentry skills, and a pile of scrap lumber you can construct a dog house fit for a king.

This dog house features a shed roof raised 5″ on the entrance side to channel the rain away from the door. The roof overhangs 5″ over all four sides to keep rain and snow from clinging to the structure. This lengthens the life of your dog house, and keeps your dog warm and dry. The roof is attached with four bolts so it can be easily removed to clean the inside.

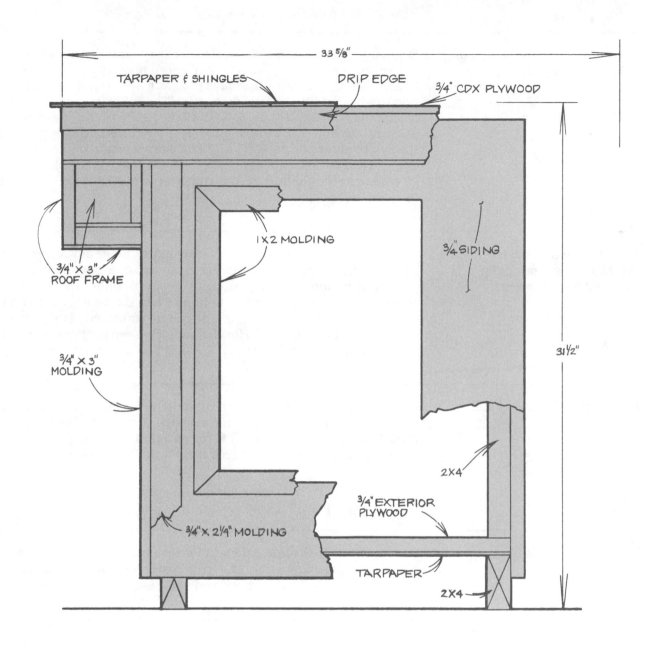

FRONT ELEVATION

Labels in diagram:
- 33 5/8"
- TARPAPER & SHINGLES
- DRIP EDGE
- 3/4" CDX PLYWOOD
- 3/4"X 3" ROOF FRAME
- 1 X 2 MOLDING
- 3/4" SIDING
- 3/4" X 3" MOLDING
- 2X4
- 3/4" EXTERIOR PLYWOOD
- 3/4"X 2 1/4" MOLDING
- TARPAPER
- 2X4
- 31 1/2"

Materials

Use 2 x 4 pressure-treated lumber for the floor frame, to prevent moisture and insect damage. Redwood or cedar will also do.

Use CDX (exterior) plywood for the floor—½″ thick plywood for small dogs, and ¾″ thick for medium-sized and large dogs.

Use 1 x 6 yellow pine siding for the walls and roof. The dog is likely to chew on these parts—especially if you have a puppy—so DO NOT use pressure-treated lumber. Much of this lumber is soaked with chemicals that could be harmful to the animal if ingested. You can also use plywood for these parts.

You'll also need some ¾″ thick #2 pine shelving for the roof frame and trim, tarpaper or plastic, drip edge, shingles, galvanized nails, roofing nails, bolts, washers, and wing nuts.

DOG HOUSE

Planning Your Dog House

1   Adjust the dimensions of the house to fit your dog.

Begin by considering the size and weight of your dog. This will determine how large—or small—you'll need to build your dog house. Generally, dogs can be comfortable in a fairly small space in proportion to what his human friends require. The rule of thumb is to design the house so it is 1½ to 2 times the length of the dog (don't measure the tail). The width should be three-fifths of the length and the height should be one-fifth higher than the height of the standing dog. (Measure the height of the dog from the top of the skull to the ground in a straight line.) As drawn here, the dog house will fit most medium-sized dogs.

2   Make further adjustments based on your pet's individual requirements.

Once you determine the size of your structure, make adjustments to suit your pet's individual needs. For example, some dogs may require only enough room to turn around and lie down, while other more active dogs may want a little extra room when the weather's too bad for playing outdoors. If you have a bitch and you breed her regularly, you may want to make room for the puppies. Keep your dog in mind when building his or her home.

3   Locate the door *above* the floor.

It's important to pay special attention to the door on your dog house. An open door design, such as the one illustrated here, should be raised about 4″ above the floor to prevent drafts from disturbing your sleeping dog. It will also keep him warmer in the winter.

Making the Floor

4   Cut and assemble the floor frame.

Cut the 2 x 4's to the proper length for the floor frame, as shown in the *Floor Frame Layout*. Set the frame parts on edge and nail them together, using 16d nails. Check and make sure the frame is square.

5   Insulate the frame and attach the floor.

Staple a vapor barrier of tarpaper or UV resistant plastic to the floor frame. (See Figure 1.) In particularly wet areas, use *two* sheets. This will help insulate the floor and help keep your dog warm and dry. Lay the plywood floor over the vapor barrier and attach it to the frame with 6d nails.

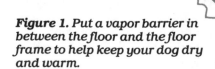

3/4″ THICK EXTERIOR PLYWOOD

TAR PAPER OR PLASTIC

FLOOR FRAME

Figure 1. Put a vapor barrier in between the floor and the floor frame to help keep your dog dry and warm.

Making the Walls

Cut the 2 x 4's to make the wall posts, mitering the tops at 82½°, as shown in the *Wall Assembly, Side View* drawing. If you're building a house for a medium-sized dog, the front edge of the front posts should be 5″ higher than the back edge of the back posts. For smaller or larger dog houses, you'll need to adjust the rise to keep the proper slope on the roof.

6 **Cut the wall posts.**

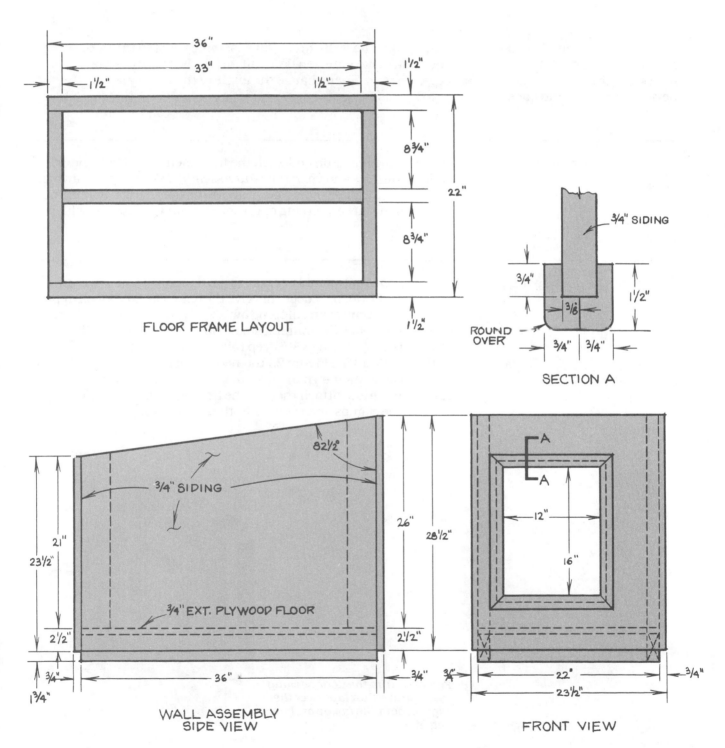

FLOOR FRAME LAYOUT

SECTION A

WALL ASSEMBLY
SIDE VIEW

FRONT VIEW

7   Cut and assemble the walls.

Because the dog house is a small structure, there is no proper frame to make. Instead cut out the side, front, and back walls from plywood or siding. Bevel the top edge of the front and back walls, and slope the side wall at 82½°, to match the posts. Nail the side walls to the wall posts with 6d nails.

8   Attach the walls to the floor frame.

Set the side walls on the floor, and nail the bottom edges to the floor frame with 6d nails. Then nail the front and back walls to the assembly.

TIP   You may find it easier to bevel and slope the walls *after* you assemble them to the floor frame. Just leave yourself plenty of extra stock and make sure you don't hit any nails. Cut the slope in the side walls first, using a circular saw or saber saw. Then change the angle of the blade and cut the bevel in the front and back.

9   Cut the door opening.

Turn the dog house on end, with the front facing up. Mark the door opening, as shown in the *Wall Assembly, Front View* drawing. Remember, the door should be 4″ above the floor. Drill a hole in the door area to start a 'piercing cut', then saw out the opening with a saber saw.

10   Frame the door opening.

The door opening should be framed, particularly if you're working with plywood. Not only does this give the door a finished look, it prevents the plywood from chipping away and protects the pooch from splinters. To make the framing stock, rip some 1 x 2's from the #2 pine and cut a ¾″ wide x ⅜″ deep rabbet in them. Miter the stock to fit the opening. (See Figure 2.) You need to make both an *inside* and *outside* frame, as shown in *Section A*. Temporarily clamp the frame parts in place and attach them to the front wall with 2d nails. Remove the clamps and round over the edges of the frame.

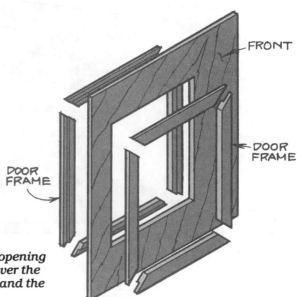

Figure 2. *Trim the door opening with a frame that laps over the edge on both the inside and the outside.*

Making the Detachable Roof

To make the front-to-back roof frame parts, rip 3″ stock from the #2 pine. Miter the ends at 82½°, as shown in the *Roof Frame, Side View*. To make the side-to-side part, bevel-rip them at the same angle and cut them off square. The overall width of the side-to-side parts should be 3⅛″. Set the outer parts on edge and assemble them with 4d nails.

11 Cut and assemble the outer roof frame.

Cut plywood sheathing to fit flush with the outside edges of the roof frame. Then nail the sheathing to the frame, using 4d nails. Turn the roof upside down on the floor of your shop, and turn the dog house upside down on top of it. Center the dog house on the roof and mark the position. Remove the dog house and assemble the inner roof frame parts to the outer frame with 4d nails, as shown in the *Roof Frame, Bottom View*. Turn the roof over and nail the roof to the inner frame.

12 Assemble the roof and inner frame.

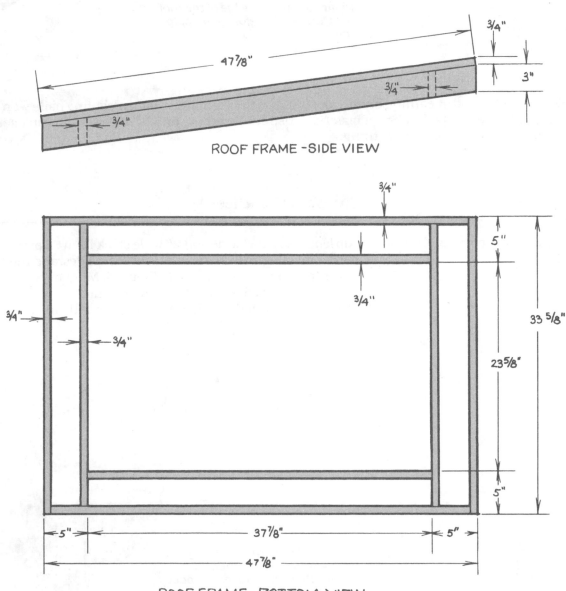

ROOF FRAME – SIDE VIEW

ROOF FRAME – BOTTOM VIEW

13   Attach the roof to the dog house.

Turn the dog house right side up and position the roof on top of it. The fit should be snug, but not too snug. Drill ¼″ holes through the inner frame, the side walls, and the wall posts, as shown in the *Finished House, Side View.* From inside the house, insert ¼″ x 3½″ carriage bolts in the holes. Secure the roof with washers and wing nuts on the outside.

14   Finish the roof.

Attach drip edge all around the edge of the roof. Staple two layers of tarpaper to the sheathing to waterproof the roof, then attach shingles or roofing felt according to manufacturer's directions. (See Figure 3.)

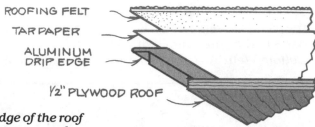

ROOFING FELT

TAR PAPER

ALUMINUM DRIP EDGE

½″ PLYWOOD ROOF

Figure 3. Line the edge of the roof with 'drip edge' before you apply the roofing materials.

TIP   The plywood sheathing on the roof must be thick enough that the roofing nails won't poke through when you attach the roof materials. Use ¾″ sheathing and ⅝″ nails. If you can't get short nails, use two layers of ½″ sheathing.

Finishing Touches

15   Attach the corner molding.

Rip lengths of 1¼″ wide and 2″ wide stock from #2 pine to make the corner molding. Attach the 2″ stock to the sides, and the 1¼″ stock to the front or back, as shown in Figure 4. Miter or bevel the top ends of these pieces at 82½°, so that they will butt up flush against the roof frame. Attach the molding to the walls with 4d nails.

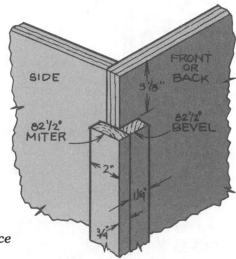

SIDE

FRONT OR BACK

3⅛″

82½° MITER

82½° BEVEL

2″

1¼″

¾″

Figure 4. The corner molding wraps around the corner, as shown. Remember to leave space at the top for the roof frame.

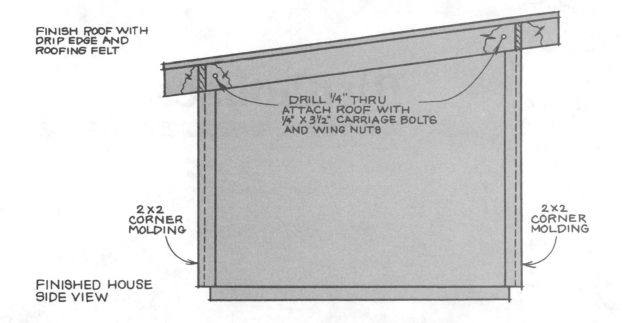

FINISH ROOF WITH
DRIP EDGE AND
ROOFING FELT

DRILL ¼" THRU
ATTACH ROOF WITH
¼" X 3½" CARRIAGE BOLTS
AND WING NUTS

2 X 2
CORNER
MOLDING

2 X 2
CORNER
MOLDING

FINISHED HOUSE
SIDE VIEW

---

16   **Paint the dog house.**

Paint the exposed wooden parts of the dog house with exterior paint. Use latex paint, particularly if your dog likes to gnaw.

---

17   **Add bedding to the floor, if you want.**

No bedding is necessary in mild weather, although you can use cedar shavings, straw, or newspaper shreds if your dog is accustomed to a "mattress." In cold weather, use at least 4″ of straw, cedar shavings or heavy carpet or blanket (nailed down).

TIP   Don't put linoleum on the floor because it's cold, tears easily and is chewable.

Location

18   **Select a spot for the house.**

Finally, put the house out where your dog can enjoy it. You may not want to put it on a permanent foundation—a 'mobile' dog house will allow you to reposition it when the weather changes. In chilly weather, the house should face east or south to get the warmth of the sun. The back of the house should be placed against the prevailing wind. You can also put it next to a fence or wall to further protect the dog from the elements. In the summer, move the house to a shady spot to keep your dog from overheating. For a mobile foundation, just prop the corners of the house up on four bricks. Make sure the house is at least 4″ off the ground so the moisture won't rot the wood. An airspace under the floor also helps keep the house cooler in the summer and warmer in the winter.

TIP   If your dog house is particularly large, consider putting it on wheels. Use a spike for an 'anchor'. The wheels make it easier to move and keep it off the ground.

Birdhouses

There's something wonderfully restful about watching the birds building nests, tending the eggs, feeding their young. If you enjoy birds, you can attract them to your backyard by building a birdhouse or two. Spend a few hours in your shop, and you can have a mini-nature preserve right outside your picture window.

However, building a birdhouse isn't the simple project you may think it is. It's a science! Each species of birds has its own nesting habits, and prefers a particular type of home. Shown here are four basic birdhouses—a nesting shelf, a 'mounted' house, a 'hanging' house, and an 'apartment' house. We've also included the necessary information so that you can adjust the design of these houses— the size of the opening, floor size, and depth of the cavity—to attract the species you'd like to watch.

MISCELLANEOUS OUTDOOR STRUCTURES

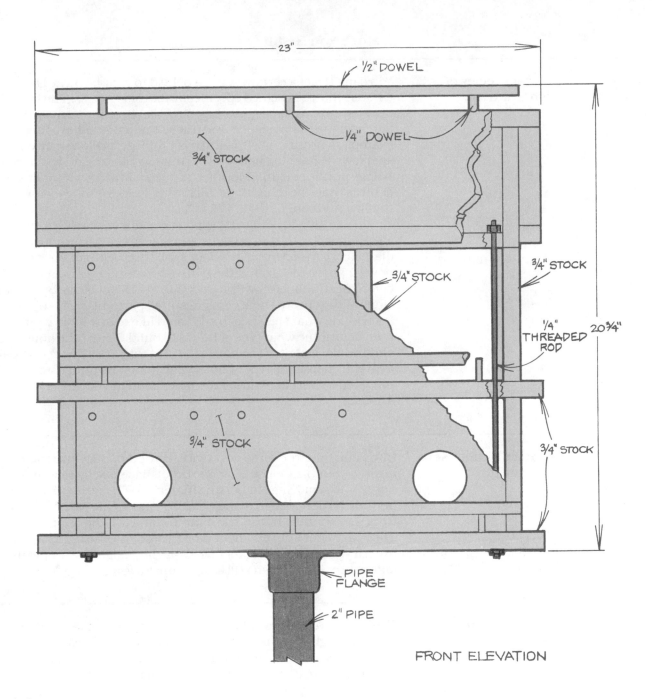

23"

1/2" DOWEL

1/4" DOWEL

3/4" STOCK

3/4" STOCK

3/4" STOCK

1/4" THREADED ROD

20 3/4"

3/4" STOCK

3/4" STOCK

PIPE FLANGE

2" PIPE

FRONT ELEVATION

Materials

Birds live in trees; trees are made of *untreated* wood, so use untreated wood to build your birdhouses. The chemicals used in pressure-treated wood may be harmful to birds. Select redwood, cedar, or cypress instead. These woods are naturally resistant to damage caused by moisture and insects. We've used ¾″ wood for the birdhouses in this chapter, but if you have some redwood, cedar, or cypress scraps in larger sizes around the house, adapt the design to make use of them and save money. Purchase dowels to serve as perches.

In addition to the wood, purchase waterproof (resorcinol) glue, non-rusting hinges, and galvanized nails. Birds will shy away from bright colors or reflective materials, so paint the hardware on your birdhouse with a flat black, green, or brown. Give the hardware two coats before you hang or mount the house.

Before You Begin

1 Decide what species of birds you wish to attract.

You can attract a certain species of bird to your backyard by building a house that is designed specifically for that species. The four types of houses shown here—nesting shelf, hanging house, mounted house, and apartment house—each attract a limited number of species. For example, a nesting 'shelf' will attract robins, swallows, phoebes, and sparrows. A hanging house will attract wrens, and an apartment house, purple martins. Mounted houses will attract a wider variety of birds than the other three, including hawks, owls, woodpeckers, finches, and songbirds.

In addition to being attracted to certain types of houses, birds also respond to the dimensions and placement of the house—the diameter of the opening, the floor space, the depth of the cavity, and how far above the ground the house is mounted. As we mentioned before, nesting shelves will attract four species of birds. But the dimensions of the one we show here are suited particularly to robins.

Decide what species of birds you'd like to attract to your yard, then determine what type of house to build. Adjust the dimensions and the placement of the house to suit that particular species. To help you choose the type of house and adjust its dimensions, we've prepared a chart for some of the more popular species in North America.

2 Make sure the birdhouse can be easily cleaned.

Birds don't use the same nest twice. However, you can attract other birds to your birdhouse after one brood has gone by cleaning out the house and removing old nesting materials. The birdhouses shown here are all designed to make cleaning easier. One side swings away from the mounted house, the bottom comes off the hanging house. The apartment house comes apart in sections after you loosen four nuts. If you change the dimensions or the design of your birdhouse, make sure that you can still clean it once a year.

3 Be sure that the bird-house is properly ventilated and drained.

Birds require good ventilation and drainage—don't make your bird-house airtight or watertight. Young birds can suffocate on a hot day in a wooden birdhouse if there are no air holes. And they can drown in a birdhouse if you don't place holes in the bottom for drainage.

4 Mount your house where you—and the birds—can enjoy it all year round.

While many people take their birdhouses inside for storage during the winter, consider leaving yours out. You may attract more winter-time birds. Birds who normally wouldn't make their nests in a house in the summer will use it for shelter in the winter. They will huddle together for protection against icy winds and sub-zero temperatures. Nesting shelves even make good wintertime tray feeders.

MISCELLANEOUS OUTDOOR STRUCTURES

Dimensions and Placement of Birdhouses

Species	Floor Size	Depth of Cavity	Diameter of Entrance	Height of Entrance	Height above ground
Dimensions in		Inches			Feet
Nesting Shelves:					
American Robin	6 x 8	8			6 to 15
Barn Swallow	6 x 6	6			8 to 12
Eastern Phoebe	6 x 6	6			8 to 12
Song Sparrow	6 x 6	6			1 to 3
Mounted Houses:					
American Kestrel	8 x 8	12 to 15	3	9 to 12	10 to 30
Eastern Bluebird	5 x 5	8	1½	6	5 to 10
Chickadee	4 x 4	8 to 10	1⅛	6 to 8	6 to 15
Downy Woodpecker	4 x 4	8 to 10	1¼	6 to 8	6 to 20
House Finch	6 x 6	6	2	4	8 to 12
Northern Flicker	7 x 7	16 to 18	2½	14 to 16	6 to 20
Nuthatch	4 x 4	8 to 10	1¼	6 to 8	12 to 20
Red-bellied Woodpecker	6 x 6	12 to 15	2½	9 to 12	12 to 20
Red-headed Woodpecker	6 x 6	12 to 15	2	9 to 12	12 to 20
Screech Owl	8 x 8	12 to 15	3	9 to 12	10 to 30
Starling	6 x 6	16 to 18	2	14 to 16	10 to 25
Titmouse	4 x 4	8 to 10	1¼	6 to 8	6 to 15
Tree Swallow	5 x 5	6	1½	1 to 5	10 to 15
Hanging Houses:					
Carolina Wren	4 x 4	6 to 8	1½	4 to 6	6 to 10
House Wren	4 x 4	6 to 8	1 to 1¼	4 to 6	6 to 10
Winter Wren	4 x 4	6 to 8	1 to 1¼	4 to 6	6 to 10
Apartment House:					
Purple Martin*	6 x 6	6	2½	1	12 to 20

*Dimensions are for one compartment (one pair of martins).

Adapted from *Home for Birds*, U.S. Department of the Interior, Fish and Wildlife Service.

Building a Nesting Shelf

1 Check the dimensions for your particular species.

As shown, this nesting shelf is designed to attract robins; however, you can easily adjust these dimensions to attract other varieties of birds. Check the chart in the beginning of this chapter. For additional information, contact your local chapter of the National Audubon Society.

2 Cut and shape all the pieces.

Cut all the pieces to size, as shown in the *Nesting Shelf, Front View* and *Nesting Shelf, Side View* drawings. Bevel the edges of the roof pieces at 60°, so they will come together in a peak. Drill ¼″ drainage holes in the corners of the floor, to allow rainwater to drain out of the shelf, as shown in the *Floor Layout* drawing. Also, drill ¼″ mounting holes in the back.

3 Assemble the nesting shelf.

Assemble the floor, the sides, and the front with 6d finishing nails and resorcinol glue. (Resorcinol is waterproof and won't dissolve or lose its grip when exposed to the weather.) Then attach the shelf assembly to the back with #10 x 2″ flathead wood screws and glue. Glue the roof pieces together, and wait for the glue to dry. When it's dry, attach the roof to the back with screws and glue, as you did with the shelf.

TIP To hold the roof pieces together while the glue dries, nail two wire staples across the joint between the pieces, one staple on either end of the roof assembly. Don't pound these staples all the way into the wood, so that you can easily remove them when the glue dries.

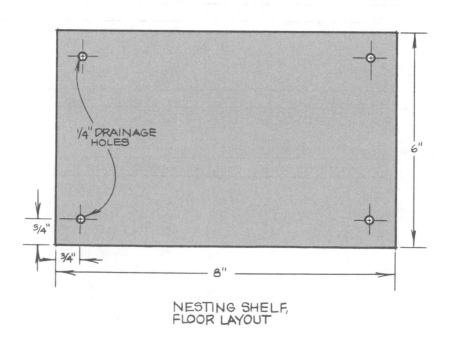

¼″ DRAINAGE HOLES

¾″

¾″

6″

8″

NESTING SHELF,
FLOOR LAYOUT

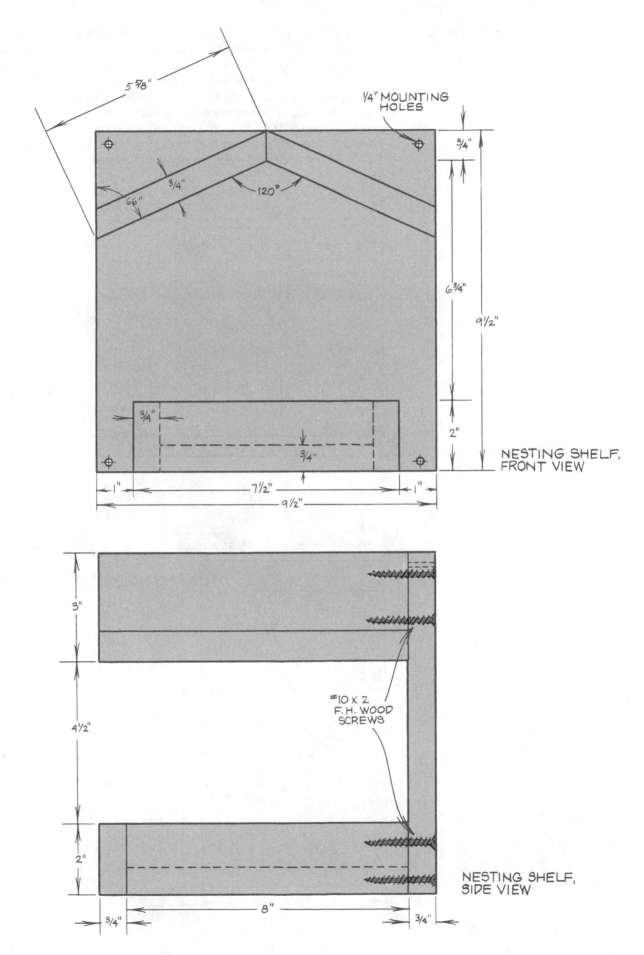

5 5/8"

1/4" MOUNTING HOLES

3/4"

3/4"

66"

120°

6 3/4"

9 1/2"

3/4"

3/4"

2"

NESTING SHELF, FRONT VIEW

1"

7 1/2"

1"

9 1/2"

3"

4 1/2"

#10 X 2 F.H. WOOD SCREWS

2"

NESTING SHELF, SIDE VIEW

3/4"

8"

3/4"

4 **Mount the nesting shelf.** Select a likely spot in a tree or under the eaves of your house where birds might come to nest. From scrap wood, cut four spacer blocks ¼"-¾" thick. Drill ¼" holes through each. With glue, secure the spacer blocks on the back of the shelf, lining up the holes in the blocks with the mounting holes. (The spacer blocks will keep water from collecting between the nesting shelf and whatever surface you've mounted it to. This, in turn, will keep the water from soaking through the wood, and the shelf will stay drier.) Mount the shelf with #12 x 2½" roundhead wood screws, as shown in Figure 1.

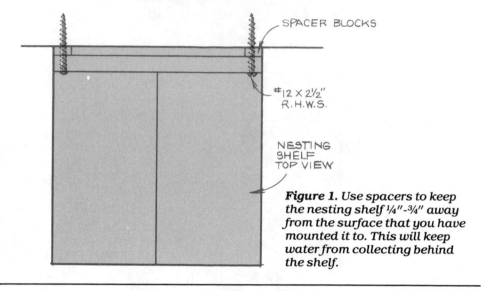

Figure 1. Use spacers to keep the nesting shelf ¼"-¾" away from the surface that you have mounted it to. This will keep water from collecting behind the shelf.

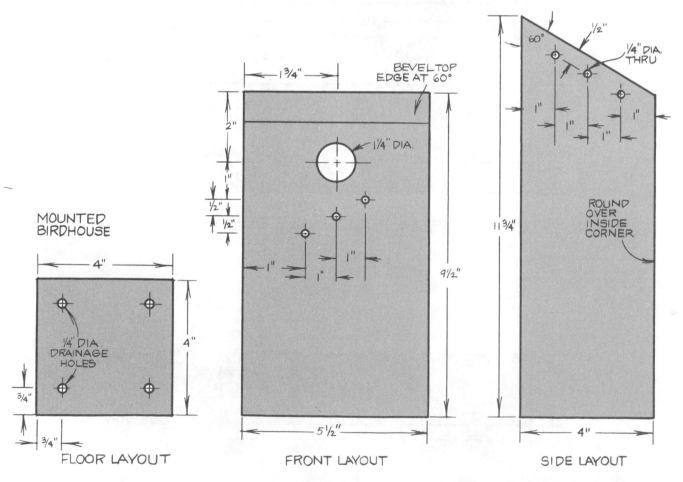

MOUNTED BIRDHOUSE

FLOOR LAYOUT

FRONT LAYOUT

SIDE LAYOUT

Building a Mounted House

As shown, this mounted house is designed to attract chickadees; however, you can easily adjust these dimensions to attract other varieties of birds. Check the chart in the beginning of this chapter. For additional information, contact your local chapter of the National Audubon Society.

1 **Check the dimensions for your particular species.**

Cut all the pieces to size, as shown in the *Mounted Birdhouse, Front View* and *Mounted Birdhouse, Side View* drawings. Bevel the edges of the roof piece at 60°, and miter the top edge of the sides to match. Drill ¼″ drainage holes in the corners of the floor, to allow rainwater to drain out of the house, as shown in the *Floor Layout* drawing. Also, drill ¼″ mounting holes in the back, ¼″ ventilation holes in the sides where shown in the *Side Layout* drawing, and ¼″ holes in the front to mount the perches where shown in the *Front Layout* drawing. Finally, drill an entrance hole in the front, above the perch holes.

2 **Cut and shape all the pieces.**

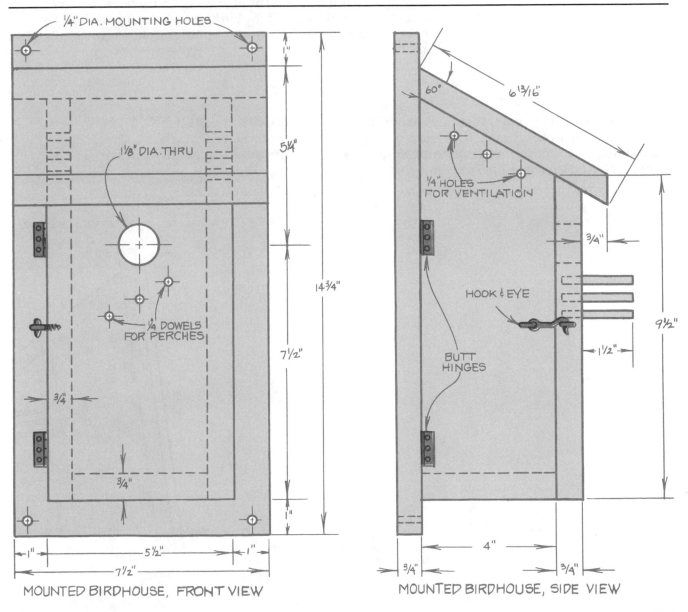

MOUNTED BIRDHOUSE, FRONT VIEW

MOUNTED BIRDHOUSE, SIDE VIEW

3 **Assemble the birdhouse.**

Assemble the back, floor, roof, front, and *one* side with 6d finishing nails and waterproof resorcinol glue. While the glue cures, paint the hardware—butt hinges and a hook-and-eye—with flat paint to dull the shiny metal surfaces. Round over the inside corner of the front edge of the remaining side. (This will allow you to open and shut the side, like a door.) Using the butt hinges, mount this side to the back, as shown in the working drawings. Keep it closed with the hook-and-eye. Finally, cut lengths of ¼" dowel and glue them in the perch holes on the front of the birdhouse. These dowels should protrude 1½".

TIP If you want to paint the outside of your birdhouse, allow it to sit for at least 3 weeks after you paint it so that the smell from the paint will dissipate. This smell will keep birds away.

---

4 **Mount the house at a slight angle.**

Select a likely spot in a tree or under the eaves of your house where birds might come to nest. From scrap wood, cut four spacer blocks ¼"-¾" thick. Drill ¼" holes through each. With glue, secure the spacer blocks on the back of the birdhouse, lining up the holes in the blocks with the mounting holes. Mount the birdhouse with #12 x 2½" roundhead wood screws.

TIP Birds seem to prefer that these mounted houses be tipped forward slightly, 5° to 15°. (See Figure 2.) This probably makes the entrance hole seem more protected from the weather and the elements. To mount the house at an angle, make the upper mounted blocks thicker than the lower blocks. Or find a surface on a tree limb that is slightly angled.

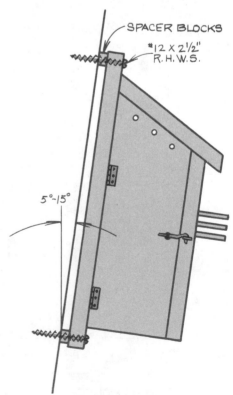

SPACER BLOCKS

#12 X 2½"
R.H.W.S.

5°-15°

Figure 2. Mount the birdhouse at a slight angle, from 5° to 15°. This keeps the house drier and makes the birds feel more protected.

MISCELLANEOUS OUTDOOR STRUCTURES

Building a Hanging House

Wrens are the only birds who consistently nest in hanging houses, and as such, this house is designed for house wrens or winter wrens. By changing the dimensions slightly, you can also attract Carolina wrens. Check the chart in the beginning of this chapter to find the proper dimensions for the species you want to attract. For additional information, contact your local chapter of the National Audubon Society.

1 Check the dimensions for your particular species.

> **TIP** If you aren't partial to wrens, you can also mount this house on a pole and use the basic design to attract many other species that would rather live in a mounted house.

Cut all the pieces to size, as shown in the *Hanging Birdhouse, Front View* and *Hanging Birdhouse, Side View* drawings. There are many bevels and miters to cut in this structure: Bevel the edges of the roof pieces at 45°, and bevel the top edge of the sides at 60°. The bottom edge of the sides and the adjoining edges of the bottom should be beveled at 75°. Miter the front and back as shown in the *Front Layout* drawing. Drill ¼″ drainage holes in the corners of the floor, to allow rainwater to drain out of the house, as shown in the *Floor Layout* drawing. Also, drill ¼″ ventilation holes in the front and back, and a ⅜″ hole in the front to mount the perch. Finally, drill an entrance hole in the front, above the perch hole.

2 Cut and shape all pieces.

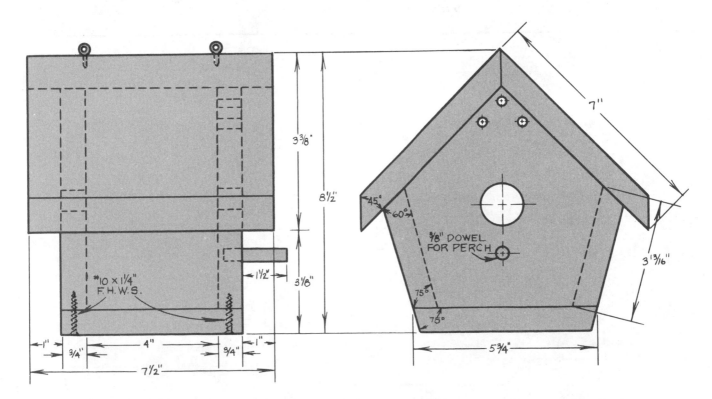

HANGING BIRDHOUSE, SIDE VIEW   HANGING BIRDHOUSE, FRONT VIEW

3 **Assemble the birdhouse.**

Assemble the roof pieces, sides, front, and back with 6d finishing nails and waterproof resorcinol glue. Screw (but *don't* glue) the floor to the birdhouse assembly, using #10 x 1¼″ flathead wood screws. This will enable you to easily disassemble the floor from the house when you want to clean it. Cut a length of ⅜″ dowel and mount it to the front, where you've drilled the perch hole. The perch should protrude 1½″.

4 **Hang the birdhouse.**

Paint four eye screws and two short lengths of chain so they won't reflect the light, then screw two of the eye screws into the peak of the roof, near the front and the back. You want to hang the birdhouse from at least *two* points for stability. (See Figure 3.) If you hang it from just one, it will wobble and twist in the breeze, making it difficult for the birds to land on the perch. Select a spot in a tree or under the eaves of your house where birds are likely to nest, and screw the other two eye screws into the underside of a limb or a soffit. Then hang the house, attaching the chains to the eye screws with S-hooks.

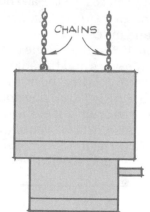

Figure 3. Hang the birdhouse from two or more chains. If you hang it from just one, the house won't be stable enough for the birds to land on the perches.

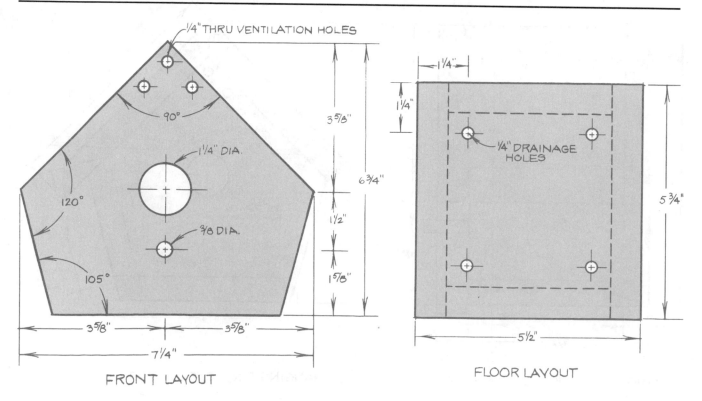

FRONT LAYOUT

FLOOR LAYOUT

MISCELLANEOUS OUTDOOR STRUCTURES

Building an Apartment House

Martins are an extremely useful bird—a single martin will eat more than 2000 mosquitoes a day! Furthermore, martins are among the few species of birds that nest in groups. A small colony will provide effective pest control for your backyard. The 'apartment' birdhouse shown here is designed specifically for purple martins, and will house up to 24 birds—12 nesting pairs.

However, to attract purple martins you not only need the right house, you need the right backyard. Make sure you have no trees or other obstructions for 40 yards around the spot where you plan to put the birdhouse. Martins swoop and glide to their nesting place, and they need plenty of open space. A nearby pond or stream, an open meadow, and telephone wires (for perches) will also help to attract the birds.

1 Be sure that your backyard environment is suited for purple martins.

Cut the floors, sides, fronts and backs, top, gable ends, and roof pieces to size, as shown in the *Apartment House, Front View* and *Apartment House, Side View* drawings. Cut the long and short dividers to size, as shown in the *Apartment House Floor Plan* drawing. Many of these parts are wider than 1 x 12 stock. For these parts, you'll have to glue up wider stock, edge to edge, using waterproof resorcinol glue.

2 Cut all the pieces to size.

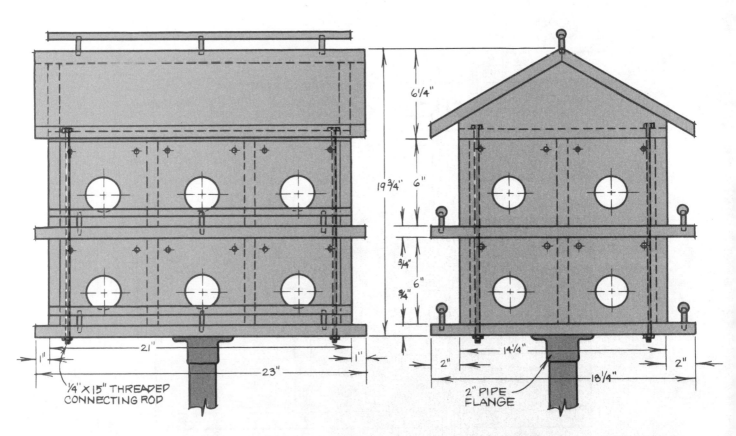

APARTMENT HOUSE, FRONT VIEW   APARTMENT HOUSE, SIDE VIEW

3 Shape and drill the parts.

Drill ¼″ holes in the floors and the top for drainage, ventilation, mounting perches, and connecting the separate assemblies of the birdhouse, as shown in the *Floor Layout* drawing. (The top won't get holes for perches, of course, but it does need ventilation and connecting holes.) Also, drill ¼″ ventilation and 2½″ entrance holes in the fronts, backs, and sides, where shown in the *Side Layout* drawing. Use a hole saw to make the entrance holes. Bevel the edges of the roof pieces at 60°, and miter the gable ends to match, as shown in the *Gable Layout* drawing. Cut notches in the short and long dividers, where shown in the *Short Divider Layout* and *Long Divider Layout* drawings.

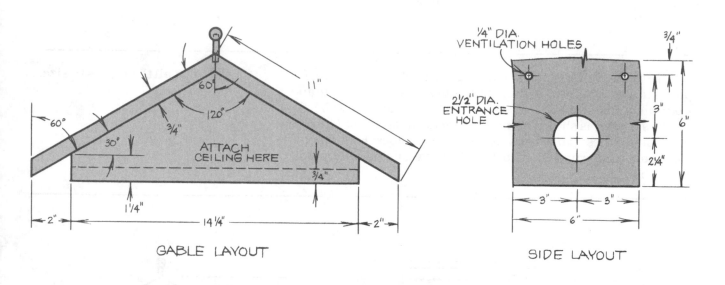

GABLE LAYOUT

SIDE LAYOUT

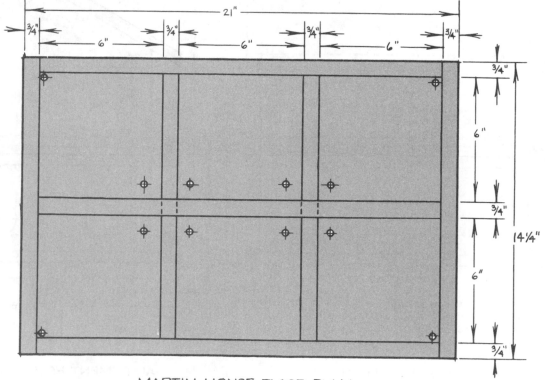

MARTIN HOUSE FLOOR PLAN

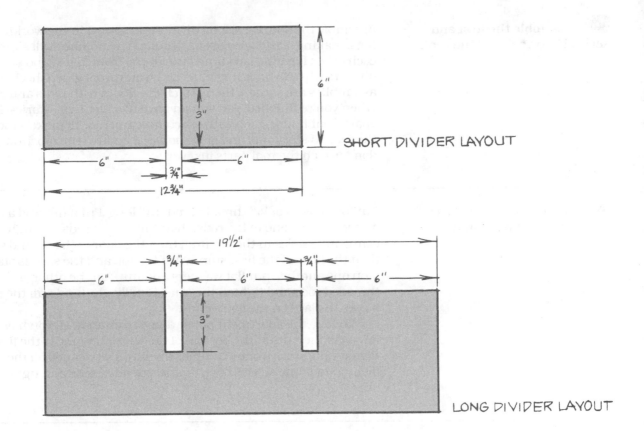

SHORT DIVIDER LAYOUT

LONG DIVIDER LAYOUT

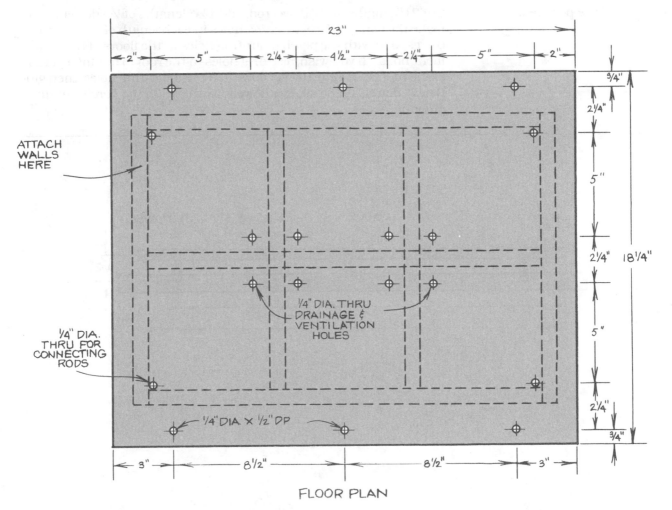

ATTACH WALLS HERE

1/4" DIA. THRU FOR CONNECTING RODS

1/4" DIA. THRU DRAINAGE & VENTILATION HOLES

1/4" DIA. X 1/2" DP

FLOOR PLAN

4 **Assemble the first and second 'stories', and the roof.**

Assemble the short and long dividers as shown in the working drawings, making a grid with six partitions. The notches will slip over each other, forming lap joints. Attach the front, back, and sides to the dividers with 6d finishing nails and waterproof glue. Check that the assemblies are square, then attach the floors with nails and glue. When you're finished, you should have two identical 'stories' for your apartment house, each with six compartments. To make a roof for the second story, nail and glue the gable ends to the top. However, don't attach the roof parts just yet.

---

5 **Connect the assemblies to make the apartment house.**

Cut four pieces of ¼″ threaded rod, 15″ long. Put a nut and a flat washer on one end of the rods, then thread the rods through the connecting holes in the top first, then the floor of the second story, then the floor of the first story. With the top and the stories stacked atop one another, put flat washers and nuts on the other ends of the threaded rods and tighten the nuts. Finally, glue and nail the roof pieces in place on the gable ends.

To take the apartment house apart to clean it, all you have to do is loosen the nuts at the bottom of the assembly, under the first floor. Remove the stories one at a time. When you've cleaned all the apartments, stack the stories back on the threaded rods and tighten the nuts.

---

6 **Install the perches.**

Cut 21″ lengths of ½″ dowel rod, and 1⅜″ lengths of ¼″ dowel rod to make the perches. Drill three ¼″ holes in each length of ½″ rod, to correspond with the perch mounting holes in the floors of the apartment house. If you wish, also drill holes in the roof to mount a perch at the roof peak. Glue the ¼″ dowels in the mounting holes, then glue the ½″ dowels to the smaller dowels, as shown in the *Perch Detail* drawing.

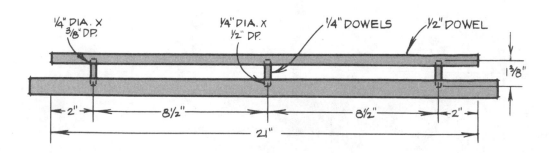

PERCH DETAIL

To mount the apartment house, purchase a 2″ pipe flange, a 3′ length of 2¼″ pipe, and a 15′ to 23′ length of 2″ pipe, threaded on one end. Dig a 3′ deep hole and place a large rock in the bottom. Set the 2¼″ pipe on the rock, and throw some gravel in the hole for drainage. Fill the rest of the hole with concrete. (See Figure 4.) Attach the pipe flange to the bottom of the apartment house with #12 x ¾″ flathead wood screws, and screw the threaded end of the 2″ pipe into the pipe flange. When the concrete cures, slip the other end of the 2″ pipe into the 2¼″ pipe, set in the ground. This arrangement will allow you to take the apartment house down from time to time for cleaning.

7 **Mount the finished apartment house.**

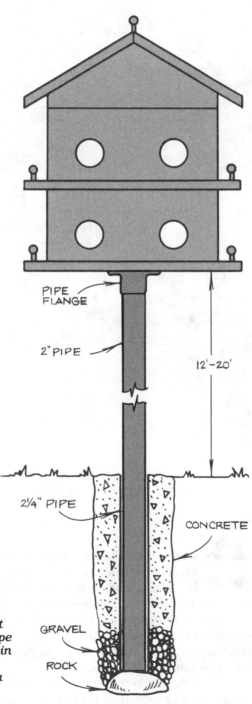

PIPE FLANGE

2″ PIPE

12′–20′

2¼″ PIPE

CONCRETE

GRAVEL

ROCK

Figure 4. Mount the apartment house on a pipe, and set the pipe into a slightly larger pipe cast in the ground. This way, you can take the house down when you need to.

Wishing Well

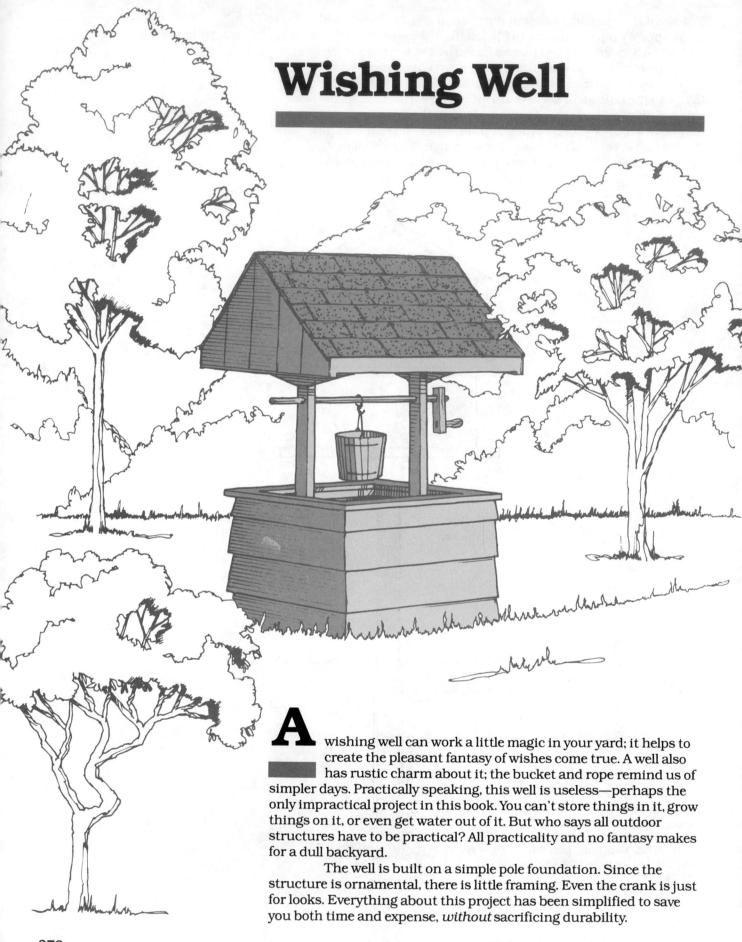

A wishing well can work a little magic in your yard; it helps to create the pleasant fantasy of wishes come true. A well also has rustic charm about it; the bucket and rope remind us of simpler days. Practically speaking, this well is useless—perhaps the only impractical project in this book. You can't store things in it, grow things on it, or even get water out of it. But who says all outdoor structures have to be practical? All practicality and no fantasy makes for a dull backyard.

The well is built on a simple pole foundation. Since the structure is ornamental, there is little framing. Even the crank is just for looks. Everything about this project has been simplified to save you both time and expense, *without* sacrificing durability.

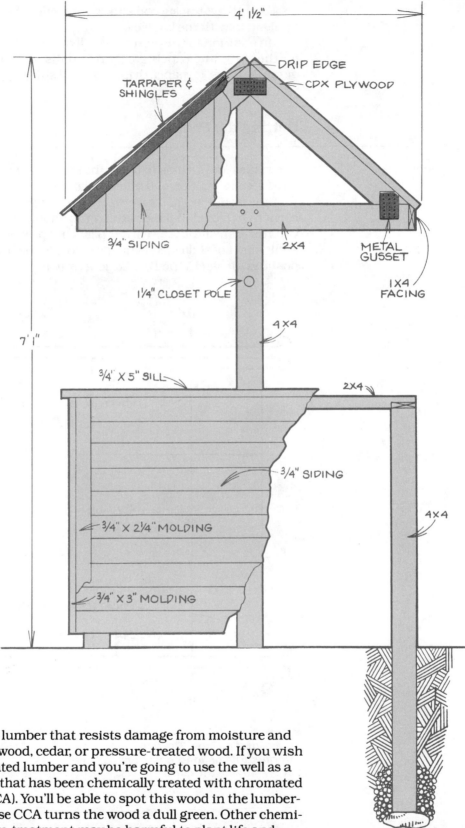

4' 1½"

DRIP EDGE

TARPAPER &
SHINGLES

CDX PLYWOOD

3/4" SIDING

2X4

METAL
GUSSET

1X4
FACING

1¼" CLOSET POLE

4X4

7' 1"

3/4" X 5" SILL

2X4

3/4" SIDING

3/4" X 2¼" MOLDING

4X4

3/4" X 3" MOLDING

SIDE ELEVATION

Materials

For this project, use lumber that resists damage from moisture and mildew, such as redwood, cedar, or pressure-treated wood. If you wish to use pressure-treated lumber and you're going to use the well as a planter, use lumber that has been chemically treated with chromated copper arsenate (CCA). You'll be able to spot this wood in the lumber-yard right off because CCA turns the wood a dull green. Other chemicals used in pressure-treatment may be harmful to plant life and could stunt the growth of sensitive flowers or vines growing on the arbor.

Use 4 x 4's for the posts, 2 x 4's for the top plate and roof frame, 'one-by' (¾″ thick) stock for the molding and sill, and ½″ thick CDX plywood for the roof sheathing. Cover the base of the well and the roof gables with ¾″ thick wood siding that will either match your home or blend in with the landscape.

In addition to this lumber, you'll also need some 1¼″ 'closet pole' for the crank and winch, galvanized nails, sheet metal nailing gussets or truss plates, tarpaper, drip edge, and shingles.

Setting the Posts

1 Lay out the posts.

Use stakes and strings to mark the location of the posts. (See Figure 1.) Mark the location of the four corner (short) posts first, then check that the layout is square by measuring diagonally, from corner to corner. Both diagonal measurements should be the same. When you're satisfied that the corner posts are properly located, put up another set of stakes and strings to locate the intermediate (long) posts, as shown in the *Post Layout* drawing.

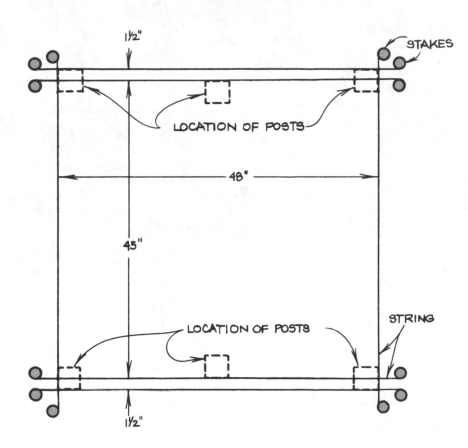

Figure 1. Locate the posts with stakes and strings. The two intermediate or 'long' posts are set 1½″ inside the corner posts.

Use a post hole digger to dig holes at least 24"-36" deep. (The bottom of the hole must be below the frost line for your area.) Make each hole at least 10 to 12 inches across to allow space for gravel and dirt fill. Put a rock about 8" in diameter at the bottom of each hole to help keep them from settling, and set the posts in the ground. Use a level to set the posts straight up and down, then hold them upright with stakes and temporary braces. Fill the holes with gravel and dirt. (See Figure 2.)

2 Set the posts in the ground.

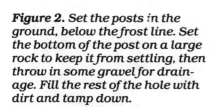

Figure 2. Set the posts in the ground, below the frost line. Set the bottom of the post on a large rock to keep it from settling, then throw in some gravel for drainage. Fill the rest of the hole with dirt and tamp down.

TAMPED EARTH

24"-36"

4"-6" GRAVEL

ROCK

TIP Do *not* tamp down the dirt right away. Wait until you've added top plates and trued up the well frame.

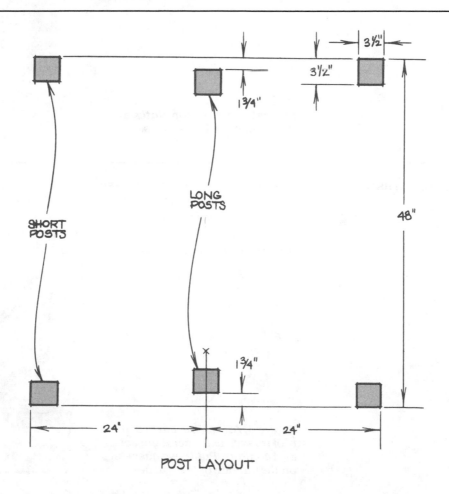

3½"

3½"

1¾"

SHORT POSTS

LONG POSTS

48"

1¾"

24"

24"

POST LAYOUT

3 **Cut the corner posts to the proper height.**

Pick a corner post and measure 34½" up from the ground. With a string and string level, find the tops of the other posts, using the first corner post as a reference. Cut the corner posts off at the proper height with a handsaw.

4 **Cut and attach the top plates to the posts.**

Cut the top plates to the dimensions shown in the *Base Frame, Side View* and *Base Frame, Top View* drawings. Cut lap joints in the ends of all four top plate pieces. (See Figure 3.) Notch two of the pieces to fit around the long posts, as shown in the top view. Check the fit of all the parts, and nail the top plate to the posts with 16d nails.

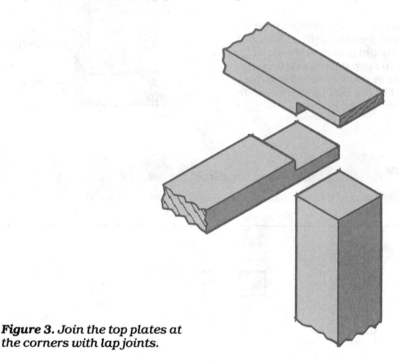

Figure 3. Join the top plates at the corners with lap joints.

5 **Build the roof trusses.**

Cut the parts for two roof trusses, as shown in the *Roof Frame, Side View* drawing. Hold the parts of each truss together with metal gussets or 'truss plates' and 4d nails, as shown in Figure 4. Put these gussets on the *outside* of the trusses only. Eventually, they will be covered over by siding and they won't show.

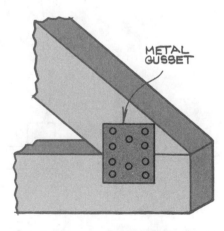

METAL GUSSET

Figure 4. Attach the parts of the roof trusses with metal gussets and 4d nails. Put these gussets on the outside of the trusses.

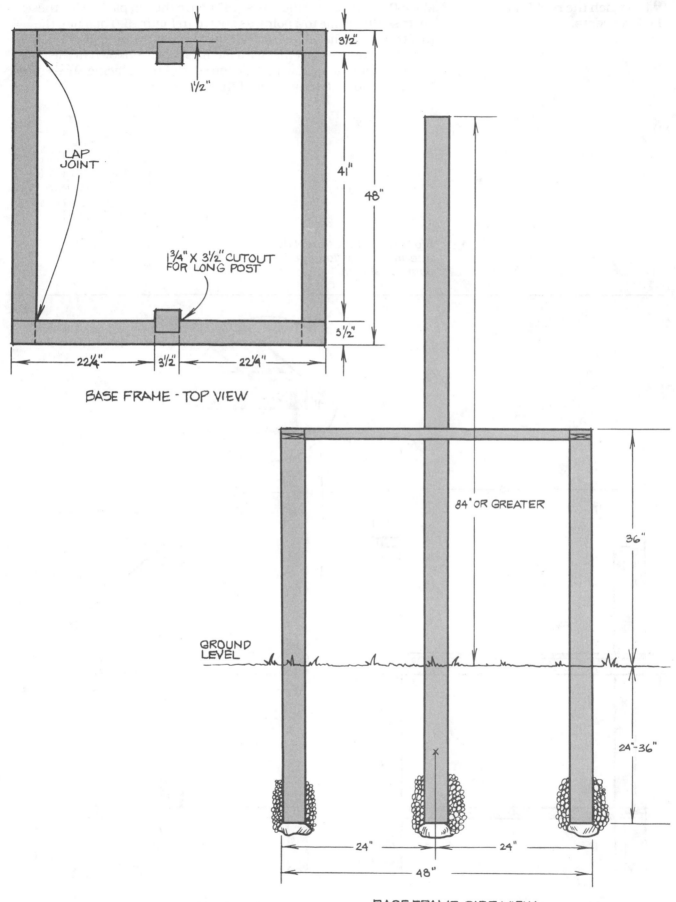

LAP JOINT

$1\frac{3}{4}$" X $3\frac{1}{2}$" CUTOUT FOR LONG POST

$3\frac{1}{2}$"

$1\frac{1}{2}$"

41"

48"

$3\frac{1}{2}$"

$22\frac{1}{4}$"

$3\frac{1}{2}$"

$22\frac{1}{4}$"

BASE FRAME - TOP VIEW

84" OR GREATER

36"

GROUND LEVEL

24"-36"

24"

24"

48"

BASE FRAME-SIDE VIEW

8 **Attach the roof frame to the long posts.**

Mark the tops of the long posts, 48″ above the top plates. Cut these posts so they come to a point, as shown in Figure 5. Then nail the roof trusses to the outside of the long posts with 16d nails. Finally, tie the two trusses together with facing strips, as shown in the *Roof Frame, Front View* drawing. The top edge of these facing strips must be mitered to match the slope of the roof.

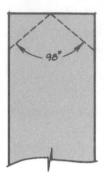

Figure 5. Cut the tops of the intermediate or 'long' posts to a point, as shown.

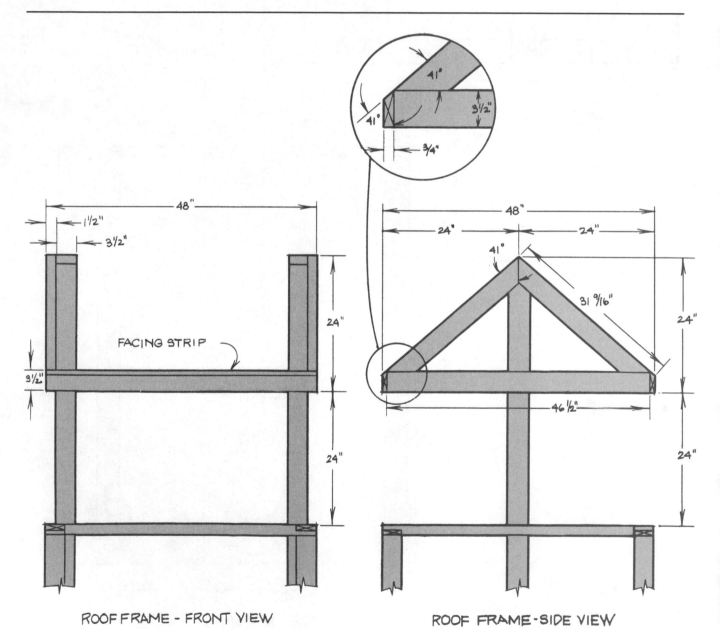

ROOF FRAME - FRONT VIEW

ROOF FRAME - SIDE VIEW

Installing the Siding and Roofing

Cover the roof frame with ½″ CDX plywood sheathing. Use 4d nails to attach the plywood to the frame. Be sure to turn the 'good' side of the plywood down.

7 **Cover the roof with plywood.**

Attach wood siding to the well base and the gable ends, as shown in the *Roofing and Siding, Front View* and *Side View* drawings. As drawn, there is horizontal siding on the base and vertical siding on the gable. This design matches many frame homes that use horizontal 'ship-lap' siding on the walls and vertical sheet siding on the gables. However, you may wish to substitute other siding materials to suit your own tastes. Whatever material you use, install it on the base so that the bottom edge of the siding comes no closer than 2″ above the ground.

8 **Cover the base and the gable ends with siding.**

TIP When the well is finished, you can hide the gap at the bottom with shrubs or other garden plants.

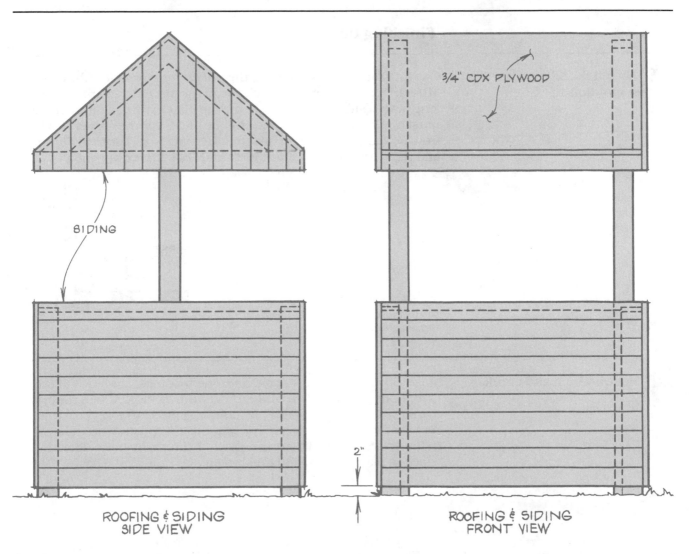

ROOFING & SIDING
SIDE VIEW

ROOFING & SIDING
FRONT VIEW

9 **Install the roofing materials.**

Cover the roof with a double layer of tarpaper. Then attach drip edge all the way around the perimeter of the roof. Attach shingles to the roof, over the tarpaper and drip edge. (See Figure 6.) Use ⅝" long roofing nails to install the shingles, so they won't poke through the bottom of the plywood sheathing.

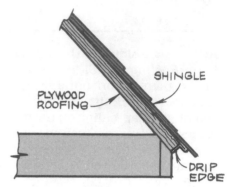

Figure 6. Cover the roof with tarpaper, then put drip edge all around the perimeter of the roof. Install shingles over the tarpaper and drip edge.

TIP If the nails still come through the plywood, or you can't find short roofing nails, use ¾" exterior plywood to sheath the roof.

Finishing Up

10 **Attach the sill and the corner molding.**

Cut the sill pieces, as shown in the *Sill Layout, Top View* drawing. Miter the ends at 45° and notch two of the pieces to fit around the long posts. Nail the sill pieces to the top plates with 6d nails. The sill must overlap the siding by ¾". Rip the pieces for the corner molding from ¾" thick stock, and nail it to the corners, as shown in Figure 7 and the *Finished Well, Front View* and *Side View* drawings.

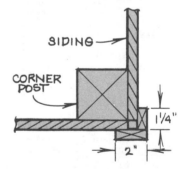

Figure 7. Finish the corners of the well base with corner molding, as shown.

11 **Make a winch and crank.**

Make a crank from 'closet pole' and 'one-by' stock, as shown in the *Crank Detail, Front View* and *Side View* drawings. Attach the crank to a piece of closet pole 49" long. This long piece of closet pole will serve as the winch. Drill 1¼" holes in the long posts, where shown in the drawings, and insert the winch through the holes. Tack the winch in place with two 8d finishing nails through the posts.

Paint or stain the exposed wood surfaces of the well to blend in with or augment the landscape. Drill a ⁷/₁₆" hole through the winch, near the middle, and thread a ⅜" rope through the hole. Tie a knot in one end of the rope so that it won't come loose. Wrap the rope around the winch several times and tie a bucket to the other end of the rope, making it look as if you've just cranked the bucket up out of the well. Finally, throw a penny in the bucket and make a wish.

12 Finish the well and add a rope and bucket.

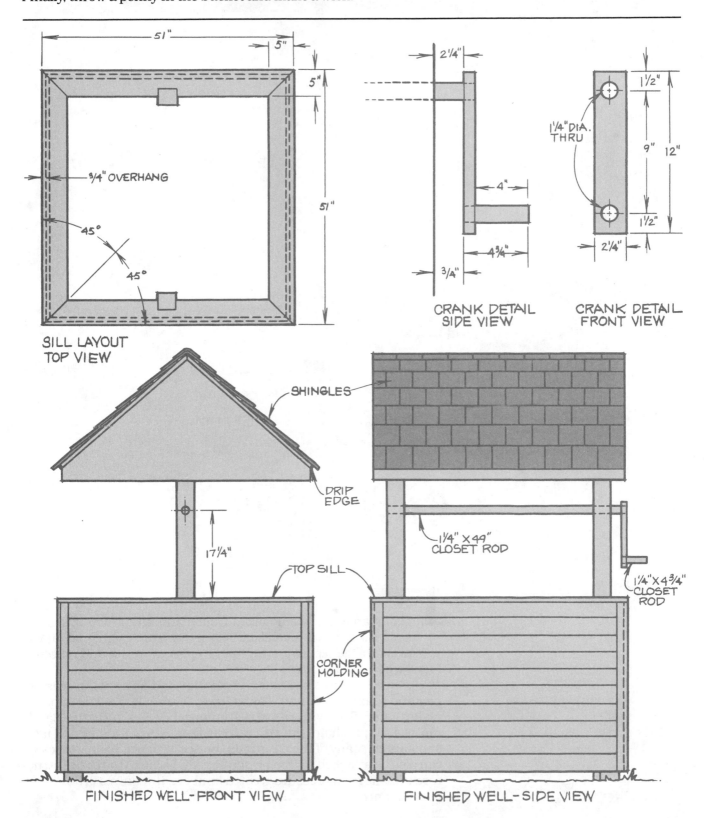

Footbridge

If your landscape would profit from it, this simple footbridge will span the gap over a creek or between two hills with grace. It will also provide secure footing—and a place to pause and reflect—for all who cross it.

The bridge uses a simple truss design and an easy, but reliable, pier and beam foundation. The walkway is actually 1 x 4 decking that makes a firm floor with lots of rustic appeal. As designed, the 'handrail' is a simple chain. But you can easily substitute something more substantial. Many of the handrails you see attached to open decks can also be attached to this bridge. Finally, because the entire structure is made from pressure-treated lumbers, it requires little maintenance.

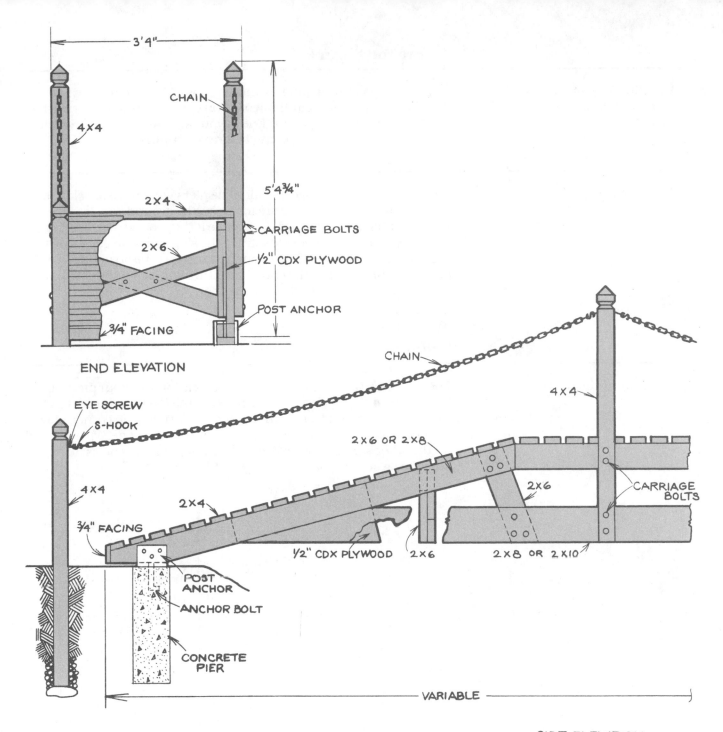

3'4"

CHAIN

4×4

5'4¾"

2×4

CARRIAGE BOLTS

2×6

½" CDX PLYWOOD

POST ANCHOR

¾" FACING

END ELEVATION

EYE SCREW

S-HOOK

CHAIN

4×4

2×6 OR 2×8

4×4

2×4

2×6

CARRIAGE BOLTS

¾" FACING

½" CDX PLYWOOD

2×6

2×8 OR 2×10

POST ANCHOR

ANCHOR BOLT

CONCRETE PIER

VARIABLE

SIDE ELEVATION

Materials

If your bridge doesn't span over 32', you can use 2 x 6's and 2 x 8's for the trusses and the braces. If it's longer, you'll have to use 2 x 8's and 2 x 10's, or build a pier to support the middle of the span. The hand-rail posts are 4 x 4's, and the walkway is made from 2 x 4's. Only use pressure-treated wood for this structure, particularly if you will be spanning a creek. Moisture will decay non-treated wood.

Besides this lumber, you'll also need some concrete, several 24" lengths of 8" stovepipe, anchor and hex bolts, eye screws, S-hooks, chain, truss plates, galvanized nails, post anchors, and some scraps of ½" exterior plywood.

Before You Begin

1 Check local codes and secure a building permit, if needed.

Check with your local building inspections office to make sure your structure meets code requirements. You can also obtain information about the location of the frost line in your area from this office. And you may need to obtain a permit before you begin construction.

2 Select a site.

Select a site for the bridge in which the piers will not be too close to the edge of the creek. If they are too close, the ground won't support them. You can place the piers on a slope if the ground is firm enough and cut the beams to compensate for the slope. Also, make sure the ground on *both* sides is firm enough to support the structure. If one side of the bridge sags, the entire structure will lean and you'll have a bridge that grows rickety before its time.

Pouring the Piers

3 Lay out the piers.

Use string and a line level to lay out the locations for your piers on each bank of the creek or gully. Stretch string between the stakes to form a rectangle, and locate the center of a pier at each corner. Make sure the piers are the same distance apart on each side. With a friend on the other side, measure diagonally from corner to corner, to be sure the layout is square. (See Figure 1.)

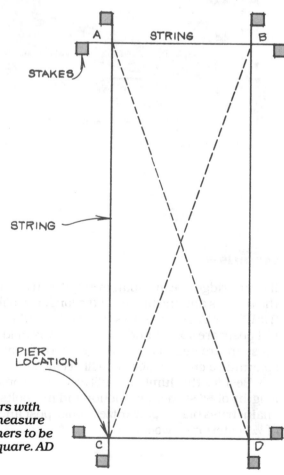

Figure 1. Lay out the piers with stakes and string, and measure diagonally from the corners to be sure that the layout is square. AD should equal BC.

MISCELLANEOUS OUTDOOR STRUCTURES

Dig holes for the piers at least 36″ deep on the slope, below the frost line for your area so ice won't force the piers to heave out of line. Use 8-inch stovepipes for the forms and set them in the holes. The tops of the forms must be flush to the string, so that the tops of the piers will be level with each other. Once you're sure the forms are level, simply mix cement and pour into the form.

4　**Pour the piers.**

TIP　Before you pour the concrete, throw 2″-3″ of gravel into the bottom of each form. This will help provide drainage.

Before the concrete sets up, position anchor bolts in the center of each pier. (See Figure 2.) Use the string to locate the precise position of the bolts. The tops of the bolts should protrude far enough to attach the truss anchors.

5　**Set the anchor bolts.**

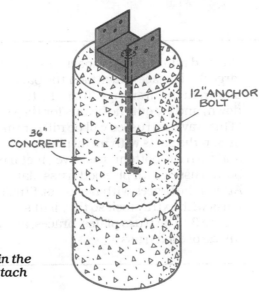

12″ ANCHOR BOLT

36″ CONCRETE

Figure 2. To make the truss anchors, set an anchor bolt in the concrete pier and use it to attach a post anchor to the pier.

TIP　To hold the anchor bolts at the proper height, drive two small stakes on either side of the piers. Wrap a wire around the end of the bolt, then wrap the ends of the wire around the stakes so that the bolt is suspended in the concrete.

Wait at least 24 hours for the concrete to cure and remove stovepipe, string, wire, and stakes. Place the truss anchors over the bolts. (Use ordinary 'post anchors' for the truss anchors.) Stretch a string from one anchor to its pair on the other side of the creek or gully. Use this string to line up the sides of the anchors. Repeat for the second set of anchors. When all anchors are properly aligned, secure them to the piers with washers and nuts.

6　**Attach the truss anchors.**

7   Plan the truss.

Carefully measure the distance between the inside edges of the truss anchors. Use this as your 'base' measurement. Sit down with a paper and pencil and carefully plan out your truss, using the angles and the basic design in the *Truss, Side View* drawing. We have purposely left out many of the measurements, because these will vary depending on the length of the bridge. As shown, our truss design will work for a bridge between fourteen and eighteen feet long. If yours is longer, you will have to lengthen the middle span. (This span is 2'-6" long on our drawing.) If it's shorter, you will have to shorten or eliminate the span. Draw carefully to scale, then measure the length of the beams and braces you need to cut.

TIP   Use an engineer's or an architect's rule to draw your plan. These rules help you scale parts down accurately.

8   Cut and assemble the parts.

Cut and miter the truss parts. Lay out the beams of one truss on a large flat surface and attach the parts with metal truss plates, as shown in the drawings. Turn the truss over (so the metal parts face down) and lay out the beams for the second truss on top of the first. (This way the trusses will be mirror images of each other.) Again, attach the parts with metal plates. Move the top truss off to one side, and turn the bottom truss over, metal plates up. Cover the plates on both trusses with plywood truss plates, as shown in the drawings. Attach the braces with hex bolts. Finally, 'beef up' the ends of each truss with extra stock so they'll fit snugly in the truss anchors. (See Figure 3.) The truss plates, braces, and extra stock all go on the *inside* of each truss.

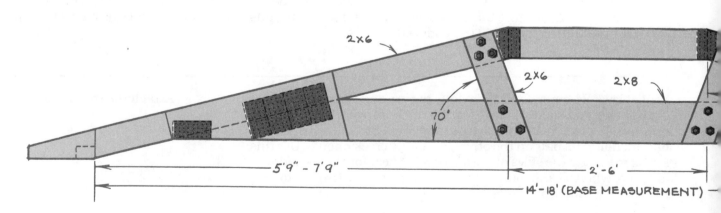

TRUSS-INSIDE VIEW

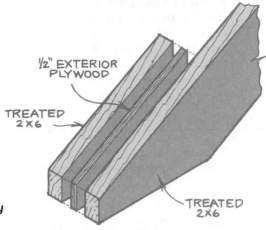

Figure 3. *Thicken the truss ends as shown so that they'll fit snugly in the truss anchors.*

½" EXTERIOR PLYWOOD

TREATED 2×6

TREATED 2×6

---

9   **Build the crossbraces.**

Cut and miter the parts for the crossbraces, as shown in the *Crossbrace, End View* drawing. Lap the boards at the proper angle and assemble them with ⅜" x 3½" carriage bolts.

TIP   If you're building a bridge longer or shorter than the one shown in the plans, you may have to adjust the angle of the miter cuts and the lap on the crossbraces. Measure your truss carefully to be sure.

Assembling the Bridge Frame

---

10   **Set the trusses in place.**

With a friend, set the trusses in place on the anchors. Remember, all the braces and the truss plates should be on the inside. With stakes and braces, temporarily hold the trusses in place so they don't fall over. Use a level to make sure that they are perfectly straight up and down.

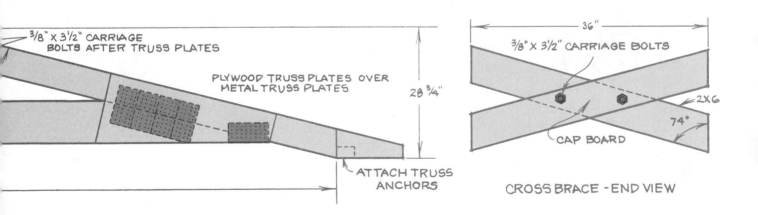

⅜" X 3½" CARRIAGE BOLTS AFTER TRUSS PLATES

PLYWOOD TRUSS PLATES OVER METAL TRUSS PLATES

28 ¾"

ATTACH TRUSS ANCHORS

36"

⅜" X 3½" CARRIAGE BOLTS

2×6

74°

CAP BOARD

CROSS BRACE - END VIEW

11 Attach the crossbraces.

Nail the crossbraces to the trusses, using 16d nails. Locate them in the approximate position shown on the *Finished Bridge, Side View* drawing. The actual position will depend on the length of your bridge. As you attach the crossbraces, continually check the trusses with a level to make sure that they have remained in the proper alignment. If any corners of the crossbraces stick up above the truss beams, trim them flush with a handsaw.

12 Attach the trusses to the truss anchors.

Nail the trusses to the truss anchors with 4d nails. When all is secure, remove the temporary braces and stakes.

Adding Decking and Handrails

13 Install the walkway.

Cut a sufficient number of 2 x 4's to the proper length to make the walkway. The two end pieces should be beveled at 76°, as shown in the *Finished Bridge, Side View* drawing. Nail 2 x 4's to trusses sides, using two 12d nails. Leave a small space between the boards to allow the lumber to swell without buckling. Do *not* attach walkway boards to the area of the truss where you will attach the handrail posts; leave that until later.

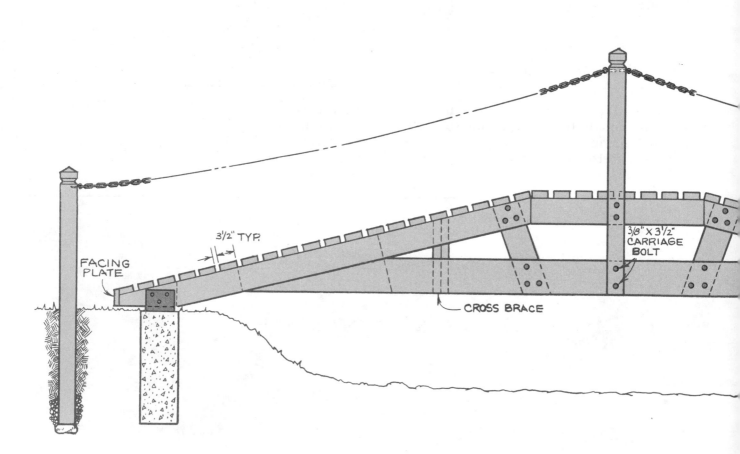

FINISHED BRIDGE, SIDE VIEW

MISCELLANEOUS OUTDOOR STRUCTURES

Cut six posts to hold the handrail. Shape the top of the posts as shown in the drawings with a table saw or radial arm saw. Carefully measure and notch the middle posts as shown in the *Middle Post Layout* drawing.

14 Cut the posts.

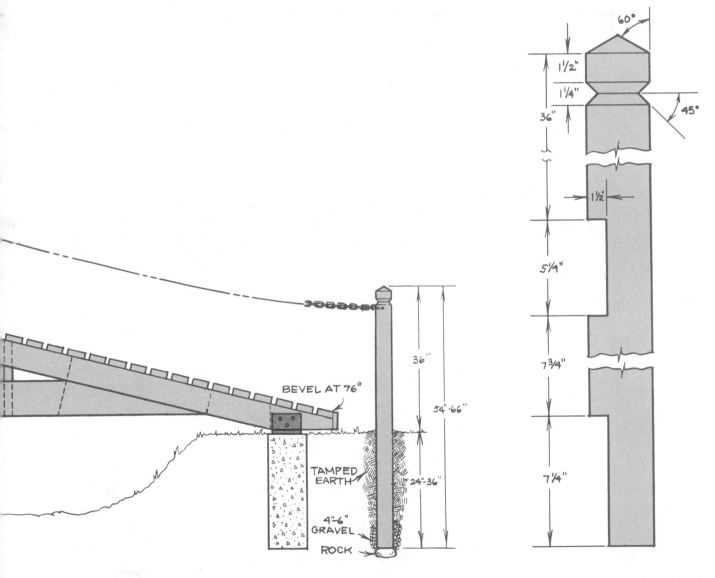

MIDDLE POST LAYOUT

15 Set the end posts.

Set the end posts in the ground, two at either end of the bridge. Locate them 8″ from the end of the bridge and 40″ apart, as shown on the *Finished Bridge, Top View* drawing. They should stand 36″ above the ground. Even them up by adding or removing dirt at the bottom of the post hole. Then pack the holes with gravel and dirt.

TIP If a post stands a little bit too high, you can lower it by using it to 'tamp' the ground at the bottom of the hole (before you fill the hole). This technique is especially useful if the rocks you put at the base of each post aren't all quite the same width—just tamp the rocks into the ground a little further.

16 Attach the middle posts and finish the walkway.

Attach the middle posts to trusses with ⅜″ x 3½″ carriage bolts. Notch the remaining 2 x 4's for the walkway to fit around the posts, and nail them in place.

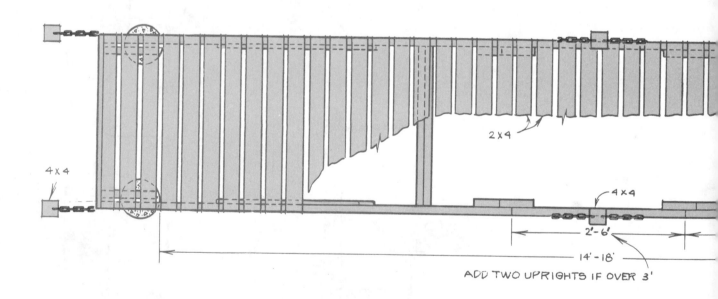

FINISHED BRIDGE, TOP VIEW

Screw eye screws into the posts about 5″ from the top, to hang the chain. Carefully measure the distance between the eye screws and have a hardware store cut the lengths of chain about 1″ longer than the measured distance. Attach the chain to the eye screws with S-hooks. (See Figure 4.) The extra length and the S-hooks will provide enough slack that the chain will hang in a graceful arc.

17 **Hang the chain between the posts.**

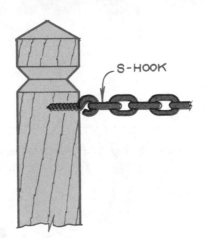

Figure 4. Attach the chain to the posts with eye screws and S-hooks.

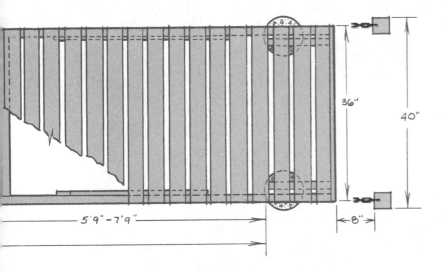

Raised Garden Beds

There's no more economical way to spruce up your landscape than to build a raised garden bed and fill it to overflowing with flowers and shrubs. These beds are not just decorative, they can also be used to line a walk, set a driveway off from the lawn, or prevent erosion on a steep slope. They're also quite a boon for gardeners. A raised garden bed provides better ventilation for roots and makes cultivating and weeding easier.

There are many different ways to design your raised garden bed. It can be a square, rectangle, triangle, hexagon—or just about anything you can imagine. The basic building steps are the same for every design. Just vary the length and arrangement of the railroad ties or landscape timbers to fit the size and shape you've selected.

MISCELLANEOUS OUTDOOR STRUCTURES

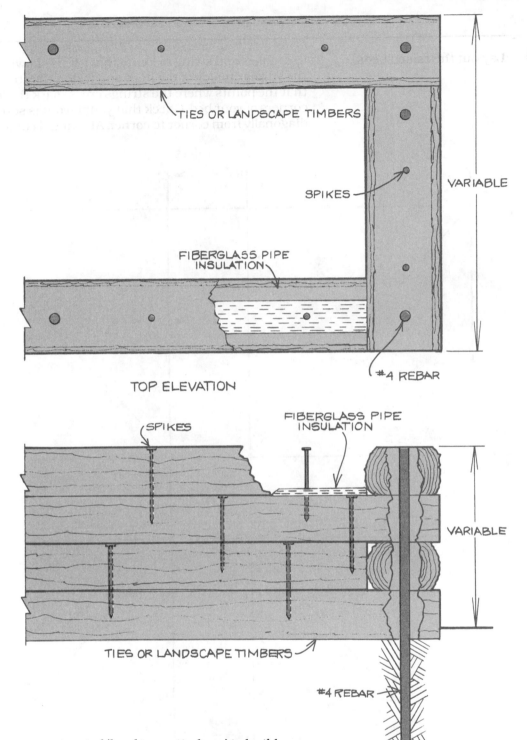

TIES OR LANDSCAPE TIMBERS

VARIABLE

SPIKES

FIBERGLASS PIPE INSULATION

#4 REBAR

TOP ELEVATION

SPIKES

FIBERGLASS PIPE INSULATION

VARIABLE

TIES OR LANDSCAPE TIMBERS

#4 REBAR

SIDE ELEVATION

Materials

Use railroad ties or pressure-treated 'landscape timbers' to build the frame for the beds. These ties and timbers are treated to prevent damage from ground moisture and insects. These woods also weather to a soft shade of gray that enhances the rustic appeal of the structure. If you work with railroad ties, make sure you get old, *used* ties. New ties are soaked in creosote, and this chemical will damage your plants. After a few years in the weather, most of the creosote leeches out of older ties.

In addition to the lumber, purchase fiberglass pipe insulation. You'll use this to seal between the ties and prevent water or dirt from seeping out of the bed. Also purchase 6″ long spikes to secure the ties to each other, and #4 reinforcing rod to anchor the ties to the ground.

1 **Lay out the raised beds.**

Use stakes and string to locate the ties, as shown in Figure 1. Place the stakes outside of the lines and fasten the string between them so that the points where the strings cross mark the location of the corners of your bed. Check that your layout is square by measuring diagonally from corner to corner. AD should equal BC.

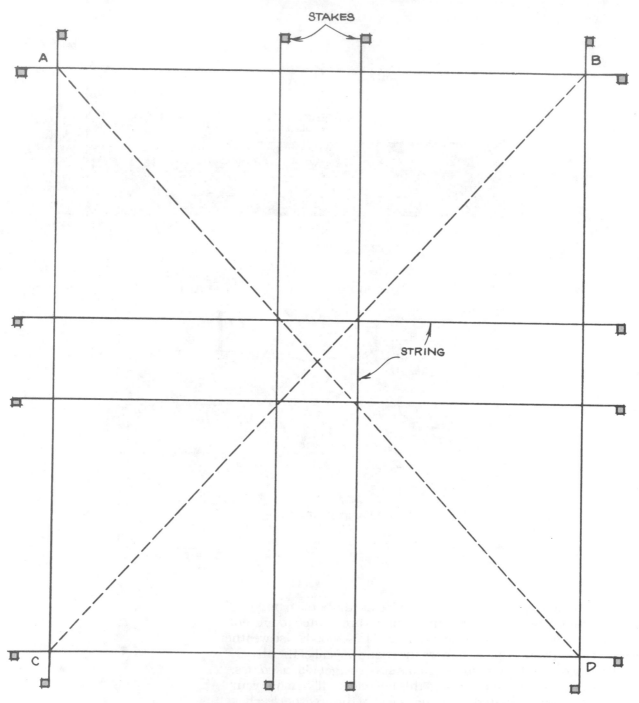

Figure 1. Use stakes and string to lay out the location of your ties or timbers. Check that the beds are square by measuring diagonally from corner to corner. AD should equal BC.

Peel up the sod and remove any large rocks, debris, or plants from the bed area. Rake the area, then level the bed with a shovel. To check if the bed is level, lay a 2 x 4 on edge across the bed. Use a carpenter's level to tell you if any portion of the bed needs to be filled or stripped.

2 Clear sod and level the bed.

Cut the wood to the proper size. Then lay the first course of ties in place. Use string and a string level to level the ties, as shown in Figure 2. If one end of a tie is low, shim it up with dirt. If it's high, remove the tie and strip away some more dirt.

3 Set the first course of ties or timbers in place.

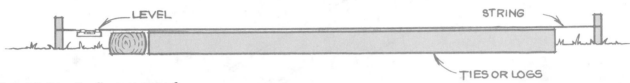

Figure 2. Lay the first course of ties or timbers in place and check that it's level, using string and a string level. Add or remove dirt, as necessary, to level the ties.

TIP Plan your beds so that you can cut the tie or timbers in multiples of 2'—2', 4', 6', 8', etc.—as shown in the *Typical Bed Layout* drawing. This will save you time and materials.

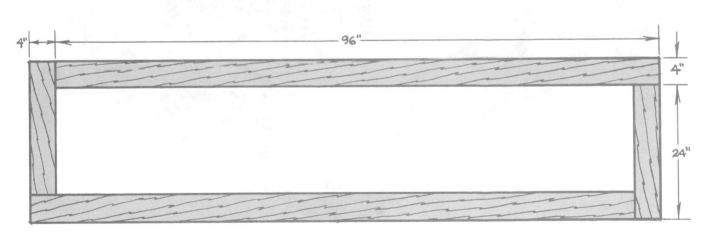

TYPICAL BED LAYOUT

4 Lay up the remaining courses of ties or timbers.

Lay a strip of fiberglass pipe insulation on the top surface of the first course of ties. (This insulation creates a 'seal' between the ties, and keeps soil from washing out of the beds.) Then lay up the second course, lapping the ends of the second-course ties over the joints between the first-course ties, as shown in Figure 3. Nail the second course in place with spikes. Follow the same procedure for the third and fourth courses, setting insulation between each course and making sure the ends of the ties lap over the joints between ties on the preceding course.

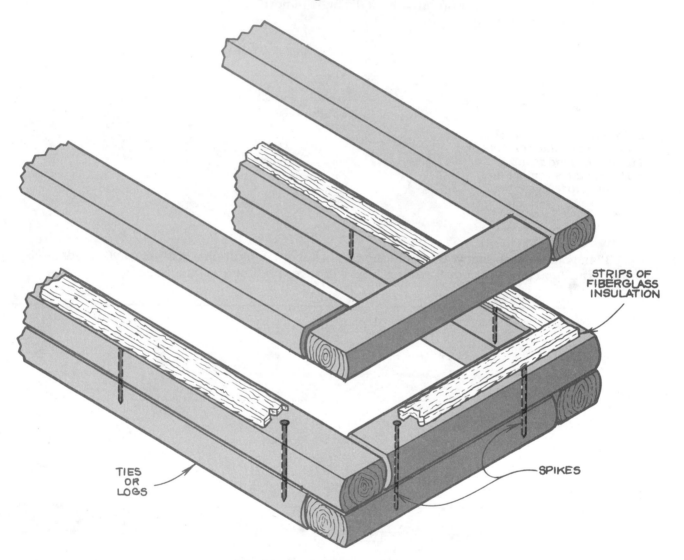

Figure 3. *Place strips of fiberglass pipe insulation between each course of ties or timbers to prevent the soil from washing away. Lay up the courses so that the ends of each course lap the joints between the ties of the preceding course. Nail each course in place with spikes.*

MISCELLANEOUS OUTDOOR STRUCTURES

To secure the assembly in place, drill ⅝″ holes down through the wood, all the way to the ground, as shown in Figure 4. Drill these holes at the corners and about every 4′. Cut rebar 12″ longer than the height of the wall, then drive the rebar rods down through the walls and into the ground. This will anchor the ties to the ground so they won't slip out of place.

5 Anchor the ties or timbers in place.

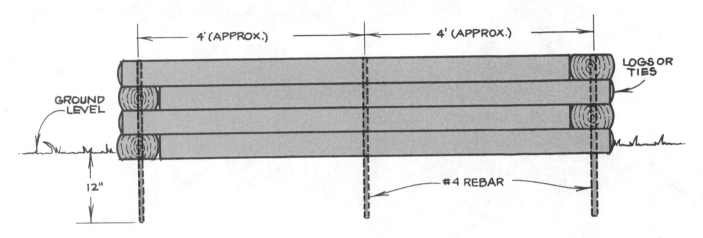

Figure 4. *Drill ⅝″ holes through the ties, then drive #4 rebar rods down through the wood and 12″ into the ground. This will anchor the beds in place.*

To keep dirt from washing out of the bed at the corners, stuff the inside corners of the ties with more fiberglass insulation. Stuff insulation in any crevice that might let the dirt escape.

6 Finish sealing the bed.

Fill the bed with topsoil and soak it with water. Let the soil settle, then top it off with more soil until the bed is filled. Generally, soil will settle the width of one tie after you first fill the bed. If you can, wait for a thunderstorm before planting your flowers or shrubs. This will settle the dirt completely, and it will show you if there are any 'leaks' in the beds.

7 Fill the bed with topsoil.

Trash Can Enclosure

If you don't like the trash cans cluttering up your yard, there's a simple way to hide them from view—but still keep them handy. Build an enclosure! This wooden structure is attractive, yet simple and relatively inexpensive to build. There are two hinged doors in front to provide easy access to the cans, and the top is open to provide plenty of ventilation.

This trash can enclosure is 3′ wide, 5′ high, and stretches nearly 6′ long, so you can store three standard-size trash cans in it. It's designed like a pole building without a roof. The walls are covered with wood siding that can either be painted to match your home or to blend in with the landscape. You can build this structure easily in a single afternoon—then never have to look at the trash again.

MISCELLANEOUS OUTDOOR STRUCTURES

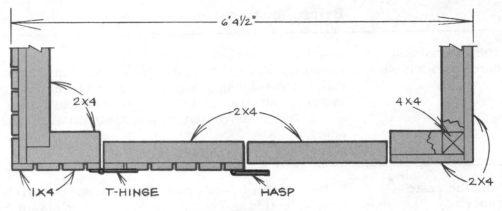

6' 4½"

2 X 4

2 X 4

4 X 4

2 X 4

1 X 4   T-HINGE   HASP

TOP ELEVATION

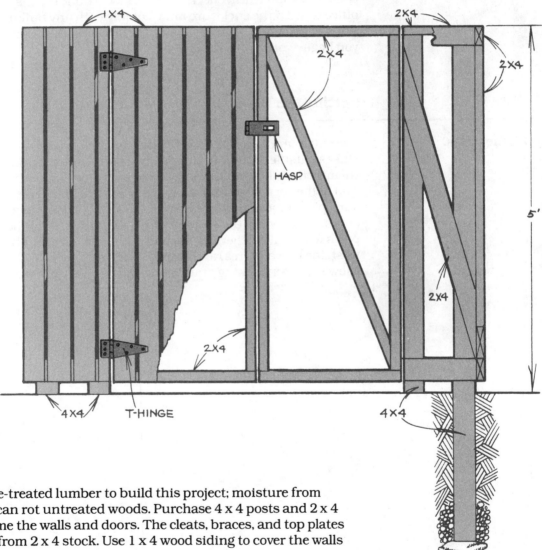

1 X 4

2 X 4

2 X 4

2 X 4

HASP

2 X 4

2 X 4

5'

4 X 4   T-HINGE   4 X 4

Materials

Use pressure-treated lumber to build this project; moisture from the ground can rot untreated woods. Purchase 4 x 4 posts and 2 x 4 studs to frame the walls and doors. The cleats, braces, and top plates are also cut from 2 x 4 stock. Use 1 x 4 wood siding to cover the walls and the doors.

In addition to these materials, purchase galvanized nails—regular nails will rust and stain your wood. For the doors, you'll need T-hinges and a hasp.

FRONT ELEVATION

Before You Begin

1   **Adjust the dimensions of the enclosure for your trash cans.**

As shown, the enclosure is 3' wide, 5' high, and 6' long. As we mentioned, it will hold three standard-size trash cans. If you have fewer cans to conceal, simply reduce the length of the enclosure. To conceal more cans, either increase the length or double the depth. You may also have to adjust the dimensions if your cans are larger than standard.

2   **Check with the waste collection agency.**

Unless your trash is collected at the curb, you should check with your waste collection agency regarding the location of your enclosure. They may not be willing to walk to the back of your lot to collect the cans. You should also check with your local building inspections office to see if the enclosure must comply with any building restrictions in your area. You may need to obtain a building permit before you begin work.

Setting the Posts

3   **Lay out the posts.**

Use stakes and string to find the locations of your posts. Place the stakes outside of the foundation lines and stretch the string between them so that the points where the strings cross mark the exact location of the corners of your enclosure. (See Figure 1.) For this enclosure, plant 6 posts—one at every corner and two to frame the front doors, as shown in the *Post Layout* drawing. Mark the location of the posts with stakes, then dig post holes 24"-36" deep (or below the frost line for your area) and twice as wide as your posts. This will allow space for packing gravel and dirt around the posts.

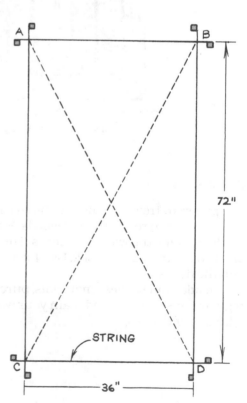

Figure 1. Lay out the location of your posts with stakes and string. Measure diagonally from corner to corner to be sure your layout is square. AD should equal BC.

MISCELLANEOUS OUTDOOR STRUCTURES

Place a large rock in the base of each hole to keep the posts from settling. These rocks should be twice the diameter of the posts—about 8″ in diameter. (See Figure 2.) Put the posts in place, then shovel gravel in the hole to a depth of 12″. This will help drain the ground water away from the posts. Finally, fill the rest of the hole with dirt and tamp lightly around the post.

4 **Set the posts in the ground.**

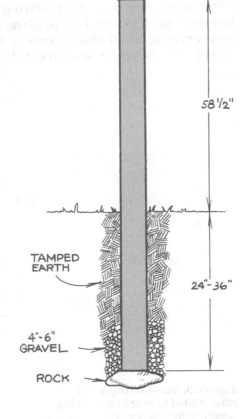

58 1/2″

TAMPED EARTH

24″-36″

4″-6″ GRAVEL

ROCK

Figure 2. Set the posts at least 2′ into the ground or below the frost line in your area. Plant each post on a large rock to keep them from settling, and fill the bottom of the post hole with several shovels full of gravel to provide drainage.

TIP Don't tamp the earth completely until after you have aligned the posts with a level and braced them so that they are straight up and down.

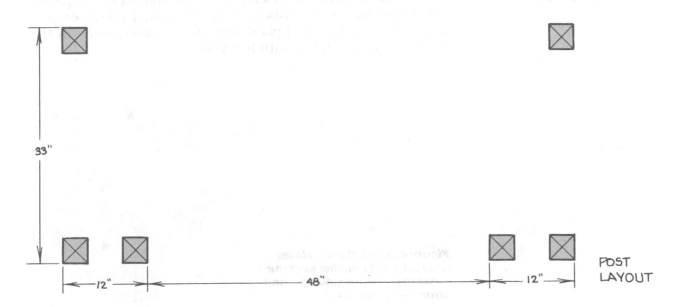

33″

12″   48″   12″

POST LAYOUT

TRASH CAN ENCLOSURE

5 Cut the posts to the proper height.

Once the posts are in the ground, brace them upright by driving stakes about 3' away and nailing scrap lumber from the post to the stakes. Use two braces per post, placing the braces at right angles to each other, as shown in Figure 3. Use a carpenter's level to make sure the posts are plumb. When the posts have been braced, cut them to the proper height. To do this, measure one post and mark the top 58½" above the ground. With a string and string level, use this as a reference mark to find the tops of the other posts. Remove the string and cut the posts off with a handsaw. Check again that the posts are still plumb, then tamp down the dirt around the posts as tight as you can.

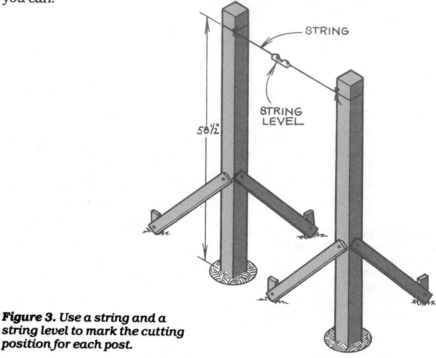

STRING

STRING LEVEL

58½"

Figure 3. Use a string and a string level to mark the cutting position for each post.

Building the Enclosure

6 Attach the top plates.

Cut top plates from 2 x 4 stock and notch the ends where the plates will fit together over the posts, as shown in Figure 4. After the top plates are cut and notched, place them over the posts, as shown. Nail the top plates to the posts with 16d nails.

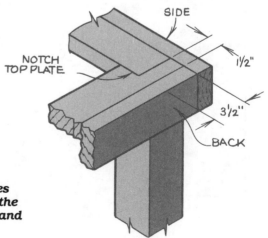

SIDE

NOTCH TOP PLATE

1½"

3½"

BACK

Figure 4. Notch the top plates where they fit together over the posts to provide a smooth—and strong—connection.

406

Cut 2 x 4 cleats and braces to size, as shown in the *Back Wall, Side Wall*, and *Front Wall* drawings. Nail the cleats to the posts and top plates with 16d nails. The bottom cleats must be at least 2″ from the ground. Next, place the braces diagonally across the wall frames, as shown, and nail them to the posts. Miter the ends of the braces so that the end grains face the side, not the top or bottom. End grains that face up absorb moisture and rot quickly.

7 **Attach the cleats and braces.**

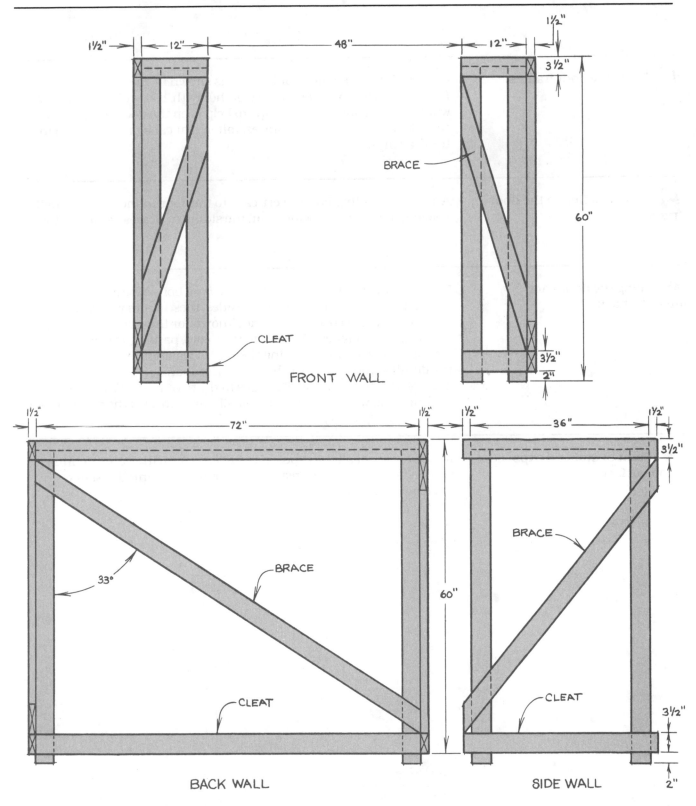

FRONT WALL

BACK WALL

SIDE WALL

8 Attach the siding.

Cut 1 x 4 siding boards to size. Nail the siding in place with 6d nails, as shown in the *Siding Installation Detail* drawing. Cover the back and the front of the structure, then the sides. The bottom edge of the siding must be flush with the bottom edge of the cleats, at least 2″ off the ground. Space the boards ½″ apart to allow the wood room to swell in warm, humid weather. At the corners, butt the siding boards together, as shown in the *Corner Detail, Top View* drawing.

Enclosing the Enclosure

9 Build the door frames.

Cut 2 x 4's to make the door frames, as shown in the *Door Frame Layout* drawing. Nail the frame together with 16d nails. You may also want to use metal framing strips to help keep the door rigid. Brace each door frame with 2 x 4 braces, mitered at each end as shown in the drawings.

10 Attach siding to the door frames.

Attach 1 x 4 siding boards vertically to the door frames with 6d nails, spacing the siding as before. Nail the siding to braces as well as the frame.

11 Hang the doors and install a hasp.

Mount two T-hinges on each door with long bolts. Use bolts to better support the weight of the doors. Besides, these doors will get a lot of heavy use—trash collectors are not known for their gentility. Attach the straps or 'tongues' of the 'T' to the doors, passing the bolts through the door and securing them with washers and nuts on the inside. After you mount the hinges to the doors, lift the doors in place and bolt the 'butts' of the 'T' hinges to the door posts. When the doors are hung and working properly, install a hasp to keep the doors closed.

12 Paint or stain all exposed wood surfaces.

If you wish, coat all the exposed wood surfaces with waterproofing paint or stain. Pressure-treated lumber can be painted just like untreated lumber.

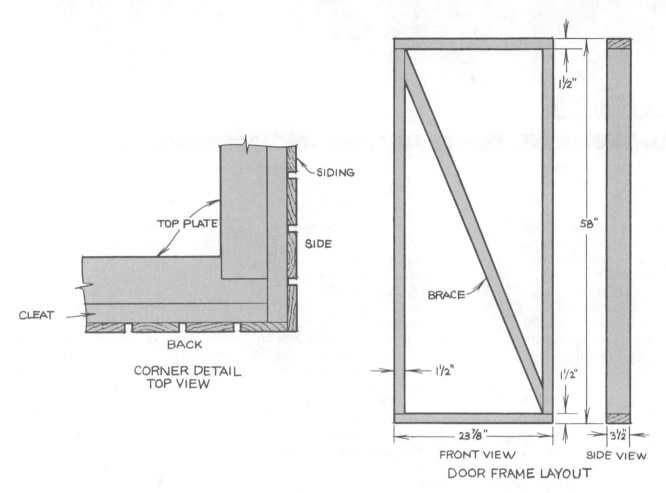

SIDING

TOP PLATE

SIDE

CLEAT

BACK

CORNER DETAIL
TOP VIEW

1 1/2"

58"

BRACE

1 1/2"

1 1/2"

23 7/8"

3 1/2"

FRONT VIEW

SIDE VIEW

DOOR FRAME LAYOUT

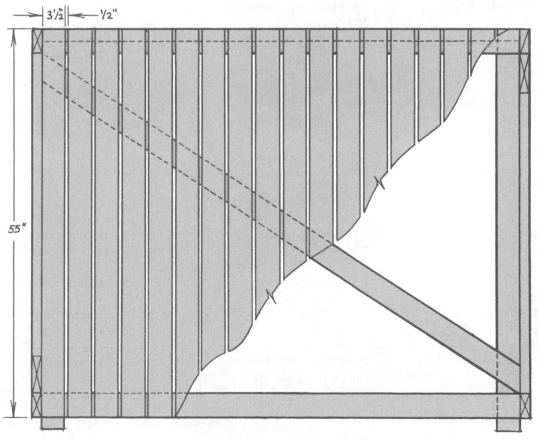

3 1/2"

1/2"

55"

SIDE INSTALLATION DETAIL

Index

METRIC EQUIVALENCY CHART

MM—MILLIMETRES   CM—CENTIMETRES

INCHES TO MILLIMETRES AND CENTIMETRES

INCHES	MM	CM	INCHES	CM	INCHES	CM
⅛	3	0.3	9	22.9	30	76.2
¼	6	0.6	10	25.4	31	78.7
⅜	10	1.0	11	27.9	32	81.3
½	13	1.3	12	30.5	33	83.8
⅝	16	1.6	13	33.0	34	86.4
¾	19	1.9	14	35.6	35	88.9
⅞	22	2.2	15	38.1	36	91.4
1	25	2.5	16	40.6	37	94.0
1¼	32	3.2	17	43.2	38	96.5
1½	38	3.8	18	45.7	39	99.1
1¾	44	4.4	19	48.3	40	101.6
2	51	5.1	20	50.8	41	104.1
2½	64	6.4	21	53.3	42	106.7
3	76	7.6	22	55.9	43	109.2
3½	89	8.9	23	58.4	44	111.8
4	102	10.2	24	61.0	45	114.3
4½	114	11.4	25	63.5	46	116.8
5	127	12.7	26	66.0	47	119.4
6	152	15.2	27	68.6	48	121.9
7	178	17.8	28	71.1	49	124.5
8	203	20.3	29	73.7	50	127.0

YARDS TO METRES

YARDS	METRES	YARDS	METRES	YARDS	METRES	YARDS	METRES	YARDS	METRES
⅛	0.11	2⅛	1.94	4⅛	3.77	6⅛	5.60	8⅛	7.43
¼	0.23	2¼	2.06	4¼	3.89	6¼	5.72	8¼	7.54
⅜	0.34	2⅜	2.17	4⅜	4.00	6⅜	5.83	8⅜	7.66
½	0.46	2½	2.29	4½	4.11	6½	5.94	8½	7.77
⅝	0.57	2⅝	2.40	4⅝	4.23	6⅝	6.06	8⅝	7.89
¾	0.69	2¾	2.51	4¾	4.34	6¾	6.17	8¾	8.00
⅞	0.80	2⅞	2.63	4⅞	4.46	6⅞	6.29	8⅞	8.12
1	0.91	3	2.74	5	4.57	7	6.40	9	8.23
1⅛	1.03	3⅛	2.86	5⅛	4.69	7⅛	6.52	9⅛	8.34
1¼	1.14	3¼	2.97	5¼	4.80	7¼	6.63	9¼	8.46
1⅜	1.26	3⅜	3.09	5⅜	4.91	7⅜	6.74	9⅜	8.57
1½	1.37	3½	3.20	5½	5.03	7½	6.86	9½	8.69
1⅝	1.49	3⅝	3.31	5⅝	5.14	7⅝	6.97	9⅝	8.80
1¾	1.60	3¾	3.43	5¾	5.26	7¾	7.09	9¾	8.92
1⅞	1.71	3⅞	3.54	5⅞	5.37	7⅞	7.20	9⅞	9.03
2	1.83	4	3.66	6	5.49	8	7.32	10	9.14